Nature-Based Solutions to 21st Century Challenges

Fully revised and updated, this second edition provides a systematic review of nature-based solutions and their potential to address current environmental challenges.

With growing recognition of the need to work with ecosystems to address global environmental challenges, there is now a move toward nature-based solutions, which involve utilizing nature's ecosystem to solve societal challenges while providing multiple co-benefits. This book systematically reviews nature-based solutions from a public policy angle, assessing policy developments which encourage the implementation of nature-based solutions to address societal challenges while simultaneously providing human well-being and biodiversity benefits. This expanded second edition delves deeper into how these solutions can foster sustainable development, enhance human well-being, and conserve biodiversity. With new case studies from the Global North and Global South, the book offers fresh insights into policy implementations and innovative approaches. Highlighting recent advancements in technology, such as AI, and policy frameworks, it provides a critical analysis of nature-based solutions' effectiveness in various contexts, including urban and rural environments. The book also explores emerging trends, such as the integration of traditional knowledge and community-led initiatives, enhancing its relevance for policymakers, researchers, and practitioners. This edition is not only a valuable resource for understanding the current landscape but also a guide for future sustainable strategies, emphasizing the importance of collaborative efforts in tackling environmental challenges. The book concludes by providing a best practice guide for those aiming to turn societal challenges into opportunities.

This book will be of great interest to policymakers, practitioners, and researchers involved in nature-based solutions, sustainable urban planning, environmental management, and sustainable development generally.

Robert C. Brears is Founder of Our Future Water, Global Climate Solutions, and Mark and Focus. He is the author of fifteen books, including *Financing Water Security and Green Growth*.

Nature-Based Solutions to 21st Century Challenges

Second Edition

Robert C. Brears

Routledge
Taylor & Francis Group
LONDON AND NEW YORK

earthscan
from Routledge

Designed cover image: Getty Images

Second edition published 2026
by Routledge
4 Park Square, Milton Park, Abingdon, Oxon OX14 4RN

and by Routledge
605 Third Avenue, New York, NY 10158

Routledge is an imprint of the Taylor & Francis Group, an informa business

First edition published by Routledge 2020

British Library Cataloguing-in-Publication Data
A catalogue record for this book is available from the British Library

ISBN: 978-1-032-85388-8 (hbk)
ISBN: 978-1-032-85386-4 (pbk)
ISBN: 978-1-003-51794-8 (ebk)

DOI: 10.4324/ 9781003517948

Typeset in Optima
by Newgen Publishing UK

Contents

Figures

Tables

Acknowledgment

First, I wish to thank Hannah Ferguson for being a visionary who enables books like mine to come to fruition. Second, I wish to thank Mum, whose deep interest in the environment and unwavering support have been invaluable throughout this writing journey. Lastly, I would like to extend a special acknowledgment to Kate, my love, who enriches my life in countless ways and has been an endless source of inspiration, encouragement, and motivation throughout the creation of this work.

1 Introduction

The Role of Nature-Based Solutions in Addressing 21st-Century Challenges

In the 21st century, society faces a range of climatic and non-climatic challenges that can trigger abrupt and, in some cases, irreversible environmental changes, negatively affecting human development. Rapid population growth and urbanization drive ecosystem degradation through excessive water extraction, eutrophication, and pollution. At the same time, climate change is intensifying the frequency and severity of disasters, resulting in increased fatalities, injuries, and substantial economic and property losses.

Instead of relying solely on conventional engineering solutions, there is increasing recognition of the need to work with natural ecosystems to enhance urban livability and support communities in coping with and recovering from disasters. Both adapt to and mitigate climate change, all while safeguarding biodiversity and ecosystem health.

Nature-based solutions (NBS) are approaches that leverage natural processes or replicate them to address environmental and societal challenges. They can be strategically and equitably implemented to address both climatic and non-climatic issues, providing multiple benefits, including lower infrastructure costs, job creation, sustainable economic growth, improved public health, and enhanced recreational opportunities.

Nature-based solutions (NBS) can mitigate risks from both sudden and slow-onset events. Coastal vegetation, sand dunes, and mangroves act as natural barriers against storm surges, while healthy coral reefs help dissipate wave energy during coastal storms. NBS can also reduce secondary impacts, such as landslides triggered by earthquakes, by stabilizing soil and vegetation. Additionally, they effectively manage gradual hazards like droughts by enhancing natural water storage in lakes and aquifers, ensuring a more reliable water supply for human and ecological needs.

Despite their proven benefits, the implementation of NBS remains limited due to a lack of understanding and the evolving nature of frameworks. Policies often fail to prioritize NBS, despite offering practical solutions. The United Nations World Water Development Report highlights that, while investment in NBS is increasing, it still represents less than 1% of total water resource management infrastructure spending.

DOI: 10.4324/9781003517948-1

Scaling up NBS requires innovative governance models that foster cross-sectoral and multi-scalar collaboration. It is essential to ensure stakeholder engagement in co-developing locally tailored solutions, enhance knowledge and confidence in NBS among policymakers and the public through targeted educational programs, and involve local communities to promote a sense of ownership and ensure proper maintenance. Additionally, increasing awareness and adoption of NBS in the private sector is crucial for the widespread implementation of this approach.

No systematic survey of the policy solutions implemented to mainstream Nature-Based Solutions (NBS) exists. While NBS are inspired by, supported by, or mimic natural processes to achieve sustainable urbanization, restore degraded ecosystems, mitigate and adapt to climate change, and reduce disaster risks, most existing literature focuses on engineering and hard science aspects of conventional solutions or specific facets of NBS.

Nature-Based Solutions to 21st Century Challenges examines the technology policies employed in leading global locations to develop and promote NBS, transforming societal challenges into sustainable and environmentally friendly growth opportunities. This book follows the UNEP-DHI classification of "technologies," categorizing them into hardware (physical infrastructure and technical equipment), software (approaches, processes, and methodologies such as planning, decision support systems, models, knowledge transfer, and capacity building), and orgware (institutional technologies, including governance structures, ownership models, and organizational arrangements).

Specifically, this book examines how multiple levels of government implement or support hardware, software, and organizational systems to advance NBS for sustainable urbanization, ecosystem restoration, climate change mitigation and adaptation, and disaster risk reduction. Through this analysis, readers will gain insights into best practices from leading global examples where NBS have been successfully applied to address pressing societal challenges while maximizing environmental, economic, and social benefits.

Following this introduction, this book's chapters are structured as follows:

Chapter 2 defines NBS and explores how they can be implemented to achieve broader development goals while enhancing the resilience of socio-ecological systems to long-term environmental and social changes.

Chapter 3 provides an overview of societal challenges that contribute to abrupt and, at times, irreversible environmental damage, negatively impacting human development.

Chapter 4 examines the effects of climate change on ecosystems, highlighting increased fragility in habitat types, reduced ecosystem resilience, and heightened extinction risks for plant and animal species.

Chapter 5 reviews different forms of environmental degradation driven by societal challenges and their consequences for biodiversity.

Chapter 6 examines the social, cultural, and environmental impacts of climate change and degradation on human society.

Chapter 7 examines the role of NBS in climate change mitigation, with a focus on enhancing carbon storage.
Chapter 8 discusses how NBS support climate change adaptation through the sustainable management, conservation, and restoration of ecosystems.
Chapter 9 examines how ecosystem restoration using NBS can enhance resilience, improve ecosystem services, and address societal challenges.
Chapter 10 explores the application of NBS in promoting sustainable urbanization.
Chapter 11 highlights the role of ecosystems in disaster risk reduction and how NBS contribute to hazard mitigation, livelihood security, and disaster resilience.
Chapter 12 emphasizes the importance of an adaptive management framework in ensuring the effectiveness of NBS.
Chapter 13 reviews the financial tools available to support NBS implementation.
Chapter 14 presents best practices for scaling up NBS, followed by the conclusion.

2 Nature-Based Solutions to Societal Challenges

Introduction

It is now widely recognized that societal challenges, such as climate change, rapid urbanization, increasing demand for food and water, and heightened disaster risks, can trigger abrupt and, in some cases, irreversible environmental changes that negatively impact human development.[1,2] One approach to addressing these challenges is through engineering solutions, which are designed to be simple to implement, easily replicable, and provide predictable outcomes. For example, large-scale physico-chemical biofiltration processes purify air and water. An alternative approach is Nature-Based Solutions (NBS), which leverages ecosystems and their services to address societal challenges in a sustainable manner.[3,4] Unlike human-made infrastructure, which often requires significant investments in materials and energy, NBS provide cost-effective and environmentally integrated solutions that work with nature rather than against it.[5]

Nature-Based Solutions

NBS involve the innovative application of ecological knowledge, drawing inspiration from, being supported by, or mimicking natural processes to address environmental, social, and economic challenges. These solutions enable societies to tackle pressing issues effectively, thereby enhancing human well-being and biodiversity.

The concept acknowledges that many societal challenges arise from human activities that overlook ecological limitations. Sustainable alternatives can be developed by looking to nature for design and process knowledge. NBS serve as an umbrella term encompassing a wide range of applications.

Examples of NBS include:

- Restoring mangrove forests to reduce coastal erosion, protect local communities and infrastructure from rising sea levels, and enhance biodiversity.
- Integrating green spaces in cities to mitigate the urban heat island effect, manage stormwater during heavy rainfall, and create wildlife habitats.

DOI: 10.4324/9781003517948-2

- Restoring wetlands to sequester carbon while improving water quality for potable and recreational use.[6,7]

When addressing societal challenges, whether from industrial, engineering, or resource management issues at local or global scales, NBS expands the range of possible responses. Exploring NBS enables a broader and more sustainable approach to problem-solving.

Operationally, the concept of an NBS can be understood by breaking it down into three key elements:

- *Nature*: Encompasses biodiversity as a whole, individual components of biodiversity, and/or ecosystem services.
- *Nature-Based*: Involves ecosystem approaches, ecosystem-based strategies, biomimicry, or alternative ways of utilizing biodiversity.
- *Solutions*: Focus on addressing a problem or challenge where a recognizable solution or a more beneficial outcome is achievable.[8]

By including the word 'solution,' NBS explicitly acknowledge that a problem exists and requires resolution. This problem-solving focus is a key characteristic that distinguishes NBS from broader, more neutral approaches to ecosystem management or policy, which may simply aim to sustain or enhance ecosystems without addressing specific challenges. Therefore, identifying problems or challenges that can be effectively addressed through NBS is fundamental to this approach.[9]

Nature-Based Solutions and Ecosystem Services

Implementing NBS can enhance or restore a range of ecosystem services. Ecosystem services refer to the benefits people derive from natural ecosystems, which can be categorized into four main types:

- *Provisioning services*: Products obtained from ecosystems.
- *Regulating services*: Benefits gained from an ecosystem's regulation of natural processes.
- *Cultural services*: Non-material benefits people receive from ecosystems.
- *Supporting services*: Essential functions that maintain other ecosystem services.

These categories are summarized in Tables 2.1 through 2.4.[10,11,12]

Nature-Based Solutions Simultaneously Meeting Wider Development Goals

While ecosystem services are often valued for their immediate benefits to human well-being and the economy, NBS focus on long-term benefits for both people and the environment. Unlike traditional biodiversity conservation and

Table 2.1 Provisioning Services

Service	*Sub-Category*	*Definition*	*Examples*
Food	Crops	Cultivated plants harvested for human or animal consumption.	Grains, vegetables, fruits
	Livestock	Domesticated animals raised for meat, dairy, labor, or other commercial uses.	Cattle, pigs, chickens
	Captured fisheries	Wild fish and seafood obtained through non-aquaculture methods such as trawling or netting.	Cod, shrimp, tuna
	Aquaculture	Fish, shellfish, and aquatic plants bred and reared in controlled water environments for harvest.	Salmon, oysters, clams
	Wild foods	Edible species gathered directly from natural ecosystems.	Wild berries, mushrooms, bushmeat
Fibre	Timber and wood fibre	Materials from trees harvested from forests, plantations, or woodlots used for construction or manufacturing.	Roundwood, wood pulp, paper
	Other natural fibres	Non-wood fibres sourced from animals or plants for various uses, excluding fuel.	Cotton, silk, hemp, flax
Biomass fuel		Organic materials from plants or animals used as energy sources.	Fuelwood, dung, ethanol from grain
Freshwater		Water from inland sources used for domestic, industrial, and agricultural purposes.	Drinking water, irrigation, hydropower
Genetic resources		Genetic material used in breeding, crop enhancement, and biotechnology.	Genes for drought-resistant crops
Biochemicals, natural medicines, and pharmaceuticals		Substances derived from ecosystems used in health care, food, and industry.	Ginseng, poppy (for codeine), garlic

management, NBS integrate sustainable solutions that address environmental change and hazards while incorporating societal factors such as human well-being, poverty alleviation, socio-economic development, and governance. This allows NBS to achieve environmental, social, and economic objectives simultaneously.[13,14]

NBS offers effective responses to significant challenges, such as climate regulation, by harnessing natural processes while also enhancing biodiversity through the sustainable management of ecosystems. Additionally, NBS respects and reinforces community rights over natural resources, placing biodiversity and local human communities at their core.[15]

Table 2.2 Regulating Services

Service	*Definition*	*Examples*
Air quality regulation	How ecosystems influence air quality by emitting pollutants (as sources) or absorbing them (as sinks).	Lakes absorb sulphur compounds; vegetation fires emit particulates and volatile organic compounds.
Climate regulation	Ecosystems' role in regulating the global climate through greenhouse gas emissions or absorption.	Forests absorb CO_2; cattle and rice fields emit methane.
	Ecosystems' influence on local or regional climate variables such as rainfall and temperature.	Forests increase regional rainfall and moderate temperatures.
Water regulation	Ecosystems' effect on the timing, quantity, and storage of water runoff, flood prevention, and aquifer recharge.	Wetlands and floodplains store runoff, permeable soils recharge aquifers.
Erosion regulation	The role of vegetation in preventing soil loss through wind or water erosion.	Grass and tree cover prevents soil erosion and landslides on slopes.
Water purification and waste treatment	The filtering and decomposition functions of ecosystems for pollutants and waste in water and soil.	Wetlands trap pollutants; soil microbes break down organic waste.
Disease regulation	How ecosystems impact the presence and spread of pathogens that affect human health.	Forests reduce standing water, limiting mosquito breeding and malaria risk.
Pest regulation	Ecosystems' control over crop and livestock pest populations.	Forest-dwelling predators (e.g. bats, toads) consume agricultural pests.
Pollination	Animal-facilitated transfer of pollen necessary for plant reproduction.	Wild bees pollinate nearby crops from forest habitats.
Natural hazard regulation	The capacity of ecosystems to mitigate the impact of natural disasters and regulate fire regimes.	Mangroves buffer storm surges; decomposers reduce wildfire fuel loads.

In the broader political discourse on jobs and economic growth, the European Commission emphasizes that NBS can transform environmental and societal challenges into innovative opportunities, thereby driving green growth and sustainable development. The European Commission identifies key areas where NBS can contribute to global goals:

- *Climate change mitigation and adaptation*: NBS enhance carbon storage and provide more resilient responses to climate challenges.
- *Restoring degraded ecosystems*: NBS improve ecosystem resilience, enabling them to sustain vital ecosystem services while addressing societal challenges.

Table 2.3 Cultural Services

Service	*Definition*	*Examples*
Ethical values	Spiritual, religious, aesthetic, intrinsic, or other meanings that people attribute to ecosystems, species, or landscapes.	Spiritual fulfilment from sacred rivers; cultural reverence for forest groves.
Existence values	The value individuals place on simply knowing that certain species or ecosystems continue to exist, regardless of direct use.	Belief in the importance of protecting endangered species like tigers or whales.
Recreation and ecotourism	Enjoyment derived from outdoor activities in natural or managed ecosystems.	Hiking in national parks, birdwatching, camping in forests, visiting coral reefs.

Table 2.4 Supporting Services

Service	*Definition*
Nutrient cycling	The process through which essential nutrients like nitrogen, phosphorus, and sulphur are extracted from mineral, aquatic, or atmospheric sources – or recycled from organic matter – and returned to ecosystems via the atmosphere, water, or soil.
Soil formation	The development of soil through the breakdown and decomposition of organic and inorganic materials over time.
Primary production	The generation of organic compounds by autotrophs (e.g. plants and algae) through the assimilation of energy and nutrients from the environment.
Photosynthesis	The biological process in which green plants and other organisms convert carbon dioxide and water into sugars and oxygen using sunlight.
Water cycling	The continuous movement of water within the biosphere in its solid, liquid, or gaseous states through processes such as precipitation, evaporation, and transpiration.

- *Sustainable urbanization*: NBS stimulates economic growth, enhances the urban environment, makes cities more attractive, and improves human well-being.
- *Disaster risk management and resilience*: NBS can provide greater benefits than conventional methods by reducing multiple risks and offering co-benefits.[16]

The European Commission views NBS as solutions to societal challenges that are inspired and supported by nature, delivering environmental, social, and economic benefits while enhancing resilience and preserving natural capital. This perspective places the economy and social well-being at the core of NBS while ensuring environmental sustainability.

As a flagship concept, NBS can incentivize governments, institutions, businesses, and citizens to develop innovative approaches that integrate natural capital into policies and planning while promoting biodiversity and human well-being.[17,18]

Natural Capital

From an economic perspective, a nation's wealth is built on four core stocks of capital: manufactured capital, which includes infrastructure, machines, and buildings; human capital, referring to people, their skills, and knowledge; social capital, encompassing norms, institutions, and social structures; and natural capital, which consists of the stock of natural assets such as geology, soil, air, water, and all living organisms. It is through the interaction of various components of natural capital that ecosystem services are generated.[19,20] For example, a forest represents a component of natural capital, while climate regulation is an ecosystem service it provides. Similarly, healthy soil is part of natural capital, while its ability to reduce flood and drought risks is an ecosystem service. In summary, natural capital refers to the stock of ecosystems and natural resources that support life and generate valuable ecosystem services essential for human well-being.

Weak versus Strong Sustainability

Weak sustainability assumes that natural and manufactured capital are interchangeable and contribute equally to human well-being. Under this view, what matters is the total aggregate stock of capital, which should be maintained or ideally increased for the benefit of future generations. Additionally, weak sustainability suggests that economic growth will continue to drive technological solutions to address environmental challenges caused by the rising levels of goods and services production.

In contrast, strong sustainability views natural capital as a complex, evolving system of biotic and abiotic components that interact to sustain ecosystem services. Proponents of strong sustainability argue that natural capital is non-substitutable because it cannot be artificially reproduced, and its destruction is often irreversible, unlike manufactured capital, which can be rebuilt or restored. Furthermore, given the limited scientific understanding of how ecosystems function, society cannot accurately predict the long-term consequences of depleting natural capital. Therefore, recognizing irreversibility and uncertainty should lead to the application of the precautionary principle when considering the use of natural capital in economic growth.[21]

Nature-Based Solutions Enhancing Resilience

NBS promote and depend on biological diversity to enhance the resilience of social-ecological systems against global changes, extreme events, and unforeseen disturbances. Productive ecosystems are essential for sustaining ecosystem services; however, ecosystems that lack resilience become vulnerable to disruptions, resulting in a decline in these services.

While some ecosystem services can be replaced by human-engineered solutions, such as using grey infrastructure (e.g., concrete barriers) instead of green infrastructure (e.g., coastal marshes) to protect coastal communities from storm surges, these replacements often come with trade-offs. For instance, coastal barriers may increase erosion in adjacent areas, exacerbating vulnerabilities along the shoreline.[22,23] To sustain the continuous flow of valuable ecosystem services amid long-term social and ecological changes, the resilience of social-ecological systems can be strengthened through key guiding principles, as outlined below.

Maintaining Biodiversity and Redundancy

Increased biodiversity enhances ecosystem resilience and stability by providing redundancy that helps buffer ecosystems against species loss and reduces the risk of compromising essential ecosystem functions. Higher biodiversity also increases the likelihood that particular species adapt well to specific conditions, ensuring ecosystems function effectively under changing environmental conditions. However, not all ecosystems require high biodiversity to be resilient. Some, like coastal wetlands, maintain strong resilience despite having low species diversity. This is because particular plant species in these environments are highly specialized to extreme conditions, such as hypoxic soils and high salinity levels.

Managing Connectivity

Connectivity refers to the structure and strength of interactions within a social-ecological system, including the movement of resources, species, or actors across patches, habitats, or social domains. It is crucial in safeguarding ecosystem services by facilitating recovery after a disturbance or preventing disturbances from spreading. For example, closely situated reef habitats can facilitate the recolonization of species that have been lost due to disturbances such as storms. Connectivity to refuge areas accelerates the restoration of disturbed ecosystems, helping to sustain critical ecosystem functions and the ecosystem services they provide.

Managing Feedback and Slowly Changing Variables

Managing feedback loops and slow variables in social-ecological systems is crucial to maintaining system stability and ensuring the continued production of essential ecosystem services. Feedback loops act as two-way connectors between variables, either reinforcing change (positive feedback) or dampening change (negative feedback). Slow variables refer to gradually changing factors that influence the long-term dynamics of a system. If a social-ecological system shifts into a different configuration, reversing the change can be extremely difficult, often leading to irreversible losses in ecosystem services and system functionality.

Economic Diversification

Communities that heavily depend on specific natural resources or industries are more vulnerable to shocks affecting those resources. In contrast, highly diversified economies can better withstand disruptions to particular sectors, like ecosystems, that can lose some species while maintaining overall system functionality.

Fostering Complex Adaptive Systems Thinking

A complex adaptive systems approach acknowledges that multiple connections co-occur at different levels within a social-ecological system. It also requires embracing unpredictability and uncertainty while considering diverse perspectives. Understanding a social-ecological system requires examining how the people within it think, specifically, how their mental models (the internal frameworks that shape reasoning, decision-making, and behavior) influence their actions.

Encouraging Learning

Social-ecological systems constantly evolve, requiring continuous knowledge revision and an ongoing commitment to learning and adaptation. Adaptive and collaborative management is crucial in building resilience by integrating diverse knowledge sources and fostering inclusive decision-making. This approach encourages a willingness to experiment and take calculated risks, ensuring that solutions remain flexible and responsive to changing conditions.

Broadening Participation

Active stakeholder participation is fundamental to building social-ecological resilience. Engaging a diverse range of stakeholders in managing social-ecological systems enhances legitimacy, broadens knowledge diversity, and strengthens the ability to detect and interpret disturbances.

Promoting Polycentric Governance

Polycentricity refers to a governance system in which multiple governing bodies interact to make and enforce rules within a specific policy arena or location. It is regarded as one of the most effective approaches for achieving collective action in response to disturbance and change. Polycentric governance enhances the resilience of ecosystem services by fostering learning and experimentation, enabling broader participation, improving connectivity, creating modularity, enhancing response diversity, and building redundancy to minimize and correct governance errors.[24,25]

A Typology Characterizing Nature-Based Solutions

A typology of NBS can be characterized along two key gradients (Figure 2.1). The first considers the extent to which biodiversity and ecosystems are engineered within an NBS. The second examines the number of ecosystem services and stakeholder groups targeted by a given NBS. According to this typology, as the number of targeted services and stakeholder groups increases, the capacity to maximize the delivery of each service and simultaneously meet the specific needs of individual stakeholders may decrease. Based on these factors, NBS can be classified into three main types:

- *Type 1*: This involves no or minimal intervention in ecosystems, aiming to maintain or enhance the delivery of multiple ecosystem services within and beyond these preserved areas. Examples include the protection of mangroves in coastal regions, which reduces risks from extreme weather events while providing local communities with food sources and other benefits. Another example is the establishment of marine protected areas to conserve biodiversity while allowing the export of biomass into fishing grounds.
- *Type 2*: This category focuses on sustainable and multifunctional ecosystem management, where landscapes are extensively or intensively managed to improve the delivery of selected ecosystem services compared to conventional approaches. Examples include innovative agricultural planning to enhance landscape multifunctionality and increase tree species and genetic diversity in forests, thereby boosting resilience to extreme events.
- *Type 3*: This involves highly intrusive ecosystem management or the creation of entirely new ecosystems to address environmental challenges. Examples include artificial ecosystems such as green roofs and walls, which help mitigate the urban heat island effect and purify polluted air. This category is closely linked to green and blue infrastructure and is often applied to restoring heavily degraded or polluted areas.[26]

Based on this classification, Type 1 aligns closely with the IUCN's definition of NBS. Types 2 and 3 also fit within this definition, provided they contribute to biodiversity conservation and promote the sustainable management or restoration of ecosystems while delivering multiple ecosystem services. Meanwhile, the European Commission often emphasizes Types 2 and

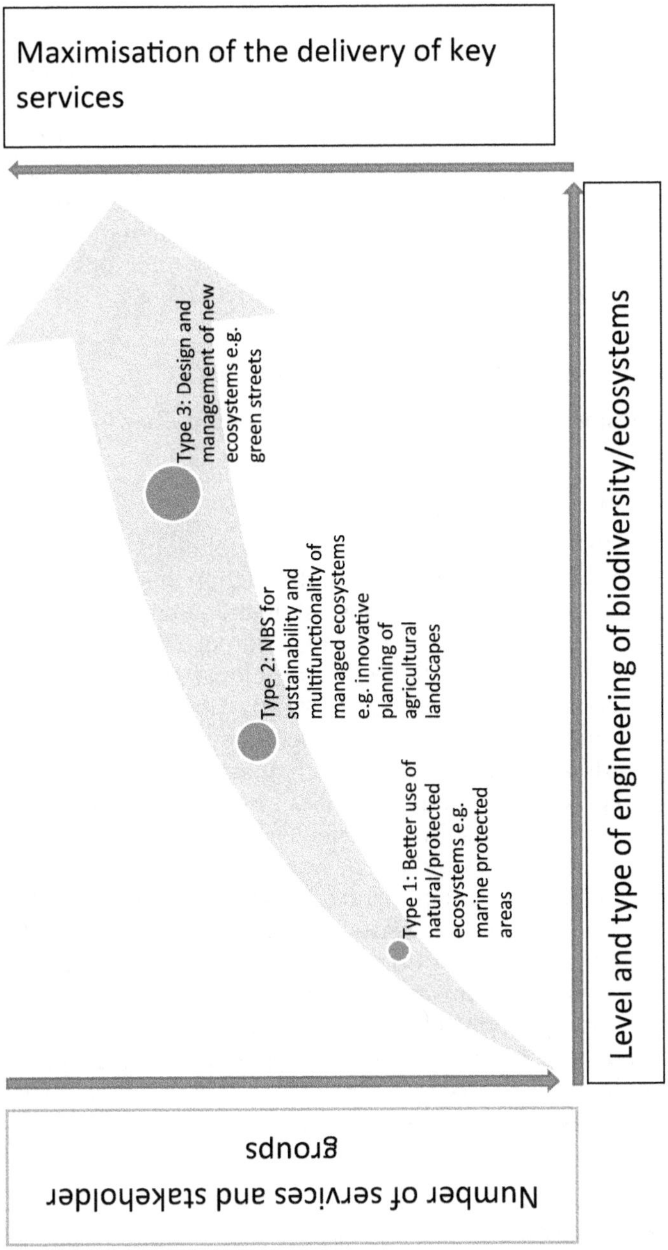

Figure 2.1 Range of Nature-Based Solution Types.

especially Type 3 for their role in transforming natural capital into a driver of green growth and sustainable development.

However, the boundaries between these types are not rigid. Hybrid solutions exist along a continuum in both space and time. For example, a combination of protected and managed areas may be necessary to achieve multifunctionality and sustainability goals. Similarly, a constructed wetland initially developed as Type 3 may, over time, evolve into a Type 1 system if it is well-established, adequately maintained, and closely monitored.

Outlier examples exist where specific ecosystems function locally and globally, engaging a broad range of stakeholders. While the design of artificial ecosystems is expected to increase over time, these systems will likely focus on specific ecosystem services and must address trade-offs between competing ecological functions.[27,28]

Considerations When Applying Nature-Based Solutions

A range of considerations must be taken into account when implementing NBS.

Multifunctionality

It is essential to clearly distinguish between solutions that qualify as NBS and those that do not. True NBS are multifunctional – they deliver multiple benefits and perform various functions within the same physical space, unlike other solutions that may serve only a single purpose.[29]

For example, a green roof that relies on clones of one or very few plant species, regardless of their biogeographical distribution, would not enhance biodiversity or improve the delivery of other ecosystem services. Furthermore, without a coordinated city-scale approach, green roof design would follow a case-by-case approach, limiting its ability to support citywide sustainability goals and enhance biodiversity and effectiveness at a meaningful scale. As a result, such an approach would not fit within the NBS framework.

In contrast, a citywide urban planning strategy that selects a diverse range of plant species for green roofs, based on biogeography and functional traits, would contribute to multiple objectives, such as reducing the urban heat island effect, capturing stormwater, limiting pollution, enhancing human well-being, and increasing biodiversity. This approach aligns with the NBS concept.

NBS expand the ecosystem services framework by actively promoting biodiversity to enhance the resistance and resilience of social-ecological systems to global changes, extreme events, and unexpected disturbances while delivering a broad range of ecosystem services.[30]

Nature-Based Solutions Can Incorporate Existing Ideas

NBS are often described as innovative, but should not be limited to entirely new solutions. While the concept creates new opportunities, it also builds on existing ideas and integrates lessons learned from the past.

NBS should balance multiple interests, particularly environmental, societal, and economic factors, while promoting sustainability. However, achieving win-win outcomes where all goals are simultaneously met is rare. Therefore, documenting and analyzing potential synergies and trade-offs between ecosystem services and stakeholder expectations is crucial in identifying and implementing NBS.

Furthermore, stakeholders and policymakers must acknowledge the complexities and uncertainties of implementing NBS. Assessing the risks associated with a given NBS should be mandatory, with alternative solutions considered and evaluated. This includes evaluating long-term and spatial impacts while accounting for environmental changes. Without careful planning, NBS could risk creating new challenges rather than delivering sustainable solutions.[31]

Key Elements for the Operationalization of Nature-Based Solutions

To mainstream NBS, they must be integrated into the existing policy mix, including biodiversity protection measures, spatial planning, environmental assessments, economic incentives, and practical demonstrations. Key decisions regarding NBS design, costs, location, scale, and management intensity will require input from a diverse range of stakeholders with different perspectives and pre-existing approaches to problem-solving. Additionally, for NBS to be widely adopted, they must be recognized as viable alternatives to conventional solutions that may not be nature-based or sustainable. To maximize the success of sustainable NBS projects, the following key elements should be considered during the design process, as summarized in Figure 2.2.

Dealing with Uncertainty and Complexity

Uncertainty is an inherent challenge in the design of NBS, as they operate within complex social-ecological systems that exhibit non-linear, heterogeneous, and often unpredictable responses to management and natural factors. One effective way to manage uncertainty and complexity is through an adaptive management framework. This approach involves implementing flexible strategies that maximize learning opportunities by treating interventions as experiments and carefully monitoring their outcomes. By continuously assessing responses, adaptive management enables refinement and adjustment to enhance the effectiveness of NBS over time.

Ensuring the Involvement of Multiple Stakeholders

Any societal challenge is likely to affect or be influenced by a diverse range of stakeholders whose involvement in the planning and implementation of NBS offers three key benefits. Substantive benefits arise as stakeholders' perspectives, knowledge, and local conditions help improve planning and decision-making. Instrumental benefits emerge when stakeholder engagement enhances understanding and acceptance of the process, leading to

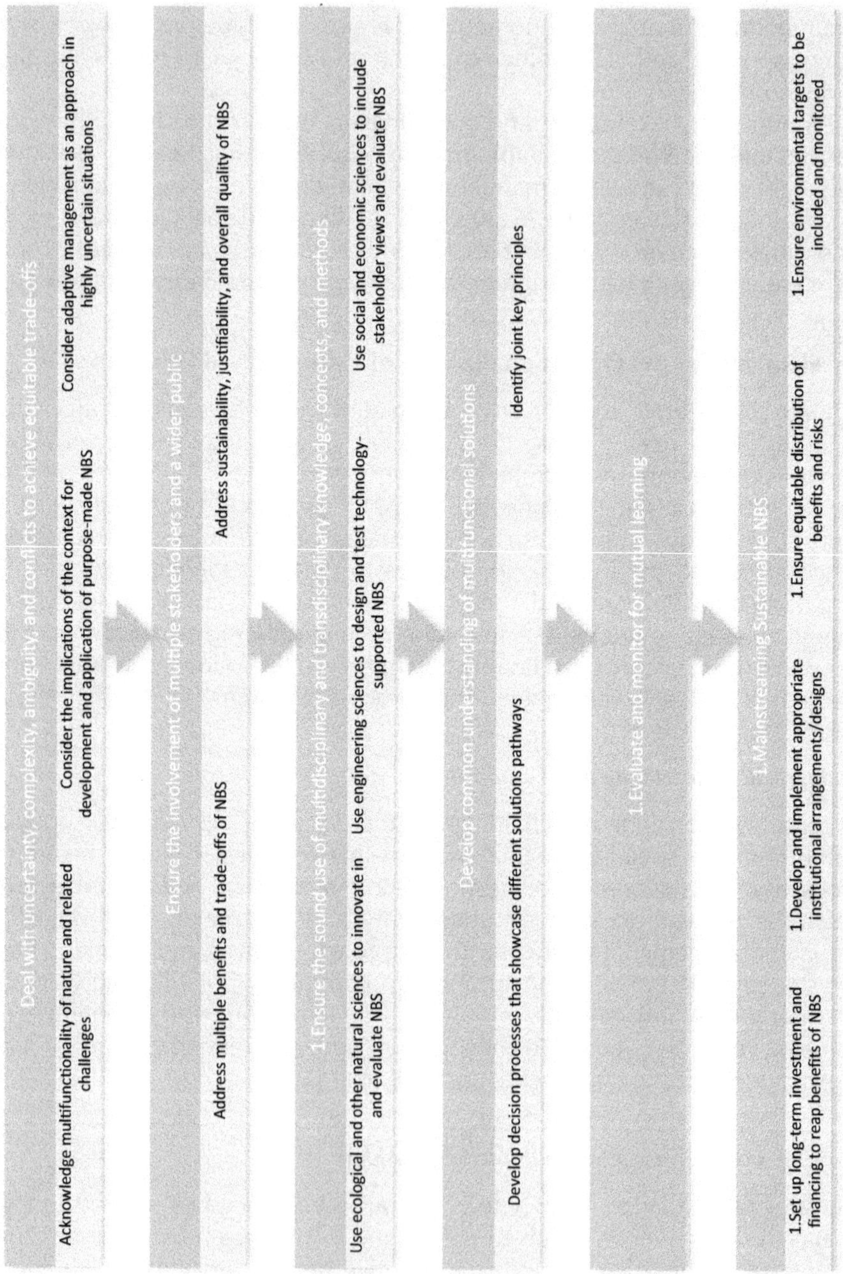

Figure 2.2 Key Elements for the Operationalization of Sustainable Nature-Based Solutions.

more substantial support for NBS. Normative benefits ensure that inclusive participation strengthens the legitimacy of decision-making and reinforces democratic principles.

Ensuring the Sound Use of Multi- and Transdisciplinary Knowledge

Social sciences play a crucial role in shaping the design, selection, implementation, and outcomes of NBS. They help identify the potential and limitations of NBS, informing their development while enhancing our broader understanding of environmental governance. Social science research offers insights into how various choices impact public support and implementation, how to promote stakeholder deliberation and empowerment, and how to identify shared social values and encourage collective action.

Developing a Common Understanding of Multifunctional Solutions, Trade-offs, and Natural Adaptation

The involvement of relevant stakeholders in NBS ensures that goals are set based on a shared understanding of available options. However, maintaining ecosystems and landscape-level ecological functions often involves trade-offs with other land-use priorities. NBS must acknowledge these trade-offs and establish goals that preserve or restore ecological structures and functions using the best available knowledge and a mutual understanding of their impacts. In addition, natural adaptation must be safeguarded by ensuring that NBS are adaptable, process-focused, and prioritize native species and local resources throughout the planning and implementation process.

Evaluating and Monitoring for Mutual Learning

Clear measures of progress and success are crucial for achieving the agreed-upon NBS goals. Indicators should be directly linked to the specific objectives of the solution, including biophysical aspects and ecosystem services. Similarly, indicators in the economic and social spheres should reflect value, capital, investment, revenue, and the impacts on health and well-being.[32]

Mainstreaming Sustainable Nature-Based Solutions

NBS provide opportunities to integrate environmental targets into policy, business, and practice sectors that may not traditionally prioritize or value the environment. To successfully mainstream sustainable NBS, the following actions are recommended:

- Establish long-term investments and financing to ensure that the benefits of NBS are realized equitably across all sectors. In this context, NBS can become a key driver of the green economy.

- Develop institutional arrangements to ensure that projects are clear, widely accepted, and aligned with key sustainability principles.
- Implement evaluation and monitoring frameworks to maximize the potential of NBS in managing environmental risks.[33]

Implementation of Nature-Based Solutions

NBS can be implemented to address specific or multiple societal challenges while simultaneously promoting human well-being and biodiversity conservation. The key implementation approaches include:

- *Restoration*: Ecological restoration, forest landscape restoration, and ecological engineering.
- *Issue-Specific Approaches*: Ecosystem-based adaptation, ecosystem-based mitigation, and ecosystem-based disaster risk reduction.
- *Infrastructure-Based Solutions*: Natural infrastructure, green infrastructure, and blue-green infrastructure.
- *Management Strategies*: Integrated Water Resources Management (IWRM), Integrated Urban Water Management (IUWM), and Integrated Coastal Zone Management (ICZM).
- *Protection Measures*: Area-based conservation, such as marine protected areas.[34,35,36,37]

Co-Benefits of Nature-Based Solutions

Implementing NBS to address societal challenges provides a range of environmental, economic, and social co-benefits, including the following.

Carbon Sequestration

Protected forests are a key NBS for reducing hazards in high-risk areas in mountainous regions. Their primary function is to shield people and infrastructure from landslides, erosion, rockfall, avalanches, and debris flows.[38]

In addition, forests store more carbon than the atmosphere, and compared to other terrestrial ecosystems, they hold some of the highest carbon stocks per unit of land area. Therefore, NBS that restore or enhance forests can significantly improve the carbon storage capacity of landscapes.[39]

Flood Peak Reduction

To reduce flood impacts, commonly implemented green infrastructure solutions include reforesting riverbanks, reconnecting rivers to floodplains, and constructing green spaces that retain and gradually release floodwater. These measures mitigate flooding and support multiple policy objectives, including enhancing wildlife habitats and providing recreational and tourism opportunities.[40]

Urban Heat Island Reduction

Green walls, also known as living walls, help prevent building envelopes from overheating, reducing the energy required for indoor cooling. In addition to lowering energy consumption, living walls can reduce air and noise pollution.[41] Living walls play a key role in mitigating urban heat island effects in dense urban areas by cooling building surfaces. One study found that living walls were up to 20°C cooler than bare walls on sunny days, significantly reducing urban heat hotspots.[42]

Health and Well-Being

Rewilding aims to restore natural processes and ecosystems within landscapes. It involves rehabilitating ecological functions, such as reducing flood risk, enhancing drought resistance, and increasing soil and plant biodiversity. Beyond environmental benefits, rewilding also supports human well-being. For example, research indicates that frequent visits to natural areas are positively associated with improved mental well-being and increased life satisfaction.[43]

Air Quality Improvements

Drylands provide essential resources, including food, medicine, wood for energy, and fodder for livestock. However, they are highly vulnerable to desertification due to climate change, land conversion, and unregulated grazing. Restoring drylands enhances the natural capital, supporting rural livelihoods while reducing sand and dust storms. Airborne dust poses significant health risks, as inhalable particles are linked to respiratory disorders, including asthma, pneumonia, and infections.[44,45]

Ecological Connectivity

The conservation and restoration of forests enhance community resilience to climate change by stabilizing land slopes, regulating water flows, and preventing flash floods and landslides as rainfall levels and intensity increase. In addition, forest restoration enhances ecological connectivity, enabling reconnected habitats and landscapes to support vital ecological processes, such as pollination, productivity, decomposition, and nutrient cycling. Increased connectivity also helps species adapt to future environmental conditions by buffering climate impacts and enhancing ecosystem resilience.[46,47,48]

Stakeholder Engagement

Due to its multifunctionality and cross-cutting nature, NBS can facilitate cooperation across sectors and promote a holistic approach to addressing societal challenges. However, collaboration is often undermined by poor information exchange and a lack of coordination, which can reduce the effectiveness of conservation initiatives. To overcome these barriers, stakeholder mapping and engagement are essential for identifying, attracting, and involving key individuals and groups in NBS initiatives. Effective engagement fosters innovative collaboration, creative problem-solving, and even more profound systemic changes in organizational culture, policies, and procedures. The greater the stakeholder buy-in, the more likely an NBS initiative will succeed.[49,50]

Reducing Vulnerability

Coastal populations face increasing risks from sea-level rise, storm surges, and other natural hazards. Restoring lost mangroves where possible and conserving remaining coastal habitats can significantly reduce coastal vulnerability and enhance protection for people and infrastructure. At the same time, local communities benefit from an increased food supply, as mangroves provide essential breeding grounds for fish and invertebrates. Further offshore, the fishing industry gains from steadily replenishing fish stocks, as mangroves continuously support offshore ecosystems by supplying new fish and invertebrates.[51]

Additional Jobs

NBS generate employment directly through construction, maintenance, and management and indirectly by boosting tourism in areas where NBS are implemented. Globally, tourism contributes over 10% of total GDP and provides jobs for one in every ten people. In Tanzania, nature-based tourism accounts for approximately 10% of GDP, while in Namibia, it supports 19% of total employment (both directly and indirectly). As such, NBS that promote tourism can serve as a vital source of income for local communities and rural households, particularly in marginal areas where economic opportunities are limited, and poverty reduction remains a challenge.[52,53]

Barriers to Nature-Based Solutions

Despite the significant environmental, economic, and social benefits of NBS, their widespread implementation still faces numerous challenges. These include economic, financial, institutional, regulatory, and infrastructural barriers, as well as limited awareness and knowledge. Examples of these challenges are outlined in Table 2.5.[54,55,56,57,58,59,60,61,62,63,64,65,66,67]

Table 2.5 Barriers to the Implementation of Nature-Based Solutions

Barrier Type	*Barrier*	*Description*
Economic	High transaction costs	Coordinating multiple stakeholders, especially across regulatory jurisdictions, is time-consuming and costly.
	High land values	NBS often require more land than grey infrastructure, making them expensive in urban areas.
	Difficulty in quantifying benefits	The benefits of NBS are hard to quantify, which skews cost-benefit analyses in favor of traditional solutions.
	Long-term horizons	NBS take longer to establish and deliver full benefits compared to conventional engineering options.
	Initial high costs	Initial investment in NBS may be high despite their relatively low ongoing maintenance costs.
Financial	Perceived higher risk	The lack of long-term financial data on NBS increases their perceived investment risk.
	Undefined financial responsibilities	Uncertainty over who should pay for NBS implementation, monitoring, and maintenance complicates financing and scaling.
	Lack of financial resources	Limited government funding and inadequate long-term investment reduce the realization of NBS benefits.
Institutional	Lock-in of traditional practices	Engineering approaches dominate planning and education, creating skepticism of NBS.
	Lack of long-term planning	Effective NBS require coordination across small-scale projects, often hindered by fragmented planning.
	Insufficient policy coherence	Policies across jurisdictions often lack integration, reducing the competitiveness of NBS.
	Lack of resources	Planning institutions may lack the capacity to implement and follow through on NBS strategies.
Regulatory	Regulatory standards	Conventional standards prioritize quantity over multifunctionality in green spaces.
	Lack of regulatory clarity	Uncertainty about how NBS comply with environmental regulations slows adoption.
	Cross-jurisdictional regulations	Inconsistent regulations across agencies and governments complicate NBS development.

(*Continued*)

Table 2.5 (Continued)

Barrier Type	*Barrier*	*Description*
Infrastructural	Lack of physical space	Urban density can limit space for NBS; retrofitting is often infeasible.
	Poor maintenance	NBS require proper upkeep to remain effective; many programs lack resources and trained personnel.
	Inadequate sizing	With extreme weather increasing, poorly sized NBS may be overwhelmed.
Awareness and knowledge	Lack of knowledge on NBS benefits	Communities may be unaware of the multiple benefits NBS offer or perceive them as costly and difficult.
	Lack of large-scale implementation data	Most existing data is from pilot projects; cost and benefit information on large-scale NBS is lacking.
	Low private sector engagement	Businesses often lack awareness of commercial benefits or fail to allocate land/resources for NBS.
	Lack of understanding	Limited understanding of multifunctional land use and ecological connectivity hampers effective planning.
	Shortage of trained professionals	Insufficient number of experts with skills to design, implement, and manage NBS.
	Short-term thinking	Long-term benefits of NBS are undervalued due to a culture of short-term planning.

Notes

1 Hilde Eggermont et al., "Nature-Based Solutions: New Influence for Environmental Management and Research in Europe," *GAIA – Ecological Perspectives for Science and Society* 24, no. 4 (2015).

2 R.C. Brears, *The Green Economy and the Water-Energy-Food Nexus* (London: Palgrave Macmillan UK, 2017).

3 IUCN, "Nature-Based Solutions to Address Global Societal Challenges," (2016), https://portals.iucn.org/library/sites/library/files/documents/2016-036.pdf.

4 Eggermont et al., "Nature-Based Solutions: New Influence for Environmental Management and Research in Europe."

5 Carsten Nesshöver et al., "The Science, Policy and Practice of Nature-Based Solutions: An Interdisciplinary Perspective," *Science of the Total Environment* 579 (2017).

6 Nigel Pontee et al., "Nature-Based Solutions: Lessons from around the World," *Proceedings of the Institution of Civil Engineers – Maritime Engineering* 169, no. 1 (2016).

7 R.C. Brears, *Climate Resilient Water Resources Management* (Cham, Switzerland: Palgrave Macmillan, 2018).

8 Conor K.M. Potschin, Roy Haines-Young, Eeva Furman, Pam Berry, and Francesc Baró, "Nature-Based Solutions," (2017), https://oppla.eu/sites/default/files/old_files/uploads/nature-based-solutions-european-and-national-policy-frameworks.pdf.

9 Potschin et al., "Nature-Based Solutions."

10 Saskia Keesstra et al., "The Superior Effect of Nature Based Solutions in Land Management for Enhancing Ecosystem Services," *Science of the Total Environment* 610–611 (2018).

11 Millennium Ecosystem Assessment, "Ecosystems and Human Well-Being: Synthesis," (2005), https://millenniumassessment.org/documents/document.356.aspx.pdf.

12 World Resources Institute, "Ecosystem Services: A Guide for Decision Makers," (2008), http://pdf.wri.org/ecosystem_services_guide_for_decisionmakers.pdf.

13 Keesstra et al., "The Superior Effect of Nature Based Solutions in Land Management for Enhancing Ecosystem Services."

14 European Commission, "Towards an Eu Research and Innovation Policy Agenda for Nature-Based Solutions & Re-Naturing Cities," (2015), https://publications.europa.eu/en/publication-detail/-/publication/fb117980-d5aa-46df-8edc-af367cddc202.

15 Eggermont et al., "Nature-Based Solutions: New Influence for Environmental Management and Research in Europe."

16 European Commission, "Towards an Eu Research and Innovation Policy Agenda for Nature-Based Solutions & Re-Naturing Cities."

17 European Commission, (n.d.), "Nature-Based Solutions," https://research-and-innovation.ec.europa.eu/research-area/environment/nature-based-solutions_en.

18 Eggermont et al., "Nature-Based Solutions: New Influence for Environmental Management and Research in Europe."

19 Pontee et al., "Nature-Based Solutions: Lessons from around the World."

20 Keesstra et al., "The Superior Effect of Nature Based Solutions in Land Management for Enhancing Ecosystem Services."

21 Brears, *The Green Economy and the Water-Energy-Food Nexus*.

22 Eggermont et al., "Nature-Based Solutions: New Influence for Environmental Management and Research in Europe."

23 Division on Earth and Life Studies National Research Council, Ocean Studies Board, Committee on the Effects of the Deepwater Horizon Mississippi Canyon-252 Oil Spill on Ecosystem Services in the Gulf of Mexico, *An Ecosystem Services Approach to Assessing the Impacts of the Deepwater Horizon Oil Spill in the Gulf of Mexico* (Washington, DC: National Academies Press, 2013).

24 GRAID, "Applying Resilience Thinking," (n.d.), https://graid.earth/projects/applying-resilience-thinking/.

25 National Research Council, *An Ecosystem Services Approach to Assessing the Impacts of the Deepwater Horizon Oil Spill in the Gulf of Mexico*.

26 Eggermont et al., "Nature-Based Solutions: New Influence for Environmental Management and Research in Europe."

27 Eggermont et al., "Nature-Based Solutions: New Influence for Environmental Management and Research in Europe."

28 Eggermont et al., "Nature-Based Solutions: New Influence for Environmental Management and Research in Europe."

29 R.C. Brears, *Blue and Green Cities: The Role of Blue-Green Infrastructure in Managing Urban Water Resources* (London: Palgrave Macmillan UK, 2018).
30 Eggermont et al., "Nature-Based Solutions: New Influence for Environmental Management and Research in Europe."
31 Eggermont et al., "Nature-Based Solutions: New Influence for Environmental Management and Research in Europe."
32 Nesshöver et al., "The Science, Policy and Practice of Nature-Based Solutions: An Interdisciplinary Perspective."
33 Nesshöver et al., "The Science, Policy and Practice of Nature-Based Solutions: An Interdisciplinary Perspective."
34 Emmanuelle Cohen-Shacham et al., "Core Principles for Successfully Implementing and Upscaling Nature-Based Solutions," *Environmental Science & Policy* 98 (2019).
35 R.C. Brears, *Urban Water Security* (Chichester and Hoboken, NJ: John Wiley & Sons, 2016).
36 Brears, *Blue and Green Cities: The Role of Blue-Green Infrastructure in Managing Urban Water Resources*.
37 Holly P. Jones, David G. Hole, and Erika S. Zavaleta, "Harnessing Nature to Help People Adapt to Climate Change," *Nature Climate Change* 2 (2012).
38 Cristian Accastello, Simone Blanc, and Filippo Brun, "A Framework for the Integration of Nature-Based Solutions into Environmental Risk Management Strategies," *Sustainability* 11, no. 2 (2019).
39 David L. Achat et al., "Forest Soil Carbon Is Threatened by Intensive Biomass Harvesting," *Scientific Reports* 5 (2015).
40 R.C. Brears, "Urban Climate Change Adaptation in Asia-Pacific: Green Jobs, Green Growth," (2015), https://ourfuturewater.com/wp-content/uploads/2020/05/pp915-urban-climate-change-adaptation.pdf.
41 Vera Enzi et al., "Nature-Based Solutions and Buildings – The Power of Surfaces to Help Cities Adapt to Climate Change and to Deliver Biodiversity," in *Nature-Based Solutions to Climate Change Adaptation in Urban Areas: Linkages between Science, Policy and Practice*, ed. Nadja Kabisch et al. (Cham: Springer International Publishing, 2017).
42 Ugo Mazzali et al., "Experimental Investigation on the Energy Performance of Living Walls in a Temperate Climate," *Building and Environment* 64 (2013).
43 Deborah F. Coldwell and Karl L. Evans, "Visits to Urban Green-Space and the Countryside Associate with Different Components of Mental Well-Being and Are Better Predictors Than Perceived or Actual Local Urbanisation Intensity," *Landscape and Urban Planning* 175 (2018).
44 FAO, "Global Guidelines for the Restoration of Degraded Forests and Landscapes in Drylands: Building Resilience and Benefiting Livelihoods," (2015), http://fao.org/3/a-i5036e.pdf.
45 Michael C. Duniway et al., "Wind Erosion and Dust from Us Drylands: A Review of Causes, Consequences, and Solutions in a Changing World," *Ecosphere* 10, no. 3 (2019).
46 Conservation Corridor, "Restoration as a Tool to Increase Landscape Connectivity," (n.d.), https://conservationcorridor.org/digests/2018/10/restoration-as-a-tool-to-increase-landscape-connectivity/#:~:text=Results%20showed%20that%20implementing%20new,existing%20corridors%20(%2B11%25).

47 UNEP, "Frontiers 2018/19: Emerging Issues of Environmental Concern," (2019), https://unenvironment.org/resources/frontiers-201819-emerging-issues-environmental-concern.

48 Brad H. McRae et al., "Where to Restore Ecological Connectivity? Detecting Barriers and Quantifying Restoration Benefits," *PLOS ONE* 7, no. 12 (2012).

49 Chantal van Ham and Helen Klimmek, "Partnerships for Nature-Based Solutions in Urban Areas – Showcasing Successful Examples," in *Nature-Based Solutions to Climate Change Adaptation in Urban Areas: Linkages between Science, Policy and Practice*, ed. Nadja Kabisch et al. (Cham: Springer International Publishing, 2017).

50 CLEVER Cities, "Clever Cities Guidance on Co-Creating Nature-Based Solutions," (2018), http://clevercities.eu/fileadmin/user_upload/Resources/D1.1_Theme_5_Co-creation_framework_FPM_12.2018.pdf.

51 Muhammad Sajjad et al., "Assessing Hazard Vulnerability, Habitat Conservation, and Restoration for the Enhancement of Mainland China's Coastal Resilience," *Earth's Future* 6, no. 3 (2018).

52 Brears, *Blue and Green Cities: The Role of Blue-Green Infrastructure in Managing Urban Water Resources*.

53 World Bank, "Ramping up Nature-Based Tourism to Protect Biodiversity and Boost Livelihoods," (n.d.), http://worldbank.org/en/news/feature/2017/05/22/ramping-up-nature-based-tourism-to-protect-biodiversity-and-boost-livelihoods.

54 Landscape Institute, "Green Infrastructure: Connected and Multifunctional Landscapes," (2009), https://landscapewpstorage01.blob.core.windows.net/www-landscapeinstitute-org/2016/03/GreenInfrastructurepositionstatement13May09.pdf.

55 U.S. EPA, "Overcoming Barriers to Green Infrastructure," (n.d.), https://epa.gov/green-infrastructure/overcoming-barriers-green-infrastructure.

56 Tony Matthews, Alex Y. Lo, and Jason A. Byrne, "Reconceptualizing Green Infrastructure for Climate Change Adaptation: Barriers to Adoption and Drivers for Uptake by Spatial Planners," *Landscape and Urban Planning* 138 (2015).

57 Kara DiFrancesco et al., "Natural Infrastructure in the Nexus," 1 (2015), https://portals.iucn.org/library/sites/library/files/documents/Nexus-001.pdf.

58 J.J. Opperman, "A Flood of Benefits: Using Green Infrastructure to Reduce Flood Risks," (2014), https://conservationgateway.org/ConservationPractices/Freshwater/HabitatProtectionandRestoration/Documents/A%20Flood%20of%20Benefits%20-%20J.Opperman%20-%20May%202014.pdf.

59 UNEP, "Green Infrastructure: Guide for Water Management. Ecosystem Based Management Approaches for Water-Related Infrastructure Projects," (2014), https://unepdhi.org/green-infrastructure-guide-for-water-management/.

60 Abby Hall, "Green Infrastructure Case Studies: Municipal Policies for Managing Stormwater with Green Infrastructure," (2010), https://scribd.com/document/514718476/Green-Infrastructure-Case-Studies.

61 Emily O'Donnell, "Overcoming Barriers to Blue-Green Infrastructure through Multiple Benefit Evaluation," in *Livable Cities Forum: Changing Climate, Changing Communities, September 13th 2016* (2016). Halifax, Nova Scotia.

62 Centre for Sustainable Community Development, "Green Municipalities: A Guide to Green Infrastructure for Canadian Municipalities," (2001), https://scribd.com/doc/315772484/A-Guide-to-Green-Infrastructure-for-Canadian-Municipalities-En.

63 Huron River Watershed Council, "Barriers Preventing Implementation of Green Infrastructure in Washtenaw County, Michigan," (2014), http://hrwc.org/wp-content/uploads/2013/02/GI_Barriers_Report_FINAL.pdf.
64 Sandra Naumann et al., "Design, Implementation and Cost Elements of Green Infrastructure Projects," (2011), http://ec.europa.eu/environment/enveco/biodiversity/pdf/GI_DICE_FinalReport.pdf.
65 Claudia Copeland, *Green Infrastructure and Issues in Managing Urban Stormwater* (Washington, DC: Congressional Research Service, 2014).
66 C.R. Thorne et al., "Overcoming Uncertainty and Barriers to Adoption of Blue-Green Infrastructure for Urban Flood Risk Management," *Journal of Flood Risk Management* 11, S2 (2015).
67 Nadja Kabisch et al., "Nature-Based Solutions to Climate Change Mitigation and Adaptation in Urban Areas: Perspectives on Indicators, Knowledge Gaps, Barriers, and Opportunities for Action," *Ecology and Society* 21, no. 2 (2016).

References

Accastello, Cristian, Simone Blanc, and Filippo Brun. "A Framework for the Integration of Nature-Based Solutions into Environmental Risk Management Strategies." *Sustainability* 11, no. 2 (2019).

Achat, David L., Mathieu Fortin, Guy Landmann, Bruno Ringeval, and Laurent Augusto. "Forest Soil Carbon Is Threatened by Intensive Biomass Harvesting." *Scientific Reports* 5 (2015/11/04): 15991.

Brears, R.C. *Blue and Green Cities: The Role of Blue-Green Infrastructure in Managing Urban Water Resources.* London: Palgrave Macmillan UK, 2018.

Brears, R.C. *Climate Resilient Water Resources Management.* Cham, Switzerland: Palgrave Macmillan, 2018.

Brears, R.C. *The Green Economy and the Water-Energy-Food Nexus.* London: Palgrave Macmillan UK, 2017.

Brears, R.C. "Urban Climate Change Adaptation in Asia-Pacific: Green Jobs, Green Growth." (2015). https://ourfuturewater.com/wp-content/uploads/2020/05/pp915-urban-climate-change-adaptation.pdf.

Brears, R.C. *Urban Water Security.* Chichester and Hoboken, NJ: John Wiley & Sons, 2016.

Centre for Sustainable Community Development. "Green Municipalities: A Guide to Green Infrastructure for Canadian Municipalities." (2001). https://scribd.com/doc/315772484/A-Guide-to-Green-Infrastructure-for-Canadian-Municipalities-En.

CLEVER Cities. "Clever Cities Guidance on Co-Creating Nature-Based Solutions." (2018). http://clevercities.eu/fileadmin/user_upload/Resources/D1.1_Theme_5_Co-creation_framework_FPM_12.2018.pdf.

Cohen-Shacham, Emmanuelle, Angela Andrade, James Dalton, Nigel Dudley, Mike Jones, Chetan Kumar, Stewart Maginnis, et al. "Core Principles for Successfully Implementing and Upscaling Nature-Based Solutions." *Environmental Science & Policy* 98 (2019/08/01): 20–29.

Coldwell, Deborah F., and Karl L. Evans. "Visits to Urban Green-Space and the Countryside Associate with Different Components of Mental Well-Being and Are Better Predictors Than Perceived or Actual Local Urbanisation Intensity." *Landscape and Urban Planning* 175 (2018/07/01): 114–22.

Conservation Corridor. "Restoration as a Tool to Increase Landscape Connectivity." (n.d.). https://conservationcorridor.org/digests/2018/10/restoration-as-a-tool-to-increase-landscape-connectivity/#:~:text=Results%20showed%20that%20implementing%20new,existing%20corridors%20(%2B11%25).

Copeland, Claudia. *Green Infrastructure and Issues in Managing Urban Stormwater.* Washington, DC: Congressional Research Service, 2014.

DiFrancesco, Kara, Todd Gartner, Suzanne Ozment, and International Water Association. "Natural Infrastructure in the Nexus." 1 (2015). https://portals.iucn.org/library/sites/library/files/documents/Nexus-001.pdf.

Duniway, Michael C., Alix A. Pfennigwerth, Stephen E. Fick, Travis W. Nauman, Jayne Belnap, and Nichole N. Barger. "Wind Erosion and Dust from Us Drylands: A Review of Causes, Consequences, and Solutions in a Changing World." *Ecosphere* 10, no. 3 (2019/03/01): e02650.

Eggermont, Hilde, Estelle Balian, José Manuel N. Azevedo, Victor Beumer, Tomas Brodin, Joachim Claudet, Bruno Fady, et al. "Nature-Based Solutions: New Influence for Environmental Management and Research in Europe." *GAIA – Ecological Perspectives for Science and Society* 24, no. 4 (2015): 243–48.

Enzi, Vera, Blanche Cameron, Péter Dezsényi, Dusty Gedge, Gunter Mann, and Ulrike Pitha. "Nature-Based Solutions and Buildings – The Power of Surfaces to Help Cities Adapt to Climate Change and to Deliver Biodiversity." In *Nature-Based Solutions to Climate Change Adaptation in Urban Areas: Linkages between Science, Policy and Practice,* edited by Nadja Kabisch, Horst Korn, Jutta Stadler, and Aletta Bonn, 159–83. Cham: Springer International Publishing, 2017.

European Commission. "Nature-Based Solutions." (n.d.). https://research-and-innovation.ec.europa.eu/research-area/environment/nature-based-solutions_en.

European Commission. "Towards an Eu Research and Innovation Policy Agenda for Nature-Based Solutions & Re-Naturing Cities." (2015). https://publications.europa.eu/en/publication-detail/-/publication/fb117980-d5aa-46df-8edc-af367cddc202.

FAO. "Global Guidelines for the Restoration of Degraded Forests and Landscapes in Drylands: Building Resilience and Benefiting Livelihoods." (2015). http://fao.org/3/a-i5036e.pdf.

GRAID. "Applying Resilience Thinking." (n.d.). https://graid.earth/projects/applying-resilience-thinking/.

Hall, Abby. "Green Infrastructure Case Studies: Municipal Policies for Managing Stormwater with Green Infrastructure." (2010). https://scribd.com/document/514718476/Green-Infrastructure-Case-Studies.

Huron River Watershed Council. "Barriers Preventing Implementation of Green Infrastructure in Washtenaw County, Michigan." (2014). http://hrwc.org/wp-content/uploads/2013/02/GI_Barriers_Report_FINAL.pdf.

IUCN. "Nature-Based Solutions to Address Global Societal Challenges." (2016). https://portals.iucn.org/library/sites/library/files/documents/2016-036.pdf.

Jones, Holly P., David G. Hole, and Erika S. Zavaleta. "Harnessing Nature to Help People Adapt to Climate Change." *Nature Climate Change* 2 (2012/06/26): 504.

Kabisch, Nadja, Niki Frantzeskaki, Stephan Pauleit, Sandra Naumann, McKenna Davis, Martina Artmann, Dagmar Haase, et al. "Nature-Based Solutions to Climate Change Mitigation and Adaptation in Urban Areas: Perspectives on Indicators, Knowledge Gaps, Barriers, and Opportunities for Action." *Ecology and Society* 21, no. 2 (2016).

Keesstra, Saskia, Joao Nunes, Agata Novara, David Finger, David Avelar, Zahra Kalantari, and Artemi Cerdà. "The Superior Effect of Nature Based Solutions in Land Management for Enhancing Ecosystem Services." *Science of the Total Environment* 610–611 (2018/01/01): 997–1009.

Landscape Institute. "Green Infrastructure: Connected and Multifunctional Landscapes." (2009). https://landscapewpstorage01.blob.core.windows.net/www-landscapeinstitute-org/2016/03/GreenInfrastructurepositionstatement13May09.pdf.

Marion Potschin, Conor Kretsch, Roy Haines-Young, Eeva Furman, Pam Berry, and Francesc Baró. "Nature-Based Solutions." (2017). https://oppla.eu/sites/default/files/old_files/uploads/nature-based-solutions-european-and-national-policy-frameworks.pdf.

Matthews, Tony, Alex Y. Lo, and Jason A. Byrne. "Reconceptualizing Green Infrastructure for Climate Change Adaptation: Barriers to Adoption and Drivers for Uptake by Spatial Planners." *Landscape and Urban Planning* 138 (2015/06): 155–63.

Mazzali, Ugo, Fabio Peron, Piercarlo Romagnoni, Riccardo M. Pulselli, and Simone Bastianoni. "Experimental Investigation on the Energy Performance of Living Walls in a Temperate Climate." *Building and Environment* 64 (2013/06/01): 57–66.

McRae, Brad H., Sonia A. Hall, Paul Beier, and David M. Theobald. "Where to Restore Ecological Connectivity? Detecting Barriers and Quantifying Restoration Benefits." *PLOS ONE* 7, no. 12 (2012): e52604.

Millennium Ecosystem Assessment. "Ecosystems and Human Well-Being: Synthesis." (2005). https://millenniumassessment.org/documents/document.356.aspx.pdf.

National Research Council, Division on Earth and Life Studies, Ocean Studies Board, Committee on the Effects of the Deepwater Horizon Mississippi Canyon-252 Oil Spill on Ecosystem Services in the Gulf of Mexico. *An Ecosystem Services Approach to Assessing the Impacts of the Deepwater Horizon Oil Spill in the Gulf of Mexico.* Washington, DC: National Academies Press, 2013.

Naumann, Sandra, M. Davis, Timo Kaphengst, Mav Pieterse, and Matt Rayment. "Design, Implementation and Cost Elements of Green Infrastructure Projects." (2011). http://ec.europa.eu/environment/enveco/biodiversity/pdf/GI_DICE_FinalReport.pdf.

Nesshöver, Carsten, Timo Assmuth, Katherine N. Irvine, Graciela M. Rusch, Kerry A. Waylen, Ben Delbaere, Dagmar Haase, et al. "The Science, Policy and Practice of Nature-Based Solutions: An Interdisciplinary Perspective." *Science of the Total Environment* 579 (2017/02/01): 1215–27.

O'Donnell, Emily. "Overcoming Barriers to Blue-Green Infrastructure through Multiple Benefit Evaluation." In *Livable Cities Forum: Changing Climate, Changing Communities, September 13th 2016*, 2016. Halifax, Nova Scotia.

Opperman, J.J. "A Flood of Benefits: Using Green Infrastructure to Reduce Flood Risks." (2014). https://conservationgateway.org/ConservationPractices/Freshwater/HabitatProtectionandRestoration/Documents/A%20Flood%20of%20Benefits%20-%20J.Opperman%20-%20May%202014.pdf.

Pontee, Nigel, Siddharth Narayan, Michael W. Beck, and Adam H. Hosking. "Nature-Based Solutions: Lessons from around the World." *Proceedings of the Institution of Civil Engineers – Maritime Engineering* 169, no. 1 (2016): 29–36.

Sajjad, Muhammad, Yangfan Li, Zhenghong Tang, Ling Cao, and Xiaoping Liu. "Assessing Hazard Vulnerability, Habitat Conservation, and Restoration for the Enhancement of Mainland China's Coastal Resilience." *Earth's Future* 6, no. 3 (2018/03/01): 326–38.

Thorne, C.R., E.C. Lawson, C. Ozawa, S.L. Hamlin, and L.A. Smith. "Overcoming Uncertainty and Barriers to Adoption of Blue-Green Infrastructure for Urban Flood Risk Management." *Journal of Flood Risk Management* 11, S2 (2015).

UNEP. "Frontiers 2018/19: Emerging Issues of Environmental Concern." (2019). https://unenvironment.org/resources/frontiers-201819-emerging-issues-environmental-concern.

UNEP. "Green Infrastructure: Guide for Water Management. Ecosystem Based Management Approaches for Water-Related Infrastructure Projects." (2014). https://unepdhi.org/green-infrastructure-guide-for-water-management/.

U.S. EPA. "Overcoming Barriers to Green Infrastructure." (n.d.). https://epa.gov/green-infrastructure/overcoming-barriers-green-infrastructure.

van Ham, Chantal, and Helen Klimmek. "Partnerships for Nature-Based Solutions in Urban Areas – Showcasing Successful Examples." In *Nature-Based Solutions to Climate Change Adaptation in Urban Areas: Linkages between Science, Policy and Practice*, edited by Nadja Kabisch, Horst Korn, Jutta Stadler, and Aletta Bonn, 275–89. Cham: Springer International Publishing, 2017.

World Bank. "Ramping up Nature-Based Tourism to Protect Biodiversity and Boost Livelihoods." (n.d.). http://worldbank.org/en/news/feature/2017/05/22/ramping-up-nature-based-tourism-to-protect-biodiversity-and-boost-livelihoods.

World Resources Institute. "Ecosystem Services: A Guide for Decision Makers." (2008). http://pdf.wri.org/ecosystem_services_guide_for_decisionmakers.pdf.

3 Environmental and Societal Challenges in the 21st Century

Introduction

The 21st century is marked by unprecedented environmental and societal challenges reshaping ecosystems, economies, and communities worldwide. Climate change, rapid urbanization, resource depletion, pollution, and biodiversity loss place immense pressure on natural systems, threatening long-term sustainability and human well-being. These challenges are deeply interconnected, often amplifying one another and leading to complex consequences such as food and water insecurity, increased disaster risks, and economic instability. Addressing these issues requires integrated and adaptive approaches that strike a balance between economic growth, environmental conservation, and social equity. Governments, businesses, and communities must work together to develop strategies that enhance resilience and promote sustainable development. This chapter examines the key global challenges of the 21st century, their far-reaching impacts, and the urgent need for coordinated policy responses to mitigate risks and foster a more sustainable future.

Climate Change

As of February 2, 2025, the global average CO_2 concentration reached 426.35 ppm, a significant increase from 422.21 ppm recorded on the same date in 2024. The annual average CO_2 concentration in 2023 was 419.3 ppm, a new record high. These rising CO_2 levels are a major driver of climate change, contributing to increasing global temperatures and more extreme weather events. The annual rate of atmospheric CO_2 increase over the past 60 years has been approximately 100 times faster than previous natural increases.[1,2] The continued buildup of CO_2 and other greenhouse gases causes the Earth's atmosphere to absorb more thermal radiation, leading to surface and lower-atmosphere warming.[3] This warming is triggering significant changes in the Earth's climate system, including the following.

DOI: 10.4324/9781003517948-3

Rising Temperatures

Global surface temperature in 2024 was 1.29°C (2.32°F) above the 20th-century average of 13.9°C (57.0°F), making it the warmest year on record. In January 2025, the Copernicus Climate Change Service reported that the global average surface air temperature reached 13.23°C (55.81°F), which is 1.75°C (3.15°F) above pre-industrial levels, marking the warmest January on record. From 2022 through 2024, each year had a global temperature departure exceeding 1.0°C (1.8°F) above the 1,880–1,900 average, a period commonly used to represent pre-industrial conditions. As of 2024, it has been the 48th consecutive year that global land and ocean temperatures were above the 20th-century average. In this century, new record highs for global land and ocean temperatures were set in 2005, 2010, 2014, 2015, 2016, and 2024. Between 1880 and 1980, a new temperature record was set on average every 13 years. However, between 1981 and 2024, a new temperature record was set on average every three years. These observations underscore the ongoing and rapid changes in our global climate system.[4,5,6,7,8,9,10,11]

Sea Level Rise

Global sea levels have been rising over the past century, with the rate of increase accelerating in recent decades. Between 1993 and 2024, global sea levels rose by approximately 111 millimeters (about 4.37 inches). If this trajectory continues, sea levels are expected to rise by more than 169 millimeters (approximately 6.65 inches) by 2050.[12] The primary drivers of global sea-level rise are thermal expansion, where ocean water expands as it warms, and the increased melting of land-based ice, including glaciers and ice sheets.[13] In the Arctic, the extent of sea ice has been declining. On September 11, 2024, Arctic sea ice likely reached its annual minimum extent of 4.28 million square kilometers (1.65 million square miles), marking the seventh-lowest extent in the nearly 46-year satellite record.[14] In the Antarctic, sea ice extent has also shown significant changes. On September 19, 2024, Antarctic sea ice reached its annual maximum extent of 17.16 million square kilometers (6.63 million square miles), the second-lowest in the 46-year satellite record, just above the record low set in 2023.[15]

Changing Precipitation Patterns

Climate change is projected to significantly alter precipitation patterns, affecting global rainfall amount, distribution, timing, and frequency. A general pattern in response to global warming is that wet regions are likely to become wetter while dry regions become drier. Specifically, in many mid-latitude and subtropical dry regions, mean precipitation is projected to decrease, while in mid-latitude wet regions, it is likely to increase. In addition, extreme precipitation events across most mid-latitude landmasses and wet tropical regions are expected to become more intense and frequent. These changes are attributed to the intensification of the water cycle due to global warming, resulting in increased atmospheric moisture and altered precipitation patterns.[16]

Floods and Storms

Warmer air holds more moisture, leading to increased precipitation. It is estimated that for every 1°C rise in temperature, the atmosphere's moisture-holding capacity increases by approximately 7%.[17] Studies worldwide have found that while extreme rainfall events are primarily influenced by natural variability, human-caused warming has played a significant role.[18] For instance, research estimates that anthropogenic climate change made the record-breaking rainfall during Hurricane Harvey (2017) up to 15 times more intense and such an event three times more likely to occur.[19] More recent studies have confirmed this trend. The 2022 Pakistan floods, which submerged one-third of the country, were significantly more likely and intense due to the effects of climate change.[20] Similarly, extreme rainfall in Eastern Spain in October 2024 was linked to fossil fuel-driven warming, highlighting the increasing frequency of these events.[21] Climate change is expected to intensify the frequency and severity of annual floods. For example, in Thailand, researchers project that an additional 60 km^2 of river basins could experience flooding with a 100-year return period.[22] Tropical storms, which derive energy from ocean heat, are also expected to intensify. By 2100, the global average intensity of tropical cyclones is expected to increase by 2%–11%. Additionally, the frequency of the most intense cyclones is projected to increase, with up to 20% more precipitation occurring within 100 kilometers of the storm center.[23]

Droughts

By 2100, meteorological drought (variability in average precipitation across different regions), hydrological drought (reduced streamflow, soil moisture, reservoir and lake levels, and groundwater recharge), and agricultural drought (insufficient water supply for crop demands due to low precipitation, timing issues, or reduced water access) are projected to become more prolonged, more frequent, or both in some regions and seasons. These changes will result from declining precipitation levels, increased evaporation, or both.[24,25] Droughts are expected to intensify in southern Europe, the Mediterranean, central Europe, central and southern North America, Central America, northeast Brazil, and southern Africa. One study estimates that if global warming reaches 3°C, the area of Europe affected by drought could expand from 13% to 26% compared to the 1971–2000 reference period.[26]

Heatwaves

Heatwaves, prolonged periods of excessively hot weather, have become more frequent, intense, and longer-lasting due to climate change, significantly impacting human health, livelihoods, and natural systems. Between the 1960s and the 2010s, the average number of heatwaves in major U.S. cities increased from two to six per year, with their average duration extending by about a day.[27] Globally, each additional 0.5°C of warming increases the intensity and frequency of temperature extremes, including heatwaves.[28] Projections indicate that heatwaves will become more frequent, prolonged, and severe by mid-century across various regions, even under scenarios with significant reductions in emissions. For instance, heatwaves in Central and South America are expected to become more frequent, longer-lasting, and more intense by mid-century under both low- and high-emission scenarios.[29]

Wildfires

Recent studies have shown that climate change contributes to longer, more intense wildfire seasons. Warmer temperatures and earlier snowmelt lead to drier soils and vegetation, increasing the likelihood of drought and extending the wildfire season. These hot, dry conditions also make wildfires more intense and longer-lasting. Climate change is altering wildfire risk patterns globally, with regions previously less affected now experiencing more frequent wildfires. For instance, in the western United States, the number of large fires has doubled between 1984 and 2015 due to warmer, drier conditions. Similarly, in the Mediterranean Basin, recent trends indicate that wildfires are becoming more frequent and severe, affecting countries that previously experienced fewer such events.[30,31]

A 1.5°C or 2°C World

Human activities have caused approximately 1.1°C of global warming above pre-industrial levels. At the current rate, human-induced warming is increasing by about 0.2°C per decade. If this trend continues, global temperatures are expected to reach 1.5°C above pre-industrial levels by the early 2030s. Climate-related risks for natural and human systems are higher at 1.5°C of global warming than present levels and increase further at 2°C. For instance, impacts on biodiversity and ecosystems, including species loss and extinction, are projected to be more severe at 2°C warming than at 1.5°C.

Recognizing the urgent threat posed by climate change, 195 countries adopted the Paris Agreement in December 2015, aiming to limit the global temperature increase to well below 2°C above pre-industrial levels and to pursue efforts to limit the increase to 1.5 °C above pre-industrial levels. To achieve the 1.5°C target with no or limited overshoot, net global carbon emissions need to decline by approximately 45% from 2010 levels by 2030 and reach net zero around 2050. This necessitates rapid and far-reaching transitions in energy, land, urban infrastructure, and industrial systems. Such transitions are unprecedented in scale and require deep emissions reductions across all sectors, a wide array of mitigation strategies, and a significant increase in investments toward these solutions.

However, recent analyses suggest that current policies and pledges are insufficient to meet these targets. If existing policies persist, global warming is projected to reach approximately 2.7°C above pre-industrial levels by the end of the century. Even with the full implementation of current Nationally Determined Contributions, the anticipated temperature rise is about 2.6°C. Incorporating binding long-term or net-zero targets could limit warming to around 2.1°C, but this still falls short of the Paris Agreement's goals.[32,33,34]

Population Growth

As of 2024, the global population has reached approximately 8.2 billion, marking an increase of over 600 million people in the past seven years. Projections indicate that the population will rise to 9.7 billion by 2050, with a peak of around 10.3 billion in the mid-2080s, before experiencing a gradual decline to 10.2 billion by 2100.[35] This rapid population growth, coupled with increased urbanization, is placing a significant strain on the ecosystem's services. Expanding urban areas lead to habitat destruction, resource depletion, and environmental pollution, resulting in degraded ecosystem functions.[36]

Rapid Urbanization

As of 2025, approximately 57% of the global population resides in urban areas, a figure projected to rise to 68% by 2050. Urbanization and overall population growth are expected to add approximately 2.5 billion people to urban areas by 2050. Currently, there are 33 cities worldwide with populations exceeding 10 million inhabitants. By 2030, this number is projected to increase to 43 megacities.[37] The impacts of urbanization include urban heat island effects, increased consumption of energy-intensive materials, lower air quality, and higher downstream water pollution. Regarding climate change, cities consume over two-thirds of the world's energy, accounting for more than 70% of global carbon emissions.[38] Between now and 2050, global urban land use is projected to increase from just over 1 million square kilometers to over 2.5 million square kilometers, resulting in the loss of food-producing areas and natural habitats, as well as increased habitat fragmentation.[39]

Rising Demand for Food

According to the Food and Agriculture Organization (FAO), feeding a projected global population of 9.1 billion people by 2050 will require a

70% increase in global food production between 2005/07 and 2050.[40] As a result, global water withdrawals for irrigation are expected to increase from 2.6 billion cubic kilometers in 2005–2007 to 2.9 billion cubic kilometers by 2050.[41] The increase in agricultural production is expected to have a significant impact on water quality due to non-point source pollution, including sediment runoff, nutrient runoff, microbial contamination, and chemical runoff from pesticides, herbicides, and other agrochemicals. This pollution will degrade surface and groundwater quality.[42] Additionally, the intensification of livestock farming could lead to a rise in the number of people affected by organic pollution, increasing from 1.1 billion in 2000 to 2.5 billion by 2050.[43] The rising demand for animal-based products, including meat and fish, drives significant shifts in global diets. In 2022, global fisheries and aquaculture production reached a record high of 223.2 million tons, with aquaculture surpassing capture fisheries for the first time. Since 1961, the annual growth in fish consumption has been twice as high as global population growth. However, one-third of the world's assessed fish stocks are now overexploited, raising concerns about the sustainability of global fisheries.[44,45]

Increased Demand for Water

According to projections, if current water consumption patterns continue, global demand for freshwater is expected to exceed supply by 40% by 2030.[46] The primary drivers of this increasing water demand include population growth, urbanization, and heightened requirements in manufacturing, energy production, and agricultural irrigation. This surge in demand is likely to lead to ecosystem degradation in many areas due to excessive water withdrawals, eutrophication, and pollution.[47,48,49] For instance, global biofuel production has seen a significant increase, with biofuel demand rising by 6% (9,100 million liters) in 2022 compared to 2021.[50] While biofuels offer renewable energy benefits, they also substantially pressure water resources. The cultivation of biofuel crops requires large volumes of water for photosynthesis and processing in biorefineries. Additionally, the intensive use of fertilizers and pesticides in growing biofuel crops, such as corn and soybeans, leads to runoff pollution, contaminating both surface and groundwater sources.[51,52]

Rising Energy Use

Rising income levels and population growth, particularly in urbanized areas, are expected to drive a nearly 50% increase in global energy consumption between 2018 and 2050, according to the U.S. Energy Information Administration (EIA). Most of this growth is expected to occur in non-OECD countries, with Asia accounting for half of the world's electricity consumption by 2025, as demand surges in China, Southeast Asia, and India.[53] The share of renewables in the global power mix is projected to increase from 29% today to 35% by 2025, driven by rapid growth in solar, wind, biomass, and hydropower. By 2040, renewables are expected to account for two-thirds of global power generation, significantly reducing reliance on fossil fuels. The United States alone is forecast to add more solar capacity than any other energy source in 2023, while China is expected to contribute nearly half of the additional renewable generation worldwide. Despite this expansion, carbon emissions from the power sector are projected to plateau by 2025, following a record high of 13.2 gigatons CO_2 in 2022. However, the electrification of heating and transportation, along with the increasing share of weather-dependent renewables in the energy mix, means that energy security and system flexibility will become critical challenges in the coming years.[54] Renewable energy sources, such as wind, solar, biomass, and hydropower, also have environmental impacts. Wind and solar farms need large land areas, which can disrupt ecosystems and wildlife habitats. Wind turbines may cause bird deaths, while concentrated solar power systems often require large amounts of water for cooling. Hydropower reservoirs, though reliable, can flood land and disrupt local ecosystems.[55]

Economic Growth

The global economy is projected to more than double in size by 2050, driven by technological advancements and productivity improvements.[56] This economic expansion is expected to significantly increase global material consumption, with estimates suggesting that material use could double to 190 billion tons by 2060, up from 92 billion tons in recent years.[57] The extraction and processing of these materials contribute substantially to greenhouse gas emissions, accounting for about 50% of total global emissions. By 2060, total emissions are projected to reach 75 gigatons of carbon dioxide equivalent, with material use in agriculture, energy supply, and industry accounting for approximately 50 gigatons. Beyond greenhouse gas emissions, increased material usage poses other environmental challenges, including eutrophication, freshwater aquatic ecotoxicity, and the release of acidifying substances into the atmosphere. These environmental impacts can compromise the ecosystems' ability to provide essential services, such as climate regulation, flood control, natural habitats, and cultural amenities, all of which are vital for sustaining human and social capital.[58]

Biodiversity Loss

Recent assessments indicate that approximately one million animal and plant species are threatened with extinction, with many facing extinction within decades. The average abundance of native species in most major land-based habitats has decreased by at least 20%, primarily since 1900.[59] Over the past 250 years, 571 plant species have been documented as extinct, more than twice that of birds, mammals, and amphibians combined. Data suggests that plant extinction is up to 500 times faster than natural background rates, mainly due to human activities. Insect biodiversity is under significant threat globally, with studies indicating that 40% of insect species may face extinction in the coming decades.[60] Vertebrate pollinators are also at risk; in Europe, 9% of bee and butterfly species are threatened, and populations are declining for 37% of bees and 31% of butterflies.[61,62] In aquatic and marine environments, more than 40% of amphibian species, nearly 33% of reef-forming corals, and over a third of marine mammals are threatened with extinction.[63] Projections estimate that for every degree Celsius of global warming, the world's oceans could lose 5% of marine animal biomass, excluding the additional impacts of overfishing.[64]

Environmental Pollution and Waste

Environmental pollution occurs when the Earth's natural systems, both physical and biological, become contaminated to a level that disrupts normal environmental processes. This occurs when natural resources are consumed at a rate faster than nature can replenish them, resulting in air, water, and land pollution. Additionally, waste disposal exceeds the environment's ability to break down or absorb pollutants, further contributing to environmental degradation.[65]

Air Pollution

Air pollution is a significant global health hazard, contributing to millions of premature deaths annually. The World Health Organization (WHO) estimates that in 2019, ambient (outdoor) air pollution was responsible for approximately 4.2 million premature deaths worldwide. These fatalities are primarily linked to ischemic heart disease, stroke, chronic obstructive pulmonary disease, lung cancer, and acute respiratory infections in children.

When considering ambient and household air pollution, the combined effects are associated with around 6.7 million premature deaths yearly. Notably, 89% of these deaths occur in low- and middle-income countries, with the highest numbers in the WHO South-East Asia and Western Pacific regions.[66]

Particulate matter (PM) is a mixture of solid particles and liquid droplets suspended in the air. Some particles, such as dust, dirt, soot, and smoke, are large enough to be seen with the naked eye, while others are so small that they require an electron microscope to detect. PM pollutants include PM10, which consists of inhalable particles 10 micrometers or smaller, and PM2.5, which are fine inhalable particles 2.5 micrometers or smaller. These particles come in different shapes and sizes and contain various chemicals. Some are directly emitted from construction sites, unpaved roads, fields, smokestacks, and fires. In contrast, others form through chemical reactions involving sulfur dioxide and nitrogen oxides, which are pollutants released by power plants, industries, and vehicles.[67]

Ozone (O_3) is a gas made up of three oxygen atoms and is found in both the upper atmosphere (stratosphere) and at ground level (troposphere). Stratospheric ozone is beneficial as it forms a protective layer that shields life on Earth from the sun's harmful ultraviolet rays. In contrast, tropospheric ozone, also known as ground-level ozone, is a harmful pollutant. It forms when nitrogen oxides (NO_x) and volatile organic compounds, released from vehicles, power plants, chemical plants, and other sources, undergo chemical reactions in the presence of sunlight. Ozone pollution is most common in urban areas on hot, sunny days, but high levels can still occur in colder months. Because ozone is transported by wind, even rural areas can experience elevated ozone levels despite being far from pollution sources.[68]

Acidifying pollutants, such as sulfur dioxide (SO_2) and nitrogen oxides (NO_x), can remain in the atmosphere for several days and travel thousands of kilometers before undergoing chemical reactions that convert them into sulfuric and nitric acids. Once these pollutants settle on land or water, they alter the chemical composition of soil and surface water, a process known as acidification. Acidification has severe environmental consequences, including forest decline and the formation of "dead" lakes where aquatic life cannot survive.[69] In the United States, 96% of the 417 National Parks assessed experience significant air pollution in at least one category, and 88% of parks have acidic pollutants deposited into their soils and waters, threatening sensitive species and ecosystems.[70]

Soil Pollution

Soil pollution occurs when chemicals or harmful substances are present in the soil at concentrations higher than usual, negatively impacting ecosystems and living organisms. Pollution can result from direct contamination, such as the release of industrial waste or oil spills, and indirect contamination, where pollutants enter the soil through water runoff or air deposition. The primary human-related sources of soil pollution include industrial activities, municipal waste, and the use of agricultural chemicals. These pollutants enter the environment accidentally, such as through oil spills or leaching from landfills, or intentionally, through practices like the use of fertilizers, irrigation with untreated wastewater, and the application of sewage sludge on land.[71,72,73,74,75] In Europe, an estimated 2.5 million sites have experienced polluting activities, with industrial production sectors contributing more to local soil pollution (60%) than service sectors (32%). The most common soil contaminants are mineral oils and heavy metals.[76] Contaminated soils can release toxic chemicals into groundwater and surface water, which can be absorbed by plants and animals, contaminate drinking water supplies, and even pollute indoor air in overlying buildings. In dry regions, soil pollution can spread further through wind-borne dust, increasing exposure risks for humans and ecosystems.[77]

Landfill Pollution

Globally, most waste is either dumped or disposed of in landfills, which vary in their treatment based on the type of waste they receive. These include mineral waste landfills, hazardous waste landfills, industrial landfills serving specific industries, and municipal waste landfills, which accept a mix of household waste, construction debris, small-scale industrial waste, and minor amounts of hazardous materials.[78] While landfills help mitigate risks to public health and safety, they pose significant environmental challenges, including leachate contamination of groundwater and the release of greenhouse gases. Leachate forms when liquid and decomposing solid waste mix with rainfall and surface runoff, creating a toxic liquid that can seep into surrounding soil and water sources. Additionally, the improper disposal of untreated industrial and hazardous chemicals in municipal landfills increases the presence of toxic and persistent contaminants, particularly heavy metals and organic compounds, which can cause severe environmental damage. Landfills also generate greenhouse gases, including carbon dioxide, methane, nitrous oxide, perfluorocarbons, sulfur hexafluoride, and hydrofluorocarbons. These gases contribute to global warming and exacerbate climate change, making proper landfill management and waste treatment essential for environmental protection.[79,80,81,82]

Freshwater Pollution

Freshwater pollution occurs when harmful substances, such as chemicals or microorganisms, contaminate streams, rivers, lakes, aquifers, and other water bodies, thereby reducing water quality and potentially making it toxic for humans and the environment. Groundwater pollution is particularly concerning because even if the pollution source is removed, contamination can persist for hundreds of years due to the slow movement and long retention time of groundwater. This makes groundwater cleanup difficult and costly, posing a long-term threat to drinking water supplies and ecosystems.[83]

Freshwater pollution occurs through point source pollution and non-point source pollution. Point source pollution originates from a single, identifiable source, such as a pipe or discharge outlet, making it easier to measure and assess. Common sources include factories, wastewater treatment plants, landfills, and storage tanks containing fuel, solvents, or industrial chemicals. Non-point source pollution originates from multiple, diffuse sources, making it more challenging to trace. One major contributor is fertilizer runoff from gardens and agricultural activities. Nitrate-based fertilizers, widely used due to their high solubility, can easily leach through soil into rivers and streams, leading to eutrophication, where excessive nutrients cause harmful algal blooms and oxygen depletion in waterways. In urban areas, rain and stormwater runoff wash heavy metals, oils, and other pollutants from roads and buildings into water bodies, further degrading water quality.[84,85]

Sediment pollution in waterways is often elevated in areas where livestock graze along riverbanks, steep farmland has been deforested, or riparian vegetation is lacking. Sediment is a physical pollutant that affects water quality in several ways. It increases turbidity, reducing sunlight penetration into the water, which limits or prevents the growth of algae and rooted aquatic plants. In spawning rivers, fine sediments settle on gravel beds, blocking fish from spawning and harming aquatic ecosystems. Excess sedimentation can also reduce river channel depth, decreasing the river's ability to carry water efficiently and increasing flood risks. Sediments can also carry pollutants, such as nutrients, bacteria, and toxic chemicals from agricultural sources, into waterways, further degrading water quality.[86]

For example, recent assessments indicate that Europe's freshwater bodies face significant pollution challenges. As of 2021, only 39.5% of the European Union's surface waters, such as lakes, rivers, and coastal waters, achieved good ecological status, and only 26.8% attained good chemical status, reflecting a decline from 33.5% in 2015. The primary factors preventing these water bodies from meeting the European Commission's minimum targets include contamination from nitrates and agricultural runoff, salt intrusion, and the leaching of hazardous chemicals from industrial sites, mining areas, and waste storage facilities. In contrast, groundwater resources in the EU are in relatively better condition. Approximately 86% of groundwater bodies were reported to be in good chemical status. However, around 23% of groundwater areas in the EU-27 have 'poor' chemical status, with a small number of pollutants, including nitrates and pesticides from agriculture, being key contributors to contamination.[87,88]

Agricultural Pollution

Agricultural activities are a significant source of marine pollution, primarily due to nutrient runoff. When excess nitrogen and phosphorus from fertilizers are applied to crops, rainfall washes these nutrients into rivers and streams, eventually reaching coastal waters and contributing to pollution and damage to the ecosystem. Around 80% of marine pollution originates from land-based sources, with agricultural runoff being a key contributor.[89,90] For example, in Bangladesh, more than 9,000 metric tons of pesticides and over two million metric tons of fertilizers are used annually, with approximately 1,800 tons of pesticide residues entering coastal waters through runoff. In addition to surface runoff, nitrogen and phosphorus can leach into groundwater, posing long-term risks of contamination.[91]

Excess nitrogen and phosphorus in aquatic ecosystems can trigger eutrophication, a process in which excess nutrients fuel rapid algal and plant growth, disrupting the natural ecological balance. This often results in dense algal blooms, which reduce water clarity and degrade water quality. The overgrowth of algae blocks sunlight, limiting the growth of submerged vegetation and altering coastal habitats. When these blooms die, microbial decomposition depletes oxygen levels, creating hypoxic or anoxic "dead zones", areas where oxygen is too low to support most marine life. These conditions can lead to widespread fish kills and significant declines in aquatic biodiversity.[92]

Hypoxic Events

Hypoxic events are common in marine coastal areas near large, nutrient-rich rivers, such as the Mississippi River and the Gulf of Mexico. Some algal blooms associated with these events produce harmful toxins, resulting in declining water quality, the destruction of valuable fisheries, and public health risks.[93] One of the largest hypoxic zones in the Western Hemisphere is the Northern Gulf of Mexico Dead Zone, caused by nutrient pollution from the Mississippi River. The primary contributors are nitrogen and phosphorus runoff, primarily from major farming states in the Mississippi Valley, including Minnesota, Iowa, Illinois, Wisconsin, Missouri, Tennessee, Arkansas, Mississippi, and Louisiana.[94] In 2024, the Gulf of Mexico's dead zone was measured at approximately 17,369 square kilometers (6,705 square miles), about the size of New Jersey. This is the 12th largest dead zone recorded in 38 years of measurements. The five-year average size of the dead zone is now 11,133 square kilometers (4,298 square miles), more than twice the target of 4,921 square kilometers (1,900 square miles) set by the Mississippi River/Gulf of Mexico Hypoxia Task Force for 2035.[95]

Sewage Pollution

Globally, approximately 80% of wastewater is discharged into the environment untreated, with rates exceeding 95% in some of the least-developed countries.[96] Untreated sewage often contains a mix of industrial waste, municipal waste, animal remains, domestic wastewater, and fecal matter. Being primarily organic, sewage undergoes bacterial decomposition, which depletes oxygen levels in water bodies. This oxygen reduction can suffocate aquatic life and release toxic compounds, such as hydrogen sulfide and ammonia.[97] Additionally, nutrient-rich sewage promotes eutrophication, leading to algal blooms and hypoxic conditions, commonly referred to as dead zones. Over the past 50 years, the ocean has lost about 2% of its oxygen, with anoxic zones more than quadrupling in size.[98] The Baltic Sea is home to one of the world's largest dead zones, primarily caused by nutrient pollution from untreated sewage and agricultural runoff.[99]

Solid Waste

As of 2023, the world generates approximately 2.1 billion tons of municipal solid waste annually, with at least 33% of it not being managed in an environmentally safe manner. This figure is projected to rise to 3.8 billion tons by 2050.[100] On average, waste generation per person per day is 0.74 kilograms, but this varies widely, ranging from 0.11 to 4.54 kilograms, depending on factors such as income level and consumption patterns. There is a positive correlation between income levels and waste generation. While accounting for a smaller share of the global population, high-income countries generate about 34% of the world's waste. With rising income levels, daily per capita waste generation in high-income countries is projected to increase by 19% by 2050. In contrast, low- and middle-income countries are expected to see an increase of approximately 40% or more in per capita waste generation during the same period.[101]

Infrastructure Expansion

By 2040, the global population is projected to increase by approximately 25%, reaching around 9.2 billion people. Urbanization is expected to continue rising, with the urban population projected to grow by 46% during this period, resulting in increased demand for infrastructure. Infrastructure investment is forecasted to reach $94 trillion by 2040 to meet these needs, with electricity and roads accounting for more than two-thirds of this requirement.[102] However, large-scale infrastructure projects, such as road networks, hydropower dams, and economic corridors, often have adverse effects on natural habitats and biodiversity. For instance, new roads in forested regions can significantly contribute to deforestation, as initial roads often lead to the development of secondary and tertiary roads, thereby expanding the extent of habitat disruption.[103]

Natural Disasters

Between 2000 and 2019, there were 7,348 major recorded disaster events, resulting in approximately 1.23 million fatalities and affecting 4.2 billion people globally. These disasters led to an estimated $2.97 trillion in economic losses. Climate-related disasters, including floods, storms, and heatwaves, have increased significantly. In the past two decades, there has been a 151% rise in direct economic losses from climate-related disasters. Between 1998 and 2018, disaster-affected countries reported direct economic losses totaling $2.9 trillion, with climate-related disasters accounting for 77% of these losses. During the same timeframe, 1.3 million people lost their lives due to various disasters, and 4.4 billion individuals were injured, displaced, rendered homeless, or required emergency assistance. Notably, earthquakes, including related tsunamis, were responsible for 56% of deaths.[104]

Natural disasters often trigger secondary hazards, which can amplify their destructive impact. For example, storms can lead to flooding, storm surges, landslides, and water contamination. Floods can cause landslides, erosion, deterioration of water quality, and sediment buildup. The aftermath of bushfires includes water pollution, soil erosion, and reduced water catchment yield. Earthquakes can trigger fires, flooding, water contamination, landslides, tsunamis, and soil liquefaction.[105]

Some disasters also cause large-scale environmental contamination. Hurricane Katrina in 2005 was a powerful storm and a major pollution event, as floodwaters in New Orleans spread hazardous chemicals. The disaster also caused the release of eight million gallons of oil, making it the second-largest oil spill in North American history at the time. Similarly, the March 11, 2011, earthquake off Japan's coast triggered a 50-foot tsunami, which severely damaged the Fukushima Daiichi nuclear power plant, leading to the release of radioactive contamination.[106]

Forced Migration

Environmental degradation, including slow-onset events such as desertification, reduced soil fertility, coastal erosion, and sea-level rise, significantly impacts livelihoods and can trigger various forms of migration, both internally and internationally. While sudden-onset natural disasters often result in mass displacement, a larger number of people are expected to migrate due to the gradual deterioration of environmental conditions. In 2020, approximately 30 million people were displaced due to weather-related disasters, surpassing the number displaced by conflicts and violence. Projections for environmental migration by 2050 vary widely, ranging from 25 million to one billion individuals, with 200 million being the most cited estimate.[107,108]

Notes

1 Daily CO2, "Daily CO_2," (n.d.), https://co2.earth/daily-co2.
2 NOAA, "Climate Change: Atmospheric Carbon Dioxide," (n.d.), https://climate.gov/news-features/understanding-climate/climate-change-atmospheric-carbon-dioxide.
3 NASA, "A Blanket around the Earth," (n.d.), https://climate.nasa.gov/causes/.
4 NOAA, "Climate Change: Global Temperature," (n.d.), https://climate.gov/news-features/understanding-climate/climate-change-global-temperature.
5 NASA, "Global Temperature," (n.d.), https://climate.nasa.gov/vital-signs/global-temperature/?intent=121.
6 WMO, "Wmo Confirms 2024 as Warmest Year on Record at About 1.55°C above Pre-Industrial Level," (n.d.), https://wmo.int/news/media-centre/wmo-confirms-2024-warmest-year-record-about-155degc-above-pre-industrial-level.
7 Copernicus, "Copernicus: January 2025 Was the Warmest on Record Globally, Despite an Emerging La Niña," (2025), https://climate.copernicus.eu/copernicus-january-2025-was-warmest-record-globally-despite-emerging-la-nina.
8 NOAA, "Global Climate Summary for July 2024," (n.d.), https://climate.gov/news-features/understanding-climate/global-climate-summary-july-2024.
9 NOAA, "Global Climate Report for April 2024," (2024), https://climate.gov/news-features/understanding-climate/global-climate-report-april-2024.

10 NOAA, "2024 Was the World's Warmest Year on Record," (2024), https://noaa.gov/news/2024-was-worlds-warmest-year-on-record.
11 NOAA, "Earth Had Its Warmest Year on Record While Upper-Ocean Heat Content Experienced a Record High," (n.d.), https://ncei.noaa.gov/news/global-climate-202413.
12 B.D. Hamlington et al., "The Rate of Global Sea Level Rise Doubled during the Past Three Decades," *Communications Earth & Environment* 5, no. 1 (2024).
13 NOAA, "Climate Change: Global Sea Level," (n.d.), https://climate.gov/news-features/understanding-climate/climate-change-global-sea-level.
14 NSIDC, "Arctic Sea Ice Extent Levels Off; 2024 Minimum Set," (2024), https://nsidc.org/sea-ice-today/analyses/arctic-sea-ice-extent-levels-2024-minimum-set.
15 NSIDC, "2024 Antarctic Sea Ice Maximum Extent Finishes at Second Lowest," https://nsidc.org/sea-ice-today/analyses/2024-antarctic-sea-ice-maximum-extent-finishes-second-lowest.
16 IPCC, "Climate Change Widespread, Rapid, and Intensifying – Ipcc," (2021), https://ipcc.ch/2021/08/09/ar6-wg1-20210809-pr/.
17 D.R. Easterling, K.E. Kunkel, J.R. Arnold, T. Knutson, A.N. LeGrande, L.R. Leung, R.S. Vose, D.E. Waliser, and M.F. Wehner,, "Precipitation Change in the United States," (2017), https://digitalcommons.unl.edu/cgi/viewcontent.cgi?article=1577&context=usdeptcommercepub.
18 Union of Concerned Scientists, "Climate Change, Extreme Precipitation and Flooding: The Latest Science," (2018), https://ucsusa.org/global-warming/global-warming-impacts/floods.
19 Geert J. van Oldenborgh et al., "Attribution of Extreme Rainfall from Hurricane Harvey, August 2017," *Environmental Research Letters* 12, no. 12 (2017).
20 Friederike E.L. Otto et al., "Climate Change Increased Extreme Monsoon Rainfall, Flooding Highly Vulnerable Communities in Pakistan," *Environmental Research: Climate* 2, no. 2 (2023).
21 World Weather Attribution, "Extreme Downpours Increasing in Southeastern Spain as Fossil Fuel Emissions Heat the Climate," (n.d.), https://worldweatherattribution.org/extreme-downpours-increasing-in-southern-spain-as-fossil-fuel-emissions-heat-the-climate/#:~:text=In%20summary%2C%20over%20the%20past,increasing%20in%20intensity%20by%2012%25.
22 Sangam Shrestha and Worapong Lohpaisankrit, "Flood Hazard Assessment under Climate Change Scenarios in the Yang River Basin, Thailand," *International Journal of Sustainable Built Environment* 6, no. 2 (2017).
23 Thomas R. Knutson et al., "Tropical Cyclones and Climate Change," *Nature Geoscience* 3 (2010).
24 Union of Concerned Scientists, "Causes of Drought: What's the Climate Connection?," (n.d.), https://ucsusa.org/global-warming/science-and-impacts/impacts/causes-of-drought-climate-change-connection.html.
25 IPCC, "Ar5 Climate Change 2014: Impacts, Adaptation, and Vulnerability," (2014), https://ipcc.ch/report/ar5/wg2/.
26 L. Samaniego et al., "Anthropogenic Warming Exacerbates European Soil Moisture Droughts," *Nature Climate Change* 8, no. 5 (2018).
27 US EPA, "Climate Change Indicators: Heat Waves," (n.d.), https://epa.gov/climate-indicators/climate-change-indicators-heat-waves.
28 WMO, "Climate Change and Heatwaves," (n.d.), https://wmo.int/content/climate-change-and-heatwaves.

29 M.V.S. Ramarao et al., "Projected Changes in Heatwaves over Central and South America Using High-Resolution Regional Climate Simulations," *Scientific Reports* 14, no. 1 (2024).
30 Center for Climate and Energy Solutions, (n.d.), "Wildfires and Climate Change," https://c2es.org/content/wildfires-and-climate-change/.
31 Jordan Richards et al., "Insights into the Drivers and Spatiotemporal Trends of Extreme Mediterranean Wildfires with Statistical Deep Learning," *Artificial Intelligence for the Earth Systems* 2, no. 4 (2023).
32 IPCC, "Climate Change Widespread, Rapid, and Intensifying – Ipcc."
33 UNFCCC, "Key Aspects of the Paris Agreement," (n.d.), https://unfccc.int/most-requested/key-aspects-of-the-paris-agreement.
34 Climate Action Tracker, "Emissions Pathways," (n.d.), https://climateactiontracker.org/global/emissions-pathways/.
35 INED, "2024: The United Nations Publishes New World Population Projections," (2024), https://ined.fr/en/everything_about_population/demographic-facts-sheets/focus-on/2024-les-nations-unies-publient-de-nouvelles-projections-de-population-mondiale/#:~:text=There%20are%208.2%20billion%20human,reaching%2010.2%20billion%20in%202100.
36 Yonghui Cheng et al., "Impact of Urbanization on Ecosystem Service Value from the Perspective of Spatio-Temporal Heterogeneity: A Case Study from the Yellow River Basin," *Land* 12, no. 7 (2023).
37 Population Division UN Department of Economic and Social Affairs, "68% of the World Population Projected to Live in Urban Areas by 2050, Says Un," (2018), https://un.org/development/desa/en/news/population/2018-revision-of-world-urbanization-prospects.html.
38 World Bank, "Urban Development," (2025).
39 UNEP, "Resource Experts Call for New Strategy to Build Better Cities," (n.d.), https://unep.org/news-and-stories/press-release/resource-experts-call-new-strategy-build-better-cities.
40 FAO, "Global Agriculture Towards 2050," (2009), https://fao.org/fileadmin/templates/wsfs/docs/Issues_papers/HLEF2050_Global_Agriculture.pdf.
41 FAO, "Towards a Water and Food Secure Future: Critical Perspectives for Policy-Makers," (2015), http://fao.org/3/a-i4560e.pdf.
42 R.C. Brears, *The Green Economy and the Water-Energy-Food Nexus* (London: Palgrave Macmillan UK, 2017).
43 Yingrong Wen, Gerrit Schoups, and Nick van de Giesen, "Organic Pollution of Rivers: Combined Threats of Urbanization, Livestock Farming and Global Climate Change," *Scientific Reports* 7 (2017).
44 FAO, "The State of World Fisheries and Aquaculture (Sofia)," (2024), https://fao.org/newsroom/detail/fao-report-global-fisheries-and-aquaculture-production-reaches-a-new-record-high/en.
45 Our World In Data, "One-Third of the World's Assessed Fish Stocks Are Overexploited," (n.d.), https://ourworldindata.org/data-insights/one-third-of-the-worlds-assessed-fish-stocks-are-overexploited#:~:text=A%20fish%20is%20defined%20as,share%20has%20increased%20over%20time.
46 UNEP, "Half the World to Face Severe Water Stress by 2030 Unless Water Use Is "Decoupled" from Economic Growth, Says International Resource Panel," (n.d.), https://unep.org/news-and-stories/press-release/half-world-face-severe-water-stress-2030-unless-water-use-decoupled?utm_source=chatgpt.com.

47 R.C. Brears, *Urban Water Security* (Chichester and Hoboken, NJ: John Wiley & Sons, 2016).
48 R.C. Brears, *The Green Economy and the Water-Energy-Food Nexus (Second Edition)* (Cham: Springer International Publishing, 2023).
49 Robert C. Brears, *Water Resources Management: Innovative and Green Solutions (Second Edition)* (Berlin and Boston, MA: De Gruyter, 2024).
50 IEA, "Transport Biofuels," (2022), https://iea.org/reports/renewables-2022/transport-biofuels.
51 Brears, *The Green Economy and the Water-Energy-Food Nexus*.
52 Xiaowei Liu, S. Kent Hoekman, and Amber Broch, "Potential Water Requirements of Increased Ethanol Fuel in the USA," *Energy, Sustainability and Society* 7, no. 1 (2017).
53 EIA, "Eia Projects Nearly 50% Increase in World Energy Usage by 2050, Led by Growth in Asia," (n.d.), https://eia.gov/todayinenergy/detail.php?id=41433.
54 World Economic Forum, "Iea: More Than a Third of the World's Electricity Will Come from Renewables in 2025," (2023), https://weforum.org/stories/2023/03/electricity-generation-renewables-power-iea/.
55 Union of Concerned Scientists, "Environmental Impacts of Renewable Energy Technologies," (n.d.), https://ucsusa.org/clean-energy/renewable-energy/environmental-impacts.
56 PwC, "The World in 2050," (n.d.).
57 UNEP, "We're Gobbling up the Earth's Resources at an Unsustainable Rate," (n.d.), https://unep.org/news-and-stories/story/were-gobbling-earths-resources-unsustainable-rate#:~:text=The%20extraction%20and%20processing%20of,increase%20by%2043%20per%20cent.
58 OECD, "Global Material Resources Outlook to 2060: Economic Drivers and Environmental Consequences," (2018), https://oecd.org/content/dam/oecd/en/publications/reports/2019/02/global-material-resources-outlook-to-2060_g1g98d7d/9789264307452-en.pdf.
59 UN, "Biodiversity – Our Strongest Natural Defense against Climate Change," (2018), https://un.org/en/climatechange/science/climate-issues/biodiversity?utm_source=chatgpt.com.
60 Aelys M. Humphreys et al., "Global Dataset Shows Geography and Life Form Predict Modern Plant Extinction and Rediscovery," *Nature Ecology & Evolution* 3, no. 7 (2019).
61 UNEP, "Insect Declines Are a Stark Warning to Humanity," (n.d.), https://unep.org/news-and-stories/story/insect-declines-are-stark-warning-humanity#:~:text=Across%20the%20world%2C%20more%20than,of%20mammals%2C%20birds%20and%20reptiles.
62 European Parliament, "Protecting Pollinators in the Eu," (2021), https://europarl.europa.eu/RegData/etudes/BRIE/2021/696175/EPRS_BRI(2021)696175_EN.pdf.
63 UNEP, "Nature's Dangerous Decline 'Unprecedented' Species Extinction Rates 'Accelerating'," (n.d.), https://unep.org/news-and-stories/press-release/natures-dangerous-decline-unprecedented-species-extinction-rates.
64 Heike K. Lotze et al., "Global Ensemble Projections Reveal Trophic Amplification of Ocean Biomass Declines with Climate Change," *Proceedings of the National Academy of Sciences* 116, no. 26 (2019).

65 Salah M. El-Haggar, "Chapter 4 – Sustainable Development and Environmental Reform," in *Sustainable Industrial Design and Waste Management*, ed. Salah M. El-Haggar (Oxford: Academic Press, 2007).
66 WHO, "Ambient (Outdoor) Air Pollution," (n.d.), https://who.int/news-room/fact-sheets/detail/ambient-(outdoor)-air-quality-and-health.
67 US EPA, "Particulate Matter (Pm) Basics," (n.d.), https://epa.gov/pm-pollution/particulate-matter-pm-basics.
68 US EPA, "Ground-Level Ozone Basics," (n.d.), https://epa.gov/ground-level-ozone-pollution/ground-level-ozone-basics.
69 EEA, "Acidification," (n.d.), https://eea.europa.eu/publications/92-826-5409-5/page031new.html.
70 National Parks Conservation Association, "Polluted Parks. How America Is Failing to Protect Our National Parks, People and Planet from Air Pollution," (2019), https://npca.s3.amazonaws.com/documents/NPCAParksReport2019.pdf.
71 FAO, "Soil Pollution: A Hidden Reality," (2018), http://fao.org/3/i9183en/i9183en.pdf.
72 Larisa Lvova and Marina Nadporozhskaya, "17 – Chemical Sensors for Soil Analysis: Principles and Applications," in *New Pesticides and Soil Sensors*, ed. Alexandru M. Grumezescu (Cambridge, Massachusetts: Academic Press, 2017).
73 Shanying He et al., "Chapter Four – Soil Biogeochemistry, Plant Physiology, and Phytoremediation of Cadmium-Contaminated Soils," in *Advances in Agronomy*, ed. Donald L. Sparks (Cambridge, Massachusetts: Academic Press, 2015).
74 Umme A. Naher et al., "Chapter 8 – Fertilizer Management Strategies for Sustainable Rice Production," in *Organic Farming*, ed. Sarath Chandran, M.R. Unni, and Sabu Thomas (Cambridge, UK: Woodhead Publishing, 2019).
75 FAO, "Soil Pollution: A Hidden Reality."
76 P. Panagos et al., "Contaminated Sites in Europe: Review of the Current Situation Based on Data Collected through a European Network," *International Journal of Environmental Research and Public Health* 2013 (2013).
77 US EPA, "Contaminated Land," (n.d.), https://epa.gov/report-environment/contaminated-land.
78 P.L. Bjerg et al., "9.16 – the Groundwater Geochemistry of Waste Disposal Facilities," in *Treatise on Geochemistry*, ed. Heinrich D. Holland and Karl K. Turekian (Oxford: Pergamon, 2003).
79 C. Postigo et al., "Groundwater Pollution: Sources, Mechanisms, and Prevention," in *Encyclopedia of the Anthropocene*, ed. Dominick A. Dellasala, and Michael I. Goldstein (Oxford: Elsevier, 2018).
80 B. Seshadri et al., "Chapter 5 – Biomass Energy from Revegetation of Landfill Sites*," in *Bioremediation and Bioeconomy*, ed. M.N.V. Prasad (Amsterdam, Netherlands: Elsevier, 2016).
81 Postigo et al., "Groundwater Pollution: Sources, Mechanisms, and Prevention."
82 Agamuthu Periathamby, "Chapter 8 – Municipal Waste Management," in *Waste*, ed. Trevor M. Letcher and Daniel A. Vallero (Boston, MA: Academic Press, 2011).
83 Brears, *Urban Water Security*.
84 Brears, *Urban Water Security*.
85 Brears, *Urban Water Security*.
86 Brears, *Urban Water Security*.

87 European Commission, "Faster Progress Needed to Protect Europe's Waters and Better Manage Flood Risks," (2021), https://commission.europa.eu/news-and-media/news/faster-progress-needed-protect-europes-waters-and-better-manage-flood-risks-2025-02-04_en.

88 EEA, "Chemicals in European Surface Water and Groundwater (Signal)," (n.d.), https://eea.europa.eu/en/european-zero-pollution-dashboards/indicators/chemicals-in-european-surface-water-and-groundwater-bodies.

89 NOAA, "What Is the Biggest Source of Pollution in the Ocean?," (n.d.), https://oceanservice.noaa.gov/facts/pollution.html#:~:text=When%20large%20tracts%20of%20land,environment%20comes%20from%20the%20land.

90 UNEP, "Agriculture, Nutrients and the Health of Fish," (n.d.), https://unep.org/news-and-stories/story/agriculture-nutrients-and-health-fish.

91 Md Shahidul Islam and Masaru Tanaka, "Impacts of Pollution on Coastal and Marine Ecosystems Including Coastal and Marine Fisheries and Approach for Management: A Review and Synthesis," *Marine Pollution Bulletin* 48, no. 7 (2004).

92 M.F. Chislock, E. Doster, R.A. Zitomer, and A.E. Wilson, "Eutrophication: Causes, Consequences, and Controls in Aquatic Ecosystems," *Nature Education Knowledge* 4, no. 4 (2013).

93 Chislock, Doster, Zitomer, and Wilson, "Eutrophication: Causes, Consequences, and Controls in Aquatic Ecosystems."

94 Carleton College Science Education Resource Center, "The Gulf of Mexico Dead Zone," (n.d.), https://serc.carleton.edu/microbelife/topics/deadzone/index.html.

95 NOAA, "Gulf of Mexico 'Dead Zone' Larger Than Average, Scientists Find," (n.d.), https://noaa.gov/news-release/gulf-of-mexico-dead-zone-larger-than-average-scientists-find?utm_source=chatgpt.com.

96 UN Water, "Water Quality and Wastewater," (2018), https://unwater.org/sites/default/files/app/uploads/2018/10/WaterFacts_water_and_watewater_sep2018.pdf.

97 Shahidul Islam and Tanaka, "Impacts of Pollution on Coastal and Marine Ecosystems Including Coastal and Marine Fisheries and Approach for Management: A Review and Synthesis."

98 Denise Breitburg et al., "Declining Oxygen in the Global Ocean and Coastal Waters," *Science* 359, no. 6371 (2018).

99 Cheryl L. Dybas, "Dead Zones Spreading in World Oceans," *BioScience* 55, no. 7 (2005).

100 UNEP, "Global Waste Management Outlook 2024," (2024), https://unep.org/resources/global-waste-management-outlook-2024?utm_source=chatgpt.com.

101 World Bank, "What a Waste 2.0: A Global Snapshot of Solid Waste Management to 2050," (2028), https://openknowledge.worldbank.org/entities/publication/d3f9d45e-115f-559b-b14f-28552410e90a.

102 World Bank, "Forecasting Infrastructure Investment Needs for 50 Countries, 7 Sectors through 2040," (n.d.), https://blogs.worldbank.org/en/ppps/forecasting-infrastructure-investment-needs-50-countries-7-sectors-through-2040#:~:text=Rural%20to%20urban%20migration%20will,demographic%20changes%20across%20the%20globe.

103 William F. Laurance et al., "Reducing the Global Environmental Impacts of Rapid Infrastructure Expansion," *Current Biology* 25, no. 7 (2015).

104 UNDRR, "The Human Cost of Disasters: An Overview of the Last 20 Years (2000-2019)," (2020), https://undrr.org/publication/human-cost-disasters-overview-last-20-years-2000-2019?utm_source=chatgpt.com.
105 Geoscience Australia, "Natural Hazards in Australia. Identifying Risk Analysis Requirements," (2007), http://ga.gov.au/webtemp/image_cache/GA10873.pdf.
106 John S. Picou, Keith Nicholls, and Rainer Guski, "Environmental Stress and Health," in *International Encyclopedia of the Social & Behavioral Sciences (Second Edition)*, ed. James D. Wright (Oxford: Elsevier, 2015).
107 IOM, "Migration and Climate Change: A Complex Nexus," (n.d.), https://iom.int/complex-nexus#estimates.
108 IFRC, "Displacement in a Changing Climate," (2021), https://ifrc.org/sites/default/files/2021-11/2021-Climate-Displacement-Report-Final.pdf.

References

Bjerg, P.L., H.J. Albrechtsen, P. Kjeldsen, T.H. Christensen, and I.M. Cozzarelli. "9.16 – the Groundwater Geochemistry of Waste Disposal Facilities." In *Treatise on Geochemistry*, edited by Heinrich D. Holland, and Karl K. Turekian, 579–612. Oxford: Pergamon, 2003.

Brears, R.C. *The Green Economy and the Water-Energy-Food Nexus*. London: Palgrave Macmillan UK, 2017.

Brears, R.C. *The Green Economy and the Water-Energy-Food Nexus (Second Edition)*. Cham: Springer International Publishing, 2023.

Brears, R.C. *Urban Water Security*. Chichester and Hoboken, NJ: John Wiley & Sons, 2016.

Brears, Robert C. *Water Resources Management: Innovative and Green Solutions (Second Edition)*. Berlin and Boston, MA: De Gruyter, 2024. https://doi.org/10.1515/9783111028101.

Breitburg, Denise, Lisa A. Levin, Andreas Oschlies, Marilaure Grégoire, Francisco P. Chavez, Daniel J. Conley, Véronique Garçon, et al. "Declining Oxygen in the Global Ocean and Coastal Waters." *Science* 359, no. 6371 (2018): eaam7240.

Center for Climate and Energy Solutions. "Wildfires and Climate Change." (n.d.), https://c2es.org/content/wildfires-and-climate-change/.

Cheng, Yonghui, Qi Kang, Kewei Liu, Peng Cui, Kaixu Zhao, Jianwei Li, Xue Ma, and Qingsong Ni. "Impact of Urbanization on Ecosystem Service Value from the Perspective of Spatio-Temporal Heterogeneity: A Case Study from the Yellow River Basin." *Land* 12, no. 7 (2023). https://doi.org/10.3390/land12071301.

Chislock, M.F., E. Doster, R.A. Zitomer, and A.E. Wilson, "Eutrophication: Causes, Consequences, and Controls in Aquatic Ecosystems." *Nature Education Knowledge* 4, no. 4 (2013): 10.

Climate Action Tracker. "Emissions Pathways." (n.d.). https://climateactiontracker.org/global/emissions-pathways/.

CO2, Daily. "Daily CO_2", (n.d.). https://co2.earth/daily-co2.

Copernicus. "Copernicus: January 2025 Was the Warmest on Record Globally, Despite an Emerging La Niña." (2025). https://climate.copernicus.eu/copernicus-january-2025-was-warmest-record-globally-despite-emerging-la-nina.

Dybas, Cheryl L. "Dead Zones Spreading in World Oceans." *BioScience* 55, no. 7 (2005): 552–57.

Easterling, D.R., K.E. Kunkel, J.R. Arnold, T. Knutson, A.N. LeGrande, L.R. Leung, R.S. Vose, D.E. Waliser, and M.F. Wehner. "Precipitation Change in the United States." (2017). https://digitalcommons.unl.edu/cgi/viewcontent.cgi?article=1577&context=usdeptcommercepub.

EEA. "Acidification." (n.d.). https://eea.europa.eu/publications/92-826-5409-5/page031new.html.

EEA. "Chemicals in European Surface Water and Groundwater (Signal)." (n.d.). https://eea.europa.eu/en/european-zero-pollution-dashboards/indicators/chemicals-in-european-surface-water-and-groundwater-bodies.

EIA. "Eia Projects Nearly 50% Increase in World Energy Usage by 2050, Led by Growth in Asia." (n.d.). https://eia.gov/todayinenergy/detail.php?id=41433.

El-Haggar, Salah M. "Chapter 4 – Sustainable Development and Environmental Reform." In *Sustainable Industrial Design and Waste Management*, edited by Salah M. El-Haggar, 125–48. Oxford: Academic Press, 2007.

European Commission. "Faster Progress Needed to Protect Europe's Waters and Better Manage Flood Risks." (2021) https://commission.europa.eu/news-and-media/news/faster-progress-needed-protect-europes-waters-and-better-manage-flood-risks-2025-02-04_en.

European Parliament. "Protecting Pollinators in the Eu." (2021). https://europarl.europa.eu/RegData/etudes/BRIE/2021/696175/EPRS_BRI(2021)696175_EN.pdf.

FAO. "Global Agriculture towards 2050." (2009). https://fao.org/fileadmin/templates/wsfs/docs/Issues_papers/HLEF2050_Global_Agriculture.pdf.

FAO. "Soil Pollution: A Hidden Reality." (2018). http://fao.org/3/i9183en/i9183en.pdf.

FAO. "The State of World Fisheries and Aquaculture (Sofia)." (2024). https://fao.org/newsroom/detail/fao-report-global-fisheries-and-aquaculture-production-reaches-a-new-record-high/en.

FAO. "Towards a Water and Food Secure Future: Critical Perspectives for Policy-Makers." (2015). http://fao.org/3/a-i4560e.pdf.

Geoscience Australia. "Natural Hazards in Australia. Identifying Risk Analysis Requirements." (2007). http://ga.gov.au/webtemp/image_cache/GA10873.pdf.

Hamlington, B.D., A. Bellas-Manley, J.K. Willis, S. Fournier, N. Vinogradova, R.S. Nerem, C.G. Piecuch, P.R. Thompson, and R. Kopp. "The Rate of Global Sea Level Rise Doubled during the Past Three Decades." *Communications Earth & Environment* 5, no. 1 (2024/10/17): 601.

He, Shanying, Zhenli He, Xiaoe Yang, Peter J. Stoffella, and Virupax C. Baligar. "Chapter Four – Soil Biogeochemistry, Plant Physiology, and Phytoremediation of Cadmium-Contaminated Soils." In *Advances in Agronomy*, edited by Donald L. Sparks, 135–225. Cambridge, Massachusetts: Academic Press, 2015.

Humphreys, Aelys M., Rafaël Govaerts, Sarah Z. Ficinski, Eimear N. Lughadha, and Maria S. Vorontsova. "Global Dataset Shows Geography and Life Form Predict Modern Plant Extinction and Rediscovery." *Nature Ecology & Evolution* 3, no. 7 (2019/07/01): 1043–47.

IEA. "Transport Biofuels." (2022). https://iea.org/reports/renewables-2022/transport-biofuels.

IFRC. "Displacement in a Changing Climate." (2021). https://ifrc.org/sites/default/files/2021-11/2021-Climate-Displacement-Report-Final.pdf.

INED. "2024: The United Nations Publishes New World Population Projections." (2024). https://ined.fr/en/everything_about_population/demographic-facts-sheets/focus-on/2024-les-nations-unies-publient-de-nouvelles-projections-de-population-mondiale/#:~:text=There%20are%208.2%20billion%20human,reaching%2010.2%20billion%20in%202100.

IOM. "Migration and Climate Change: A Complex Nexus." (n.d.). https://iom.int/complex-nexus#estimates.

IPCC. "Ar5 Climate Change 2014: Impacts, Adaptation, and Vulnerability." (2014). https://ipcc.ch/report/ar5/wg2/.

IPCC. "Climate Change Widespread, Rapid, and Intensifying – Ipcc." (2021). https://ipcc.ch/2021/08/09/ar6-wg1-20210809-pr/.

Knutson, Thomas R., John L. McBride, Johnny Chan, Kerry Emanuel, Greg Holland, Chris Landsea, Isaac Held, et al. "Tropical Cyclones and Climate Change." *Nature Geoscience* 3 (2010/02/21): 157.

Laurance, William F., Anna Peletier-Jellema, Bart Geenen, Harko Koster, Pita Verweij, Pitou Van Dijck, Thomas E. Lovejoy, Judith Schleicher, and Marijke Van Kuijk. "Reducing the Global Environmental Impacts of Rapid Infrastructure Expansion." *Current Biology* 25, no. 7 (2015/03/30): R259–R62.

Liu, Xiaowei, S.K. Hoekman, and Amber Broch. "Potential Water Requirements of Increased Ethanol Fuel in the USA." *Energy, Sustainability and Society* 7, no. 1 (2017/06/21): 18.

Lotze, Heike K., Derek P. Tittensor, Andrea Bryndum-Buchholz, Tyler D. Eddy, William W.L. Cheung, Eric D. Galbraith, Manuel Barange, et al. "Global Ensemble Projections Reveal Trophic Amplification of Ocean Biomass Declines with Climate Change." *Proceedings of the National Academy of Sciences* 116, no. 26 (2019/06/25): 12907–12.

Lvova, Larisa, and Marina Nadporozhskaya. "17 – Chemical Sensors for Soil Analysis: Principles and Applications." In *New Pesticides and Soil Sensors*, edited by Alexandru M. Grumezescu, 637–78. Cambridge, Massachusetts: Academic Press, 2017.

Naher, Umme A., M.N. Ahmed, M. Imran, U. Sarkar, Jatish C. Biswas, and Qurban A. Panhwar. "Chapter 8 – Fertilizer Management Strategies for Sustainable Rice Production." In *Organic Farming*, edited by Sarath Chandran, M.R. Unni, and Sabu Thomas, 251–67. Cambridge, UK: Woodhead Publishing, 2019.

NASA. "A Blanket around the Earth." (n.d.). https://climate.nasa.gov/causes/.

NASA. "Global Temperature." (n.d.). https://climate.nasa.gov/vital-signs/global-temperature/?intent=121.

National Parks Conservation Association. "Polluted Parks. How America Is Failing to Protect Our National Parks, People and Planet from Air Pollution." (2019). https://npca.s3.amazonaws.com/documents/NPCAParksReport2019.pdf.

NOAA. "2024 Was the World's Warmest Year on Record." (2024). https://noaa.gov/news/2024-was-worlds-warmest-year-on-record.

NOAA. "Climate Change: Atmospheric Carbon Dioxide." (n.d.). https://climate.gov/news-features/understanding-climate/climate-change-atmospheric-carbon-dioxide.

NOAA. "Climate Change: Global Sea Level." (n.d.). https://climate.gov/news-features/understanding-climate/climate-change-global-sea-level.

NOAA. "Climate Change: Global Temperature." (n.d.). https://climate.gov/news-features/understanding-climate/climate-change-global-temperature.

NOAA. "Earth Had Its Warmest Year on Record While Upper-Ocean Heat Content Experienced a Record High." (n.d.). https://ncei.noaa.gov/news/global-climate-202413.

NOAA. "Global Climate Report for April 2024." (2024). https://climate.gov/news-features/understanding-climate/global-climate-report-april-2024.

NOAA. "Global Climate Summary for July 2024." (n.d.). https://climate.gov/news-features/understanding-climate/global-climate-summary-july-2024.

NOAA. "Gulf of Mexico 'Dead Zone' Larger Than Average, Scientists Find." (n.d.). https://noaa.gov/news-release/gulf-of-mexico-dead-zone-larger-than-average-scientists-find?utm_source=chatgpt.com.

NOAA. "What Is the Biggest Source of Pollution in the Ocean?" (n.d.). https://oceanservice.noaa.gov/facts/pollution.html#:~:text=When%20large%20tracts%20of%20land,environment%20comes%20from%20the%20land.

NSIDC. "2024 Antarctic Sea Ice Maximum Extent Finishes at Second Lowest." (2024). https://nsidc.org/sea-ice-today/analyses/2024-antarctic-sea-ice-maximum-extent-finishes-second-lowest.

NSIDC. "Arctic Sea Ice Extent Levels Off; 2024 Minimum Set." (2024). https://nsidc.org/sea-ice-today/analyses/arctic-sea-ice-extent-levels-2024-minimum-set.

OECD. "Global Material Resources Outlook to 2060: Economic Drivers and Environmental Consequences." (2018). https://oecd.org/content/dam/oecd/en/publications/reports/2019/02/global-material-resources-outlook-to-2060_g1g98d7d/9789264307452-en.pdf.

Otto, Friederike E.L., Mariam Zachariah, Fahad Saeed, Ayesha Siddiqi, Shahzad Kamil, Haris Mushtaq, T. Arulalan, et al. "Climate Change Increased Extreme Monsoon Rainfall, Flooding Highly Vulnerable Communities in Pakistan." *Environmental Research: Climate* 2, no. 2 (2023/03/17): 025001.

Our World in Data. "One-Third of the World's Assessed Fish Stocks Are Overexploited." (n.d.). https://ourworldindata.org/data-insights/one-third-of-the-worlds-assessed-fish-stocks-are-overexploited#:~:text=A%20fish%20is%20defined%20as,share%20has%20increased%20over%20time.

Panagos, P., M. Van Liedekerke, Y. Yigini, and L. Montanarella. "Contaminated Sites in Europe: Review of the Current Situation Based on Data Collected through a European Network." [In eng]. *Journal of Environmental and Public Health* 2013 (2013): 158764.

Periathamby, Agamuthu. "Chapter 8 – Municipal Waste Management." In *Waste*, edited by Trevor M. Letcher, and Daniel A. Vallero, 109–25. Boston, MA: Academic Press, 2011.

Picou, John S., Keith Nicholls, and Rainer Guski. "Environmental Stress and Health." In *International Encyclopedia of the Social & Behavioral Sciences (Second Edition)*, edited by James D. Wright, 804–08. Oxford: Elsevier, 2015.

Postigo, C., D.E. Martinez, S. Grondona, and K.S.B. Miglioranza. "Groundwater Pollution: Sources, Mechanisms, and Prevention." In *Encyclopedia of the Anthropocene*, edited by Dominick A. Dellasala, and Michael I. Goldstein, 87–96. Oxford: Elsevier, 2018.

PwC. "The World in 2050," (n.d.). www.pwc.com/gx/en/research-insights/economy/the-world-in-2050.html

Ramarao, M.V.S., Saravanan Arunachalam, Brisa N. Sánchez, Leah H. Schinasi, Maryia Bakhtsiyarava, Waleska T. Caiaffa, Iryna Dronova, et al. "Projected Changes in Heatwaves over Central and South America Using High-Resolution Regional Climate Simulations." *Scientific Reports* 14, no. 1 (2024/10/04): 23145.

Richards, Jordan, Raphaël Huser, Emanuele Bevacqua, and Jakob Zscheischler. "Insights into the Drivers and Spatiotemporal Trends of Extreme Mediterranean Wildfires with Statistical Deep Learning." [In English]. *Artificial Intelligence for the Earth Systems* 2, no. 4 (2023/10/01): e220095.

Samaniego, L., S. Thober, R. Kumar, N. Wanders, O. Rakovec, M. Pan, M. Zink, et al. "Anthropogenic Warming Exacerbates European Soil Moisture Droughts." *Nature Climate Change* 8, no. 5 (2018/05/01): 421–26.

Science Education Resource Center, Carleton College. "The Gulf of Mexico Dead Zone." (n.d.). https://serc.carleton.edu/microbelife/topics/deadzone/index.html.

Seshadri, B., N.S. Bolan, R. Thangarajan, U. Jena, K.C. Das, H. Wang, and R. Naidu. "Chapter 5 – Biomass Energy from Revegetation of Landfill Sites*." In *Bioremediation and Bioeconomy*, edited by M.N.V. Prasad, 99–109. Amsterdam, Netherlands: Elsevier, 2016.

Shahidul Islam, Md, and Masaru Tanaka. "Impacts of Pollution on Coastal and Marine Ecosystems Including Coastal and Marine Fisheries and Approach for Management: A Review and Synthesis." *Marine Pollution Bulletin* 48, no. 7 (2004/04/01): 624–49.

Shrestha, Sangam, and Worapong Lohpaisankrit. "Flood Hazard Assessment under Climate Change Scenarios in the Yang River Basin, Thailand." *International Journal of Sustainable Built Environment* 6, no. 2 (2017/12/01): 285–98.

UN. "Biodiversity – Our Strongest Natural Defense against Climate Change." (2018). https://un.org/en/climatechange/science/climate-issues/biodiversity?utm_source=chatgpt.com.

UN Department of Economic and Social Affairs, Population Division. "68% of the World Population Projected to Live in Urban Areas by 2050, Says Un." (2018). https://un.org/development/desa/en/news/population/2018-revision-of-world-urbanization-prospects.html.

UN Water. "Water Quality and Wastewater." (2018). https://unwater.org/sites/default/files/app/uploads/2018/10/WaterFacts_water_and_watewater_sep2018.pdf.

UNDRR. "The Human Cost of Disasters: An Overview of the Last 20 Years (2000-2019)." (2020). https://undrr.org/publication/human-cost-disasters-overview-last-20-years-2000-2019?utm_source=chatgpt.com.

UNEP. "Agriculture, Nutrients and the Health of Fish." (n.d.). https://unep.org/news-and-stories/story/agriculture-nutrients-and-health-fish.

UNEP. "Global Waste Management Outlook 2024." (2024). https://unep.org/resources/global-waste-management-outlook-2024?utm_source=chatgpt.com.

UNEP. "Half the World to Face Severe Water Stress by 2030 Unless Water Use Is "Decoupled" from Economic Growth, Says International Resource Panel." (n.d.). https://unep.org/news-and-stories/press-release/half-world-face-severe-water-stress-2030-unless-water-use-decoupled?utm_source=chatgpt.com.

UNEP."Insect Declines Are a Stark Warning to Humanity." (n.d.). https://unep.org/news-and-stories/story/insect-declines-are-stark-warning-humanity#:~:text=Across%20the%20world%2C%20more%20than,of%20mammals%2C%20birds%20and%20reptiles.

UNEP. "Nature's Dangerous Decline 'Unprecedented' Species Extinction Rates 'Accelerating'." (n.d.). https://unep.org/news-and-stories/press-release/natures-dangerous-decline-unprecedented-species-extinction-rates.

UNEP. "Resource Experts Call for New Strategy to Build Better Cities." (n.d.). https://unep.org/news-and-stories/press-release/resource-experts-call-new-strategy-build-better-cities.

UNEP. "We're Gobbling up the Earth's Resources at an Unsustainable Rate." (n.d.). https://unep.org/news-and-stories/story/were-gobbling-earths-resources-unsustainable-rate#:~:text=The%20extraction%20and%20processing%20of,increase%20by%2043%20per%20cent.

UNFCCC. "Key Aspects of the Paris Agreement." (n.d.). https://unfccc.int/most-requested/key-aspects-of-the-paris-agreement.

Union of Concerned Scientists. "Causes of Drought: What's the Climate Connection?" (n.d.). https://ucsusa.org/global-warming/science-and-impacts/impacts/causes-of-drought-climate-change-connection.html.

Union of Concerned Scientists. "Climate Change, Extreme Precipitation and Flooding: The Latest Science." (2018). https://ucsusa.org/global-warming/global-warming-impacts/floods.

Union of Concerned Scientists. "Environmental Impacts of Renewable Energy Technologies." (n.d.), https://ucsusa.org/clean-energy/renewable-energy/environmental-impacts.

US EPA. "Climate Change Indicators: Heat Waves." (n.d.). https://epa.gov/climate-indicators/climate-change-indicators-heat-waves.

US EPA. "Contaminated Land." (n.d.). https://epa.gov/report-environment/contaminated-land.

US EPA. "Ground-Level Ozone Basics." (n.d.). https://epa.gov/ground-level-ozone-pollution/ground-level-ozone-basics.

US EPA. "Particulate Matter (Pm) Basics." (n.d.). https://epa.gov/pm-pollution/particulate-matter-pm-basics.

van Oldenborgh, Geert Jan, Karin van der Wiel, Antonia Sebastian, Roop Singh, Julie Arrighi, Friederike Otto, Karsten Haustein, et al. "Attribution of Extreme Rainfall from Hurricane Harvey, August 2017." *Environmental Research Letters* 12, no. 12 (2017/12/01): 124009.

Wen, Yingrong, Gerrit Schoups, and Nick van de Giesen. "Organic Pollution of Rivers: Combined Threats of Urbanization, Livestock Farming and Global Climate Change." *Scientific Reports* 7 (2017/02/23): 43289.

WHO. "Ambient (Outdoor) Air Pollution." (n.d.). https://who.int/news-room/fact-sheets/detail/ambient-(outdoor)-air-quality-and-health.

WMO. "Climate Change and Heatwaves." (n.d.). https://wmo.int/content/climate-change-and-heatwaves.

WMO. "WMO Confirms 2024 as Warmest Year on Record at About 1.55°C above Pre-Industrial Level." (n.d.). https://wmo.int/news/media-centre/wmo-confirms-2024-warmest-year-record-about-155degc-above-pre-industrial-level.

World Bank. "Forecasting Infrastructure Investment Needs for 50 Countries, 7 Sectors through 2040." (n.d.). https://blogs.worldbank.org/en/ppps/forecasting-infrastructure-investment-needs-50-countries-7-sectors-through-2040#:~:text=Rural%20to%20urban%20migration%20will,demographic%20changes%20across%20the%20globe.

World Bank. "Urban Development." (2025).

World Bank. "What a Waste 2.0: A Global Snapshot of Solid Waste Management to 2050." (2028). https://openknowledge.worldbank.org/entities/publication/d3f9d45e-115f-559b-b14f-28552410e90a.

World Economic Forum. "Iea: More Than a Third of the World's Electricity Will Come from Renewables in 2025." (2023). https://weforum.org/stories/2023/03/electricity-generation-renewables-power-iea/.

World Weather Attribution. "Extreme Downpours Increasing in Southeastern Spain as Fossil Fuel Emissions Heat the Climate." (n.d.). https://worldweatherattribution.org/extreme-downpours-increasing-in-southern-spain-as-fossil-fuel-emissions-heat-the-climate/#:~:text=In%20summary%2C%20over%20the%20past,increasing%20in%20intensity%20by%2012%25.

4 Climate Change Impacts on Habitats, Plants, and Animals

Introduction

Climate change will affect habitats in multiple ways. Rising temperatures will extend growing seasons, boosting biological production. Increased and shifting precipitation patterns will raise nutrient levels in water bodies, leading to excessive plant growth and oxygen depletion, which can harm aquatic life. Rising sea levels will accelerate coastal erosion and flooding, particularly in low-lying coastal areas, tidal zones, and river valleys. As a result, many habitats will become more fragile, and some species will face a higher risk of extinction if they cannot migrate or adapt quickly enough. Over time, ecosystems may lose resilience, making them more vulnerable to future climate shifts and increasing the likelihood of irreversible environmental changes.[1,2] Also, climate change will directly and indirectly impact numerous plant and animal species, with some facing a heightened risk of extinction if global temperatures continue to rise unchecked.[3]

Climate Change Impacts on Habitat Types

Climate change will impact habitat types in various ways, including the following.

Freshwater

Climate change will impact freshwater quantity and quality as follows.

Water Quantity

Changes in winter precipitation patterns will have direct effects on streamflow. As average temperatures rise, more precipitation will fall as rain instead of snow, and the snow that does accumulate will melt earlier in the season. This early snowmelt will lower river flows in late summer, stressing water supplies for both human and natural users downstream.[4] Higher evaporation rates from lakes and reservoirs, combined with lower aquifer recharge rates,

DOI: 10.4324/9781003517948-4

will further reduce available water resources. Summer precipitation levels are projected to decline, while rising temperatures will lead to more frequent and intense summer droughts, further reducing river flows. In Europe, river flow droughts are expected to become more frequent and severe in southern and southeastern Europe, the United Kingdom, France, the Benelux region, and western Germany over the coming decades.[5] Similarly, studies in southern Australia predict climate change will reduce average river flows by 10%–20%.[6] Droughts will also impact groundwater supplies, as prolonged dry periods will increase reliance on aquifers.[7] In some cases, long-term depletion may become irreversible. For instance, California's San Joaquin aquifer has permanently shrunk by approximately 5.25% following severe droughts between 2007 and 2009 and 2012 and 2015.[8]

Water Quality

Climate and hydrology changes will have direct and cascading effects on the quality of surface water and groundwater. More frequent high-flow events will increase the transport of sediment, nutrients, and microbial pathogens into streams, lakes, and estuaries, degrading water quality. Sediment runoff reduces light penetration in water, limiting photosynthesis for aquatic plants. Fine sediments can also clog fish gills, leading to disease and mortality in fish populations.[9] During summer months or droughts, reduced river and lake flows can raise water temperatures and lower dissolved oxygen levels, placing stress on aquatic life.[10] The risk of harmful algal blooms is also expected to increase due to more extended periods of warm water temperatures. Episodic nutrient spikes from heavier winter rainfall and summer storms can fuel algal growth, creating toxic conditions that harm ecosystems and water supplies.[11]

Groundwater supplies are at risk of contamination due to floods, droughts, and saltwater intrusion caused by sea-level rise and storm surges.[12,13] Researchers found high concentrations of heavy metals in groundwater in the Brahmaputra floodplain in Assam, India. Approximately 80% of samples contained iron levels exceeding the World Health Organization's recommended limits, while 22.5% exceeded safe levels for both manganese and lead.[14] During California's 2014 drought, rural communities dependent on wells faced increased risks of contamination, as reduced water availability led to higher concentrations of pollutants. Many of the state's groundwater basins were already contaminated due to the overuse of nitrogen fertilizers, industrial chemicals, and natural contaminants such as arsenic.[15] Saltwater intrusion is another growing concern. In Florida's Broward County, projections indicate that 41% of its coastal well field capacity could be lost within 50 years, threatening freshwater supplies.[16]

Forests

Climate change will directly and indirectly impact forest growth and productivity through rising temperatures, altered rainfall patterns, extreme weather events, and other environmental shifts. Warming temperatures will generally lengthen the growing season and shift the geographic ranges of tree species. Some tree habitats will migrate to new areas, while others may struggle to survive if their current range becomes unsuitable for them.[17,18] Changing precipitation patterns are already influencing tree migration. A study analyzing 86 tree species across the Eastern United States (1980–2015) found that 73% of species shifted westward, following increased moisture, while 62% moved poleward toward cooler climates.[19] As the climate warms, longer snow-free periods and more frequent and intense droughts will increase the risk of forest fires. For instance, in Canada, the total area burned by wildfires could double by the end of the century.[20] Drought stress will also reduce trees' ability to produce sap, which is crucial for protecting them from harmful insects. Rising temperatures may accelerate insect development, alter seasonal lifecycles, and expand insect ranges, leading to more severe pest outbreaks.[21] Additionally, climate change may favor invasive plant species that are better adapted to shifting environmental conditions, further disrupting native ecosystems.[22] Forests will face an increased risk of permanent damage with more intense storms and heavy rainfall events. Large, slow-growing trees, which provide critical habitats for birds and other wildlife, will be particularly vulnerable to breakage and loss.[23]

Drylands

Climate change is expected to cause drylands to expand by 23% compared to the 1961–90 baseline, covering 56% of the Earth's total land surface.[24] At the same time, increasing frequency and intensity of extreme climate events are making dryland ecosystems more vulnerable. Currently, 15 out of 24 key ecosystem services in drylands are already in decline.[25] Drylands are categorized into subtropical regions (e.g., the Sahara Desert, Australia's Outback) and temperate regions (e.g., the Central Asian steppes, western United States, and parts of the Mediterranean). Research suggests that by the end of the 21st century, the current extent of temperate drylands could shrink by up to 30%, with a significant portion of this lost area likely to transition into subtropical drylands. As a result, vegetation in these newly converted drylands will rely more on moisture from the upper soil layers, favoring plant species with shallow root systems. This shift will lead to significant ecosystem changes, altering the composition of plants, biodiversity, and water availability in these regions.[26]

Climate change is expected to worsen water scarcity in drylands. A study of the semi-arid drainage basin in southeast Arizona found a long-term decline in heavy rainfall events, with a corresponding increase in smaller storms that deliver less rainfall.[27] Droughts will also have varying impacts on vegetation, depending on whether annual or perennial plants dominate the ecosystem. Research analyzing 500 drought events in drylands found that centennial-scale droughts could reduce plant biomass production by approximately 45% in perennial plant-dominated ecosystems and 72% in annual plant-dominated ecosystems. This suggests that drylands dominated by annual plants will suffer more severe losses during prolonged droughts.[28] The global expansion of drylands due to climate change will also contribute to lower carbon sequestration, intensifying regional warming and further disrupting ecosystem stability.[29]

Wetlands

Climate change will significantly impact wetlands through changes in hydrology, rising temperatures, and sea-level rise. In lower latitudes, where precipitation levels are expected to decline, freshwater marshes will likely experience lower water depths and shorter periods of inundation. These hydrological shifts may alter vegetation composition, with xeric (drought-tolerant) species becoming more dominant while hydric (water-dependent) species decline. In extreme drought conditions, entire wetland systems could dry up completely.[30] Conversely, higher phosphorus and nitrogen levels in runoff may lead to more eutrophic conditions in regions where precipitation is expected to increase. This will promote the growth of floating plants and phytoplankton, reducing the presence of submerged aquatic plants and altering wetland ecosystems.

Temperature changes will significantly affect water depth and the duration of inundation in wetlands. For example, rising temperatures could lower water levels and increase the presence of xeric species in semi-permanent prairie wetlands in North America. Higher temperatures may also lead to reductions in particular plant species and shifts in vegetation composition. One predicted impact is the replacement of salt marshes by mangroves in many areas, as fewer freezing events will allow mangroves to expand into regions where cold temperatures previously limited their growth.[31]

Coastal wetland flora and fauna are sensitive to small, permanent changes in water levels, but their ability to adapt depends on whether they can migrate to suitable areas. As sea levels rise, wetlands will need to shift inland. However, this migration may be blocked by human land use or occur too rapidly for coastal plant species to adapt or relocate. Additionally, changes in freshwater runoff will pose further challenges. Reduced freshwater inflows can decrease the sediment supply, limiting the ability of wetlands to build up soil and keep pace with rising sea levels, increasing the risk of wetland loss.[32]

Coastal and Marine

Rising greenhouse gas concentrations in the atmosphere have led to global warming, including a significant increase in ocean temperatures. Simultaneously, oceans have become more acidic as they absorb excess carbon dioxide from human activities. Key effects include the following.

Ocean Warming and Oxygen Levels

In 2023, global ocean heat content reached its highest recorded levels, surpassing previous records. The upper 2,000 meters of the ocean absorbed an additional 16 ± 10 zettajoules of heat compared to 2022, continuing a long-term trend of warming.[33] Oceans absorb over 90% of the excess heat trapped by greenhouse gases, making ocean heat content a key indicator of global warming.[34] A direct consequence of ocean warming is deoxygenation, which means that less oxygen is available for marine life. Since the 1950s, the ocean has lost approximately 2% of its oxygen, and if current trends persist, levels could decline by an additional 3%–4% by 2100.[35] Data show that ocean warming and deoxygenation have been closely linked since the mid-1980s, with the rate of increase accelerating in recent decades.[36] These changes impact marine ecosystems, fisheries, and climate patterns, underscoring the importance of monitoring and reducing greenhouse gas emissions to mitigate oceanic changes.

Ocean Carbon Sink Loss

Not all carbon dioxide emitted from the combustion of fossil fuels remains in the atmosphere. The ocean absorbs carbon dioxide in two steps. First, carbon dioxide dissolves in surface waters. Second, ocean currents and mixing processes transport the dissolved carbon dioxide into the deep ocean, where it accumulates over time. Between 1994 and 2007, the ocean absorbed approximately 34 gigatons of human-generated carbon dioxide, representing about 31% of all anthropogenic emissions. While the percentage of carbon dioxide absorbed by the ocean has remained relatively stable over the past 200 years, the absolute amount has increased significantly.

Warmer ocean temperatures, however, reduce the amount of organic carbon transported into the deep ocean, creating a feedback loop where less carbon is stored as global temperatures rise.[37,38] Over the past 30 years, as global temperatures have increased, the amount of carbon removed and stored in the deep ocean has decreased by 1.5%. Around 50 billion tons of new plankton grow yearly in the surface ocean, while six billion tons of dead plankton sink to deeper waters. The 1.5% decline means that approximately 100 million tons of additional plankton remain in the surface ocean each year, allowing more carbon to return to the atmosphere.[39]

Phytoplankton absorb carbon dioxide from seawater through photosynthesis and convert it into organic carbon. Some carbon sinks occur when phytoplankton die or are consumed by organisms that later sink to the bottom. However, phytoplankton and other organic materials in warmer waters are more likely to break down in the upper ocean before sinking. As they dissolve, the carbon they contain is released back into surface waters, where it is more easily re-emitted into the atmosphere.[40,41] Additionally, absorbing carbon dioxide by seawater leads to ocean acidification, further altering marine ecosystems.[42,43]

Ocean Acidification

When carbon dioxide dissolves in seawater, it reacts to form carbonic acid, which releases hydrogen ions. These hydrogen ions bond with other molecules, making the water more acidic and lowering its pH. This process is known as ocean acidification.[44] Since the Industrial Revolution, the pH of surface ocean waters has dropped by 0.1 units. Because the pH scale is logarithmic, this represents a 30% increase in ocean acidity. Projections suggest that as carbon dioxide emissions continue, the oceans absorb even more carbon dioxide, increasing acidity. Under a business-as-usual emissions scenario, the acidity of surface waters is projected to rise by nearly 150% by the end of the 21st century. This would bring ocean pH levels to conditions not seen in over 20 million years, posing significant risks to marine ecosystems.[45]

Climate Change Impacts on Plants and Animals

Many plant and animal species will be directly and indirectly impacted by climate change:

Plants

Climate change will significantly impact plant diversity, with many species unable to migrate or adapt quickly enough to keep pace with shifting climatic conditions. Species with long life cycles or slow dispersal rates will be particularly vulnerable to extinction. A study on 236 grass species found that the rate of climate change is approximately 5,000 times faster than the speed at which grasses can adapt to new climatic niches, making survival increasingly difficult.[46]

Isolated species face a higher risk of extinction as they may have no viable migration pathways. Alpine and island endemic species are especially vulnerable, as rising temperatures force species to shift to higher altitudes. In Australia, research suggests that many alpine plant species will lose their suitable habitat as temperatures rise. Similarly, species migrating latitudinally rather than altitudinally may encounter geographical barriers, such as coastal boundaries, which limit their ability to expand into new habitats.[47]

Climate change may also alter plant genetic composition due to new selection pressures. Species with long-distance seed dispersal may have a better chance of reaching climatically suitable areas; however, this advantage may not guarantee survival. Modelling in the eastern United States suggests a low likelihood of successful colonization beyond 20 kilometers. Relying on long-distance dispersal may lead to smaller populations and reduced genetic diversity, ultimately lowering adaptive potential.[48]

Climate change could also increase the spread of invasive species, as native plants become less suited to their environment while exotic species thrive. The Mexican daisy (Erigeron karvinskianus) is a significant invasive weed in New Zealand, threatening native forest restoration by forming dense mats that choke out native plants. The species is currently illegal to propagate, sell, or spread. As temperatures rise by around two degrees Celsius, climate projections suggest that the South Island will become more suitable for the daisy, allowing it to spread along both coasts, with its range likely reaching its peak by 2050.[49,50]

Animals

Climate change will affect food availability, shelter, and nesting sites in the short term. For example, reduced plant productivity during droughts will lead to food shortages and impact seed dispersal by small mammals and birds. In the long term, shifts in habitat distribution will disrupt ecosystems as species adjust their ranges, adapt, or face extinction. Some species will relocate to more suitable environments, but those with low dispersal ability may struggle to reach new habitats. Migratory species will experience varying environmental changes across their migration routes, making adaptation more challenging.

A species' ability to survive and reproduce depends on its physiological tolerance to temperature and moisture levels. As climate conditions shift, species will adapt to new environmental conditions, experience restricted activity, or be exposed to extreme events such as wildfires and storms. Species that rely on seasonal temperature or moisture cues to regulate life events may struggle in a changing climate. Survival and reproduction rates will decline if breeding, migration, or hibernation become out of sync with the availability of food or ideal environmental conditions. Species at higher risk of timing mismatches include long-distance migrators, obligate hibernators, and those dependent on short-lived (ephemeral) resources.[51,52]

Insects

Insects comprise nearly half of all known organisms and play essential roles in pollination, nutrient cycling, and food chains. However, climate change disrupts insect populations by altering their life cycles and distribution.[53,54] Many insects emerge earlier than usual, affecting their synchronization with host plants. For example, aphids – which reproduce rapidly – now emerge one month earlier on average than in 1965, while butterflies are emerging one to two weeks earlier.[55] This shift threatens specialist insects that rely on precise timing with plant life cycles for survival.[56] A decline in insect populations also affects predators that depend on them. In Puerto Rico, the Puerto Rican tody, a bird that feeds exclusively on insects, has experienced a 90% decline in its catch rate, indicating a sharp reduction in available prey.[57]

As temperatures rise, many insect species expand their ranges to higher latitudes and altitudes, where conditions are more favorable for them.[58] In Connecticut, over 10,000 insect species are already present, with about 10 new butterfly and moth species arriving each year due to climate change. However, species adapted to colder climates are disappearing, and some newly arrived species carry diseases. For instance, the Lone Star tick, which transmits Ehrlichiosis and other tick-borne illnesses, has now established itself in the state.[59]

Climate change is also expected to increase crop losses due to insect activity. Rising temperatures accelerate insect metabolism and reproductive rates, resulting in increased feeding activity in crop-growing regions. For every 1°C increase in global temperatures, losses of rice, corn, and wheat are projected to rise by 10%–25%. A 2°C rise would result in an estimated annual loss of 213 million tons across these three staple crops, which would significantly impact global food security.[60,61]

Birds

Climate change alters the timing of key seasonal behaviors in birds, including earlier migrations, shifts in breeding seasons, and changes in nesting patterns. Some species that traditionally migrate are even ceasing migration altogether. These timing disruptions can be harmful when birds' life cycle events fall out of sync with the availability of the plants and insects they rely on for survival. In the United States, seed-eating house finches breed earlier in response to rising temperatures. For every 1°C increase, these birds lay eggs about 4.5 days earlier. While they may be adjusting to seed availability in the short term, long-term changes in food resources or mismatches in timing could make breeding more challenging. House finches are already expanding into new areas of North America, but their populations are declining in parts of their native range.[62,63]

Birds are also shifting their geographical ranges as temperatures rise. In the Andes, researchers estimate that if global temperatures increase between 2.6°C and 4.8°C this century, tropical bird species will need to move 500–900 meters higher in elevation to find suitable conditions. However, birds already living at the highest elevations have nowhere left to go and are at risk of disappearing. In Peru, eight bird species that previously lived near a mountain peak have already vanished from the area.[64,65]

Migratory bird populations are declining faster than those of non-migratory species, as climate change makes migration longer and more challenging. A study on the Eurasian-African migratory flyway predicts that for over 80% of long-distance migratory species, the distance and duration of migration will increase significantly. One of the most significant projected increases is for the thrush nightingale, which may have to travel an additional 773 kilometers between breeding and non-breeding grounds by 2070, extending the journey by at least five extra days. Longer migration routes increase mortality rates, as migration exposes birds to higher predation risk, food scarcity, and extreme energy demands. Any further strain on migratory journeys is expected to contribute to population declines in many species.[66]

Mammals

A study assessing the effects of climate change on wildlife estimates that 47% of 873 threatened terrestrial mammal species have already been negatively affected in at least part of their range or population.[67] In the short term, climate change will reduce food availability, shelter, and breeding sites, making survival more difficult for many species. In the long term, shifting habitats will disrupt ecosystems as species move, adapt, or decline in response to changing conditions. Research on European mammals has found that up to 47% of species are threatened by climate change and possess traits that limit their ability to relocate and establish in new, suitable areas.[68] Particular species face particularly severe threats. A study on Bengal tigers in the Bangladesh Sundarbans, the world's largest remaining mangrove habitat, predicts that by 2070, no suitable habitat will remain for tigers in the region due to the combined effects of climate change and sea-level rise.[69]

Freshwater Fish

Climate change is expected to cause higher water temperatures, lower oxygen levels, and increased toxicity of pollutants in freshwater ecosystems. These changes will have significant physiological and ecological effects on freshwater fish, as outlined in Table 4.1.[70]

Corals

Coral reefs support the highest marine ecosystem biodiversity, hosting over one-quarter of all marine fish species despite covering less than 0.1% of the ocean floor.[71] However, climate change is causing rapid coral loss, with many reefs having lost over 50% of their coral cover in the past 30–40 years, and some experiencing losses exceeding 90%[72]

Ocean acidification is weakening coral reefs by reducing the ability of reef-building corals to form their skeletons.[73] Ocean acidification weakens coral reefs by corroding existing structures and slowing the growth of new skeletons. This makes reefs more fragile and more vulnerable to damage from storm waves and coral-eating organisms.[74] Many coral species struggle to adapt to these conditions, which threatens the ecological function of coral reef ecosystems in the future.[75]

Rising ocean temperatures have also led to mass coral bleaching events. Corals rely on a symbiotic relationship with microscopic algae called zooxanthellae, which live in their tissues. These algae photosynthesize and provide corals with nutrients, giving them their distinctive colors. In return,

Table 4.1 Physiological and Ecological Impacts of Climate Change on Freshwater Fishes

Issue	*Impact*
Metabolic rate and energetic requirements for growth and reproduction	Fish expend more energy maintaining metabolic balance when water temperatures move outside their tolerance range, leaving less energy for growth and reproduction.
Development time	Mismatch between larval hatching and food availability due to altered temperature, rainfall, and flow patterns can increase larval mortality.
Reproduction	Seasonal temperature and water level changes drive breeding cycles. Climate change may disrupt these cues and reduce reproductive success. Storm-driven sediment can smother spawning grounds.
Migration	Altered environmental cues such as water temperature and flow may delay or prevent migration, disrupting reproduction and survival.
Disease	Changes in temperature and water conditions can lower disease resistance and increase pathogen exposure, reducing fish health and yield.
Exposure to temperature stress	Extreme temperatures and low water levels can impair growth and cause mortality. Warmer, shallower waters may concentrate pollutants, reducing water quality.
Prey availability	Warmer water increases metabolic demand for food, while altered habitats may reduce prey availability or quality, threatening fish nutrition.
Mortality and predation risk	Higher water temperatures are linked to greater fish mortality and changes in predator-prey dynamics, increasing predation risks.
Distributional shifts, colonization, extinction	Climate change may render habitats unsuitable for some species, forcing migration, enabling new colonization, or causing local/global extinctions.

corals release ammonium, which the algae use as a nutrient.[76] However, higher ocean temperatures and pollution stress corals, causing them to expel the algae, turning them pale or white. Without these algae, corals lose their primary food source, weaken, and become more vulnerable to disease.[77]

Marine Fishes

Ocean acidification affects marine fish by altering the pH of their blood, a condition known as acidosis. When seawater becomes more acidic, fish absorb carbonic acid from their surroundings, disrupting their internal balance. Even a slight decrease in ocean pH forces fish to use additional energy to expel excess acid through their gills, kidneys, and intestines. This diverts energy from essential functions such as digestion, escaping predators, reproduction, and growth, ultimately affecting survival.[78]

Tropical fish are particularly vulnerable because they already live near their thermal limits.[79] For example, ocean waters around the Pacific Islands are expected to become warmer, less oxygenated, more acidic, and have lower plankton production, which will impact the marine food web. If current climate trends continue, 50%–80% of fish species in Pacific Island waters are expected to disappear by the end of the 21st century.[80]

Notes

1 Ministry of the Environment and Food of Denmark / Environmental Protection Agency, "Climate Change Impact on Nature," (n.d.), https://en.klimatilpasning.dk/sectors/nature/climate-change-impact-on-nature.

2 Scottish Natural Heritage, "Impacts on People and Our Economy," (n.d.), https://nature.scot/climate-change/climate-change-impacts-scotland/impacts-people-and-our-economy.

3 European Commission, "Climate Change Consequences," (n.d.), https://ec.europa.eu/clima/change/consequences_en.

4 United States Department of Agriculture Forest Service, "Flows of the Future – How Will Climate Change Affect Streamflows in the Pacific Northwest?," (2016), https://fs.fed.us/pnw/sciencef/scifi187.pdf.

5 European Environment Agency, "River Flow Drought," (n.d.), https://eea.europa.eu/data-and-maps/indicators/river-flow-drought-1/assessment.

6 Francis Chiew and Ian Prosser, "Water and Climate," in *Water: Science and Solutions for Australia*, ed. Ian Prosser (Clayton: CSIRO, 2011).

7 IGRAC, "Groundwater & Climate Change," (n.d.), https://un-igrac.org/areas-expertise/groundwater-climate-change.

8 Chandrakanta Ojha, Susanna Werth, and Manoochehr Shirzaei, "Groundwater Loss and Aquifer System Compaction in San Joaquin Valley during 2012–2015 Drought," *Journal of Geophysical Research: Solid Earth* 124, no. 3 (2019).

9 R.C. Brears, *Urban Water Security* (Chichester and Hoboken, NJ: John Wiley & Sons, 2016).

10 Brears, *Urban Water Security*.

11 I Brears, *Urban Water Security*.

12 U.S. Global Change Research Program, "Impacts, Risks, and Adaptation in the United States: Fourth National Climate Assessment, Volume Ii," (2018), https://nca2018.globalchange.gov/.

13 Nigel W. Arnell et al., "The Implications of Climate Change for the Water Environment in England," *Progress in Physical Geography: Earth and Environment* 39, no. 1 (2015).

14 Nabanita Haloi and H.P. Sarma, "Heavy Metal Contaminations in the Groundwater of Brahmaputra Flood Plain: An Assessment of Water Quality in Barpeta District, Assam (India)," *Environmental Monitoring and Assessment* 184, no. 10 (2012).

15 Reuters, "Health Experts Warn of Water Contamination from California Drought," (2014), https://reuters.com/article/us-usa-california-drought/health-experts-warn-of-water-contamination-from-california-drought-idUSBREA1I06P20140219.

16 Sun-Sentinel, "Invading Seawater Jeopardizes South Florida's Drinking Water, But We Can Lessen the Threat | Editorial," (2018), https://sun-sentinel.com/opinion/editorials/fl-op-editorial-sea-level-rise-drinking-water-20180601-story.html.

17 U.S. EPA, "Climate Impacts on Forests," (2018), https://19january2017snapshot.epa.gov/climate-impacts/climate-impacts-forests_.html
18 Craig M. Costion et al., "Will Tropical Mountaintop Plant Species Survive Climate Change? Identifying Key Knowledge Gaps Using Species Distribution Modelling in Australia," *Biological Conservation* 191 (2015).
19 Songlin Fei et al., "Divergence of Species Responses to Climate Change," *Science Advances* 3, no. 5 (2017).
20 Andrei P. Kirilenko and Roger A. Sedjo, "Climate Change Impacts on Forestry," *Proceedings of the National Academy of Sciences* 104, no. 50 (2007).
21 U.S. EPA, "Climate Impacts on Forests."
22 Cory Merow et al., "Climate Change Both Facilitates and Inhibits Invasive Plant Ranges in New England," *Proceedings of the National Academy of Sciences* 114, no. 16 (2017).
23 María Uriarte, Jill Thompson, and Jess K. Zimmerman, "Hurricane María Tripled Stem Breaks and Doubled Tree Mortality Relative to Other Major Storms," *Nature Communications* 10, no. 1 (2019).
24 Jianping Huang et al., "Accelerated Dryland Expansion under Climate change," *Nature Climate Change* 6 (2015).
25 CGIAR, "Drylands and Climate Change," (n.d.), https://iucn.org/resources/issues-brief/drylands-and-climate-change
26 Daniel R. Schlaepfer et al., "Climate Change Reduces Extent of Temperate Drylands and Intensifies Drought in Deep Soils," *Nature Communications* 8 (2017).
27 Michael B. Singer and Katerina Michaelides, "Deciphering the Expression of Climate Change within the Lower Colorado River Basin by Stochastic Simulation of Convective Rainfall," *Environmental Research Letters* 12, no. 10 (2017).
28 Jan C. Ruppert et al., "Quantifying Drylands' Drought Resistance and Recovery: The Importance of Drought Intensity, Dominant Life History and Grazing Regime," *Global Change Biology* 21, no. 3 (2015).
29 Huang et al., "Accelerated Dryland Expansion under Climate Change."
30 Frederick T. Short et al., "Impacts of Climate Change on Submerged and Emergent Wetland Plants," *Aquatic Botany* 135 (2016).
31 Short et al., "Impacts of Climate Change on Submerged and Emergent Wetland Plants."
32 José Lucas and Pérez Lloréns, "Impacts of Climate Change on Wetland Ecosystems," (2008), https://zaragoza.es/contenidos/medioambiente/cajaAzul/10S3-P2-Perez%20LloresnACC.pdf.
33 Lijing Cheng et al., "Ocean Heat Content in 2023," *Nature Reviews Earth & Environment* 5, no. 4 (2024).
34 NASA, "Ocean Warming," (n.d.), https://climate.nasa.gov/vital-signs/ocean-warming/?intent=121.
35 IUCN, "Ocean Deoxygenation," (n.d.), https://iucn.org/resources/issues-brief/ocean-deoxygenation.
36 WMO, "Multi-Agency Report Highlights Increasing Signs and Impacts of Climate Change in Atmosphere, Land and Oceans," (n.d.), https://wmo.int/news/media-centre/multi-agency-report-highlights-increasing-signs-and-impacts-of-climate-change-atmosphere-land-and.
37 Chris M. Marsay et al., "Attenuation of Sinking Particulate Organic Carbon Flux through the Mesopelagic Ocean," *Proceedings of the National Academy of Sciences* (2015).

38 CarbonBrief, "Warming Oceans Less Able to Store Organic Carbon, Study Suggests," (n.d.), https://carbonbrief.org/warming-oceans-less-able-to-store-organic-carbon-study-suggests.
39 B.B. Cael, Kelsey Bisson, and Michael J. Follows, "How Have Recent Temperature Changes Affected the Efficiency of Ocean Biological Carbon Export?," *Limnology and Oceanography Letters* 2, no. 4 (2017).
40 Marsay et al., "Attenuation of Sinking Particulate Organic Carbon Flux through the Mesopelagic Ocean."
41 CarbonBrief, "Warming Oceans Less Able to Store Organic Carbon, Study Suggests."
42 ScienceDaily, "Ocean Sink for Human-Made Carbon Dioxide Measured," (2019).
43 Nicolas Gruber et al., "The Oceanic Sink for Anthropogenic CO_2 from 1994 to 2007," *Science* 363, no. 6432 (2019).
44 Smithsonian Ocean, (n.d.), https://ocean.si.edu/ocean-life/invertebrates/ocean-acidification.
45 NOAA, "What Is Ocean Acidification?," (n.d.), www.pmel.noaa.gov/co2/story/What+is+Ocean+Acidification%253.
46 F.A. Cang, A.W. Ashley, and J.W. John, "Climate Change Is Projected to Outpace Rates of Niche Change in Grasses," *Biology Letters* 12, no. 9 (2016).
47 Matthew J. Christmas, Martin F. Breed, and Andrew J. Lowe, "Constraints to and Conservation Implications for Climate Change Adaptation in Plants," *Conservation Genetics* 17, no. 2 (2016).
48 Christmas, Breed, and Lowe, "Constraints to and Conservation Implications for Climate Change Adaptation in Plants."
49 Stuff, "Climate Change Will Move New Zealand Weeds," (n.d.), https://stuff.co.nz/environment/climate-news/110593583/climate-change-will-move-new-zealand-weeds.
50 Lauren Hannah, Glenn Aguilar, and Dan Blanchon, "Spatial Distribution of the Mexican Daisy, Erigeron Karvinskianus, in New Zealand under Climate Change," *Climate* 7, no. 2 (2019).
51 Megan M. Friggens, Mary I. Williams, Karen E. Bagne, Tosha T. Wixom, and Samuel A. Cushman, "Effects of Climate Change on Terrestrial Animals," in *Climate Change Vulnerability and Adaptation in the Intermountain Region [Part 2]*, ed. Jessica E. Halofsky, David L. Peterson, Joanne J. Ho, Natalie J. Little, and Linda A. Joyce (Fort Collins, CO: U.S. Department of Agriculture, Forest Service, Rocky Mountain Research Station, 2018).
52 Friggens, Williams, Bagne, Wixom, and Cushman, "Effects of Climate Change on Terrestrial Animals."
53 Francisco Sánchez-Bayo and Kris A.G. Wyckhuys, "Worldwide Decline of the Entomofauna: A Review of Its Drivers," *Biological Conservation* 232 (2019).
54 Gary D. Powney et al., "Widespread Losses of Pollinating Insects in Britain," *Nature Communications* 10, no. 1 (2019).
55 James R. Bell et al., "Spatial and Habitat Variation in Aphid, Butterfly, Moth and Bird Phenologies over the Last Half Century," *Global Change Biology* (2019).
56 T. Cornelissen, "Climate Change and Its Effects on Terrestrial Insects and Herbivory Patterns," *Neotropical Entomology* 40 (2011).
57 Bradford C. Lister and Andres Garcia, "Climate-Driven Declines in Arthropod Abundance Restructure a Rainforest Food Web," *Proceedings of the National Academy of Sciences* 115, no. 44 (2018).

58 Cornelissen, "Climate Change and Its Effects on Terrestrial Insects and Herbivory Patterns."
59 University of Connecticut, "Insects Coping with Climate Change," (2018), https://today.uconn.edu/2018/09/insects-coping-climate-change/#.
60 Curtis A. Deutsch et al., "Increase in Crop Losses to Insect Pests in a Warming Climate," *Science* 361, no. 6405 (2018).
61 University of Washington, "Climate Change Projected to Boost Insect Activity and Crop Loss, Researchers Say," (2018), https://washington.edu/news/2018/08/30/climate-change-insects-crops/.
62 Heather E. Watts et al., "Temperature-Correlated Shifts in the Timing of Egg-Laying in House Finches Haemorhous Mexicanus," *Ibis* 161, no. 2 (2019).
63 Phys.org, "Climate Change Affects Breeding Birds," (2018), https://phys.org/news/2024-07-climate-affects-birds.pdf.
64 Benjamin G. Freeman et al., "Climate Change Causes Upslope Shifts and Mountaintop Extirpations in a Tropical Bird Community," *Proceedings of the National Academy of Sciences* 115, no. 47 (2018).
65 BBC, "Climate Change Is 'Escalator to Extinction' for Mountain Birds," (2018), https://bbc.com/news/science-environment-46020959.
66 Christine Howard et al., "Flight Range, Fuel Load and the Impact of Climate Change on the Journeys of Migrant Birds," *Proceedings of the Royal Society B: Biological Sciences* 285, no. 1873 (2018).
67 Michela Pacifici et al., "Species' Traits Influenced Their Response to Recent Climate Change," *Nature Climate Change* 7 (2017).
68 Lisbeth Morrison, Alba Estrada, and Regan Early, "Species Traits Suggest European Mammals Facing the Greatest Climate Change Are Also Least Able to Colonize New Locations," *Diversity and Distributions* 24, no. 9 (2018).
69 Sharif A. Mukul et al., "Combined Effects of Climate Change and Sea-Level Rise Project Dramatic Habitat Loss of the Globally Endangered Bengal Tiger in the Bangladesh Sundarbans," *Science of The Total Environment* 663 (2019).
70 FAO, "Impacts of Climate Change on Fisheries and Aquaculture: Synthesis of Current Knowledge, Adaptation and Mitigation Options," (2018), http://fao.org/3/I9705EN/i9705en.pdf.
71 IUCN, "Coral Reefs and Climate Change," (2017), https://iucn.org/resources/issues-brief/coral-reefs-and-climate-change.
72 Andrew C Baker, "Climate Change: Many Ways to Beat the Heat for Reef Corals," *Current Biology* 24, no. 24 (2014).
73 NOAA, "What Is Ocean Acidification?"
74 Smithsonian Ocean, "Ocean Acidification," https://ocean.si.edu/ocean-life/invertebrates/ocean-acidification.
75 S. Comeau et al., "Resistance to Ocean Acidification in Coral Reef Taxa Is Not Gained by Acclimatization," *Nature Climate Change* 9, no. 6 (2019).
76 Katarzyna Frankowiak et al., "Photosymbiosis and the Expansion of Shallow-Water Corals," *Science Advances* 2, no. 11 (2016).
77 NOAA, "What Is Coral Bleaching?," (n.d.), https://oceanservice.noaa.gov/facts/coral_bleach.html.
78 Smithsonian Ocean, "Ocean Acidification."
79 Lise Comte and Julian D. Olden, "Climatic Vulnerability of the World's Freshwater and Marine Fishes," *Nature Climate Change* 7 (2017).

80 Rebecca G. Asch, William W.L. Cheung, and Gabriel Reygondeau, "Future Marine Ecosystem Drivers, Biodiversity, and Fisheries Maximum Catch Potential in Pacific Island Countries and Territories under Climate Change," *Marine Policy* 88 (2018).

References

Arnell, Nigel W., Sarah J. Halliday, Richard W. Battarbee, Richard A. Skeffington, and Andrew J. Wade. "The Implications of Climate Change for the Water Environment in England." *Progress in Physical Geography: Earth and Environment* 39, no. 1 (2015): 93–120.

Asch, Rebecca G., William W.L. Cheung, and Gabriel Reygondeau. "Future Marine Ecosystem Drivers, Biodiversity, and Fisheries Maximum Catch Potential in Pacific Island Countries and Territories under Climate Change." *Marine Policy* 88 (2018/02/01): 285–94.

Baker, Andrew C. "Climate Change: Many Ways to Beat the Heat for Reef Corals." *Current Biology* 24, no. 24 (2014/12/15): R1166–R68.

BBC. "Climate Change Is 'Escalator to Extinction' for Mountain Birds." (2018). https://bbc.com/news/science-environment-46020959.

Bell, James R., Marc S. Botham, Peter A. Henrys, David I. Leech, James W. Pearce-Higgins, Chris R. Shortall, Tom M. Brereton, Jon Pickup, and Stephen J. Thackeray. "Spatial and Habitat Variation in Aphid, Butterfly, Moth and Bird Phenologies over the Last Half Century." *Global Change Biology* (2019/02/14).

Brears, R.C. *Urban Water Security*. Chichester and Hoboken, NJ: John Wiley & Sons, 2016.

Cael, B.B., Kelsey Bisson, and Michael J. Follows. "How Have Recent Temperature Changes Affected the Efficiency of Ocean Biological Carbon Export?" *Limnology and Oceanography Letters* 2, no. 4 (2017/08/01): 113–18.

Cang, F.A., A.W. Ashley, and J.W. John. "Climate Change Is Projected to Outpace Rates of Niche Change in Grasses." *Biology Letters* 12, no. 9 (2016/09/30): 20160368.

CarbonBrief. "Warming Oceans Less Able to Store Organic Carbon, Study Suggests." (n.d.). https://carbonbrief.org/warming-oceans-less-able-to-store-organic-carbon-study-suggests.

CGIAR. "Drylands and Climate Change." (n.d.). https://iucn.org/resources/issues-brief/drylands-and-climate-change.

Cheng, Lijing, Karina von Schuckmann, Audrey Minière, Maria Z. Hakuba, Sarah Purkey, Gavin A. Schmidt, and Yuying Pan. "Ocean Heat Content in 2023." *Nature Reviews Earth & Environment* 5, no. 4 (2024/04/01): 232–34.

Chiew, Francis, and Ian Prosser. "Water and Climate." Chap. 3 In *Water: Science and Solutions for Australia*, edited by Ian Prosser. Clayton: CSIRO, 2011.

Christmas, Matthew J., Martin F. Breed, and Andrew J. Lowe. "Constraints to and Conservation Implications for Climate Change Adaptation in Plants." *Conservation Genetics* 17, no. 2 (2016/04/01): 305–20.

Comeau, S., C.E. Cornwall, T.M. DeCarlo, S.S. Doo, R.C. Carpenter, and M.T. McCulloch. "Resistance to Ocean Acidification in Coral Reef Taxa Is Not Gained by Acclimatization." *Nature Climate Change* 9, no. 6 (2019/06/01): 477–83.

Comte, Lise, and Julian D. Olden. "Climatic Vulnerability of the World's Freshwater and Marine Fishes." *Nature Climate Change* 7 (2017/09/11): 718.

Cornelissen, T. "Climate Change and Its Effects on Terrestrial Insects and Herbivory Patterns." *Neotropical Entomology* 40 (2011): 155–63.

Costion, Craig M., Lalita Simpson, Petina L. Pert, Monica M. Carlsen, W.J. Kress, and Darren Crayn. "Will Tropical Mountaintop Plant Species Survive Climate Change? Identifying Key Knowledge Gaps Using Species Distribution Modelling in Australia." *Biological Conservation* 191 (2015/11/01): 322–30.

Deutsch, Curtis A., Joshua J. Tewksbury, Michelle Tigchelaar, David S. Battisti, Scott C. Merrill, Raymond B. Huey, and Rosamond L. Naylor. "Increase in Crop Losses to Insect Pests in a Warming Climate." *Science* 361, no. 6405 (2018): 916.

European Commission. "Climate Change Consequences." (n.d.). https://ec.europa.eu/clima/change/consequences_en.

European Environment Agency. "River Flow Drought." (n.d.). https://eea.europa.eu/data-and-maps/indicators/river-flow-drought-1/assessment.

FAO. "Impacts of Climate Change on Fisheries and Aquaculture: Synthesis of Current Knowledge, Adaptation and Mitigation Options." (2018). http://fao.org/3/I9705EN/i9705en.pdf.

Fei, Songlin, Johanna M. Desprez, Kevin M. Potter, Insu Jo, Jonathan A. Knott, and Christopher M. Oswalt. "Divergence of Species Responses to Climate Change." *Science Advances* 3, no. 5 (2017): e1603055.

Frankowiak, Katarzyna, Xingchen T. Wang, Daniel M. Sigman, Anne M. Gothmann, Marcelo V. Kitahara, Maciej Mazur, Anders Meibom, and Jarosław Stolarski. "Photosymbiosis and the Expansion of Shallow-Water Corals." *Science Advances* 2, no. 11 (2016): e1601122.

Freeman, Benjamin G., Micah N. Scholer, Viviana Ruiz-Gutierrez, and John W. Fitzpatrick. "Climate Change Causes Upslope Shifts and Mountaintop Extirpations in a Tropical Bird Community." *Proceedings of the National Academy of Sciences* 115, no. 47 (2018): 11982–87.

Friggens, Megan M., Mary I. Williams, Karen E. Bagne, Tosha T. Wixom, Samuel A. Cushman. "Effects of Climate Change on Terrestrial Animals." In *Climate Change Vulnerability and Adaptation in the Intermountain Region [Part 2]*, edited by Jessica E. Peterson, David L. Halofsky, Joanne J. Ho, Natalie J. Little, Linda A. Joyce, 264–315. Fort Collins, CO: U.S. Department of Agriculture, Forest Service, Rocky Mountain Research Station, 2018.

Gruber, Nicolas, Dominic Clement, Brendan R. Carter, Richard A. Feely, Steven van Heuven, Mario Hoppema, Masao Ishii, et al. "The Oceanic Sink for Anthropogenic CO_2 from 1994 to 2007." *Science* 363, no. 6432 (2019): 1193.

Haloi, Nabanita, and H.P. Sarma. "Heavy Metal Contaminations in the Groundwater of Brahmaputra Flood Plain: An Assessment of Water Quality in Barpeta District, Assam (India)." *Environmental Monitoring and Assessment* 184, no. 10 (2012/10/01): 6229–37.

Hannah, Lauren, Glenn Aguilar, and Dan Blanchon. "Spatial Distribution of the Mexican Daisy, Erigeron Karvinskianus, in New Zealand under Climate Change." *Climate* 7, no. 2 (2019).

Howard, Christine, A.S. Philip, A.T. Joseph, Catherine Sheard, H.M.B. Stuart, and G.W. Stephen. "Flight Range, Fuel Load and the Impact of Climate Change on the Journeys of Migrant Birds." *Proceedings of the Royal Society B: Biological Sciences* 285, no. 1873 (2018/02/28): 20172329.

Huang, Jianping, Haipeng Yu, Xiaodan Guan, Guoyin Wang, and Ruixia Guo. "Accelerated Dryland Expansion under Climate Change." *Nature Climate Change* 6 (2015/10/26): 166.

IGRAC. "Groundwater & Climate Change." (n.d.). https://un-igrac.org/areas-expertise/groundwater-climate-change.

IUCN. "Coral Reefs and Climate Change." (2017). https://iucn.org/resources/issues-brief/coral-reefs-and-climate-change.

IUCN. "Ocean Deoxygenation." (n.d.). https://iucn.org/resources/issues-brief/ocean-deoxygenation.

Kirilenko, Andrei P., and Roger A. Sedjo. "Climate Change Impacts on Forestry." *Proceedings of the National Academy of Sciences* 104, no. 50 (2007): 19697.

Lister, Bradford C., and Andres Garcia. "Climate-Driven Declines in Arthropod Abundance Restructure a Rainforest Food Web." *Proceedings of the National Academy of Sciences* 115, no. 44 (2018): E10397.

Lucas, José, and Pérez Lloréns. "Impacts of Climate Change on Wetland Ecosystems." (2008). https://zaragoza.es/contenidos/medioambiente/cajaAzul/10S3-P2-Perez%20LloresnACC.pdf.

Marsay, Chris M., Richard J. Sanders, Stephanie A. Henson, Katsiaryna Pabortsava, Eric P. Achterberg, and Richard S. Lampitt. "Attenuation of Sinking Particulate Organic Carbon Flux through the Mesopelagic Ocean." *Proceedings of the National Academy of Sciences* (2015): 201415311.

Merow, Cory, Sarah T. Bois, Jenica M. Allen, Yingying Xie, and John A. Silander. "Climate Change Both Facilitates and Inhibits Invasive Plant Ranges in New England." *Proceedings of the National Academy of Sciences* 114, no. 16 (2017): E3276.

Ministry of the Environment and Food of Denmark / Environmental Protection Agency. "Climate Change Impact on Nature." (n.d.). https://en.klimatilpasning.dk/sectors/nature/climate-change-impact-on-nature.

Morrison, Lisbeth, Alba Estrada, and Regan Early. "Species Traits Suggest European Mammals Facing the Greatest Climate Change Are Also Least Able to Colonize New Locations." *Diversity and Distributions* 24, no. 9 (2018/09/01): 1321–32.

Mukul, Sharif A., Mohammed Alamgir, Md Shawkat I. Sohel, Petina L. Pert, John Herbohn, Stephen M. Turton, Md Saiful I. Khan, et al. "Combined Effects of Climate Change and Sea-Level Rise Project Dramatic Habitat Loss of the Globally Endangered Bengal Tiger in the Bangladesh Sundarbans." *Science of the Total Environment* 663 (2019/05/01): 830–40.

NASA. "Ocean Warming." (n.d.). https://climate.nasa.gov/vital-signs/ocean-warming/?intent=121.

NOAA. "What Is Coral Bleaching?" (n.d.). https://oceanservice.noaa.gov/facts/coral_bleach.html.

NOAA. "What Is Ocean Acidification?" (n.d.). www.pmel.noaa.gov/co2/story/What+is+Ocean+Acidification%253.

Ojha, Chandrakanta, Susanna Werth, and Manoochehr Shirzaei. "Groundwater Loss and Aquifer System Compaction in San Joaquin Valley during 2012–2015 Drought." *Journal of Geophysical Research: Solid Earth* 124, no. 3 (2019/03/01): 3127–43.

Pacifici, Michela, Piero Visconti, Stuart H.M. Butchart, James E.M. Watson, Francesca M Cassola, and Carlo Rondinini. "Species' Traits Influenced Their Response to Recent Climate Change." *Nature Climate Change* 7 (2017/02/13): 205.

Phys.org. "Climate Change Affects Breeding Birds." (2018). https://phys.org/news/2024-07-climate-affects-birds.pdf.

Powney, Gary D., Claire Carvell, Mike Edwards, Roger K.A. Morris, Helen E. Roy, Ben A. Woodcock, and Nick J.B. Isaac. "Widespread Losses of Pollinating Insects in Britain." *Nature Communications* 10, no. 1 (2019/03/26): 1018.

Reuters. "Health Experts Warn of Water Contamination from California Drought." (2014). https://reuters.com/article/us-usa-california-drought/health-experts-warn-of-water-contamination-from-california-drought-idUSBREA1I06P20140219.

Ruppert, Jan C., Keith Harmoney, Zalmen Henkin, Hennie A. Snyman, Marcelo Sternberg, Walter Willms, and Anja Linstädter. "Quantifying Drylands' Drought Resistance and Recovery: The Importance of Drought Intensity, Dominant Life History and Grazing Regime." *Global Change Biology* 21, no. 3 (2015/03/01): 1258–70.

Sánchez-Bayo, Francisco, and Kris A.G. Wyckhuys. "Worldwide Decline of the Entomofauna: A Review of Its Drivers." *Biological Conservation* 232 (2019/04/01): 8–27.

Schlaepfer, Daniel R., John B. Bradford, William K. Lauenroth, Seth M. Munson, Britta Tietjen, Sonia A. Hall, Scott D. Wilson, et al. "Climate Change Reduces Extent of Temperate Drylands and Intensifies Drought in Deep Soils." *Nature Communications* 8 (2017/31): 14196.

ScienceDaily. "Ocean Sink for Human-Made Carbon Dioxide Measured." (2019). www.sciencedaily.com/releases/2019/03/190314151648.htm

Scottish Natural Heritage. "Impacts on People and Our Economy." (n.d.). https://nature.scot/climate-change/climate-change-impacts-scotland/impacts-people-and-our-economy.

Short, Frederick T., Sarian Kosten, Pamela A. Morgan, Sparkle Malone, and Gregg E. Moore. "Impacts of Climate Change on Submerged and Emergent Wetland Plants." *Aquatic Botany* 135 (2016/11/01): 3–17.

Singer, Michael B., and Katerina Michaelides. "Deciphering the Expression of Climate Change within the Lower Colorado River Basin by Stochastic Simulation of Convective Rainfall." *Environmental Research Letters* 12, no. 10 (2017/10/01): 104011.

Smithsonian Ocean. Ocean Acidificatio. (n.d.). https://ocean.si.edu/ocean-life/invertebrates/ocean-acidification.

Stuff. "Climate Change Will Move New Zealand Weeds." (n.d.). https://stuff.co.nz/environment/climate-news/110593583/climate-change-will-move-new-zealand-weeds.

Sun-Sentinel. "Invading Seawater Jeopardizes South Florida's Drinking Water, But We Can Lessen the Threat | Editorial." (2018). https://sun-sentinel.com/opinion/editorials/fl-op-editorial-sea-level-rise-drinking-water-20180601-story.html.

United States Department of Agriculture Forest Service. "Flows of the Future – How Will Climate Change Affect Streamflows in the Pacific Northwest?" (2016). https://fs.fed.us/pnw/sciencef/scifi187.pdf.

University of Connecticut. "Insects Coping with Climate Change." (2018). https://today.uconn.edu/2018/09/insects-coping-climate-change/#.

University of Washington. "Climate Change Projected to Boost Insect Activity and Crop Loss, Researchers Say." (2018). https://washington.edu/news/2018/08/30/climate-change-insects-crops/.

Uriarte, María, Jill Thompson, and Jess K. Zimmerman. "Hurricane María Tripled Stem Breaks and Doubled Tree Mortality Relative to Other Major Storms." *Nature Communications* 10, no. 1 (2019/03/25): 1362.

U.S. EPA. "Climate Impacts on Forests." (2018). https://19january2017snapshot.epa.gov/climate-impacts/climate-impacts-forests_.html.

U.S. Global Change Research Program. "Impacts, Risks, and Adaptation in the United States: Fourth National Climate Assessment, Volume II." (2018). https://nca2018.globalchange.gov/.

Watts, Heather E., Daniela Jimenez, Veronica Pacheco, and Tauras P. Vilgalys. "Temperature-Correlated Shifts in the Timing of Egg-Laying in House Finches Haemorhous Mexicanus." *Ibis* 161, no. 2 (2019/04/01): 428–34.

WMO. "Multi-Agency Report Highlights Increasing Signs and Impacts of Climate Change in Atmosphere, Land and Oceans." (n.d.). https://wmo.int/news/media-centre/multi-agency-report-highlights-increasing-signs-and-impacts-of-climate-change-atmosphere-land-and.

5 Environmental Degradation and Impacts on Biodiversity

Introduction

Human activities are driving environmental degradation, which includes the depletion of natural resources, ecosystem destruction, wildlife extinction, and pollution.[1,2] Over the past 50 years, the rate of environmental change has been unprecedented in human history. According to the Intergovernmental Science-Policy Platform on Biodiversity and Ecosystem Services, the most significant direct drivers of environmental change are land and sea-use changes, overexploitation of organisms, climate change, pollution, and the introduction of invasive species. Terrestrial and freshwater ecosystems have been most affected by land-use changes, followed by overexploitation through activities such as harvesting, logging, hunting, and fishing. The most significant impacts on marine ecosystems have been overfishing, followed by changes in land and sea use.[3] Indirect factors, including consumption patterns, population growth, economic expansion, technological advancements, and governance systems, influence these direct drivers. The rate and intensity of these environmental changes vary across regions and countries, influenced by local and global economic and policy conditions.[4]

Environmental Degradation

Societal challenges have resulted in various types of environmental degradation, as discussed below.

Land Degradation and Desertification

Land degradation refers to the deterioration of land resources that support life, caused by soil erosion, desertification, salinization, and acidification.[5] It results from natural forces, such as droughts and extreme weather, as well as human activities that pollute or degrade soil quality, impacting food production, livelihoods, and essential ecosystem services.[6] Agriculture is the most extensive human-driven land use, covering 50% of the world's habitable land. Livestock production alone accounts for 77% of all agricultural land,

DOI: 10.4324/9781003517948-5

including land used for growing animal feed, grazing, and pasture. This heavy land use contributes to soil degradation, which in turn affects long-term food security and the stability of ecosystems.[7]

Soil Erosion

Soil erosion is a natural process that affects all landforms. In agriculture, it refers to the removal of topsoil by water, wind, and human activities, such as monocropping and tillage in industrial farming.[8,9,10] Erosion rates in arable land and intensively grazed areas are estimated to be 100–1,000 times higher than natural erosion rates, significantly exceeding the rate of soil formation. This poses a significant challenge to agricultural productivity. When topsoil, rich in nutrients and organic matter, is lost, soil fertility declines, making it harder for crops to grow. In some cases, erosion also alters soil structure, increasing its susceptibility to drought.[11,12] Farmers rely on fertilizers to compensate for nutrient loss, which contributes to environmental issues such as waterway eutrophication and groundwater contamination. Additionally, lost soil pollutes water bodies, with about 60% of eroded soil ending up in rivers, streams, and lakes. This soil contains pesticides, synthetic fertilizers, and other pollutants, which further degrade aquatic ecosystems, reduce biodiversity, and contaminate groundwater supplies.[13,14]

Soil Acidification

Soil acidification is the gradual decline in soil pH over time. While it can occur naturally over thousands of years, especially in high-rainfall regions, it can also happen rapidly within a few years due to intensive agriculture. Certain farming practices accelerate this process, including the excessive application of ammonium-based nitrogen fertilizers beyond plant needs, the leaching of nitrate nitrogen from fertilizers out of the root zone, and the continuous removal of alkaline plant and animal products, which depletes the soil's natural buffering capacity. As soil acidity increases, plants sensitive to low pH levels experience reduced growth and productivity, allowing weeds to spread or leading to a loss of soil cover, which in turn increases the risk of erosion.[15,16]

Desertification

Desertification is a form of land degradation that occurs in arid, semi-arid, and dry sub-humid regions, resulting in reduced biological productivity and creating conditions similar to those of a desert. It is not the natural expansion of existing deserts, but rather a gradual process driven by human activities, such as agricultural expansion, poor irrigation practices, deforestation, overgrazing, and climate variations, including prolonged droughts and floods.[17,18,19] Infrastructure development, including roads, human settlements, and irrigation systems, can alter water availability, triggering sudden shifts in hydrology and accelerating desertification.[20] Desertification affects approximately 250 million people and one-third of the Earth's land surface, covering more than 4 billion hectares. The issue is global, not limited to Africa, where two-thirds of the continent consists of deserts or drylands. In the United States, 30% of the land is impacted, while one-quarter of Latin America and the Caribbean is also affected. In Europe, one-fifth of Spain's land is at risk of desertification.[21]

Deforestation, Forest Degradation, and Forest Fragmentation

Deforestation, forest degradation, and fragmentation threaten the survival of species and weaken forests' ability to provide essential ecosystem services, such as clean air and water, healthy soils for agriculture, and climate regulation.[22,23] According to the Global Forest Resources Assessment 2020 by the Food and Agriculture Organization (FAO), the rate of net forest loss has declined over the past decades. Specifically, the net loss decreased from 7.8 million hectares per year from 1990 to 2000 to 5.2 million hectares per year from 2000 to 2010, and then to 4.7 million hectares per year from 2010 to 2020.[24] Human activities have degraded over 75% of the world's remaining tropical forests, resulting in reduced biodiversity, increased carbon emissions, and disrupted ecosystem stability.[25]

Deforestation

Deforestation refers to the permanent loss of forest cover or its conversion to another land use, resulting in a reduction of tree canopy cover to below 10%. This includes forests cleared for agriculture, pasture, reservoirs, and urban development.[26] Globally, agriculture and timber extraction are the leading causes of deforestation, with commercial farming responsible for 40% and subsistence activities for 33%. Other contributors include mining, infrastructure expansion, and urbanization.[27] Deforestation is a significant contributor to land degradation, resulting in severe soil erosion, increased flooding, and the loss of fertile soil. Eastern Africa has experienced moderate to severe land degradation, particularly in northern Ethiopia, Sudan, and parts of Somalia. In Ethiopia, forest cover has declined from 65% to just 2.2% of the country's total area. This deforestation has reduced the soil's water-holding capacity and lowered stream base flows, affecting residual soil moisture for agricultural and domestic water resources.[28]

Forest Degradation and Fragmentation

Forest degradation refers to the long-term decline in the benefits that forests provide, such as timber, biodiversity, carbon storage, oxygen production, and water regulation, without a significant reduction in total forest area. The primary global drivers of forest degradation include unsustainable logging, fuelwood collection, charcoal production, overgrazing, uncontrolled fires, and defaunation (the loss of wildlife populations).[29] Forest fragmentation occurs when forests are converted to agriculture, urban areas, or other land uses, resulting in smaller, isolated patches of forest scattered across the landscape. This creates more forest edges, less interior forest habitat, and fewer corridors, disrupting the movement of wildlife, plant species, and water flows.[30] Sudan is considered a hotspot for forest degradation, experiencing a 17% increase in bare land and a 5% expansion of large-scale mechanized agriculture between 1973 and 2015. During this period, dense forest cover declined by 37%, while light forest cover increased by 20%. The leading causes of forest degradation and fragmentation in Sudan include the expansion of rain-fed agriculture, tree felling, woodcutting, overgrazing, and infrastructure development.[31]

Loss of Natural Grassland

Grasslands provide essential ecosystem services, including mitigating droughts and floods, preventing soil erosion, detoxifying waste, dispersing seeds, protecting waterways, and supporting wildlife habitats.[32,33] However, they have been heavily impacted by changes in grazing intensity, the introduction of non-native grasses, the use of agrochemicals, invasive species, and air pollution. Between 1998 and 2013, approximately 19% of the world's grasslands experienced persistent degradation, often cleared for soy, oil palm, cotton, wood pulp, and biofuels.[34] The global area of biofuel crops was estimated at 45 million hectares in 2010 and is projected to double by 2030, covering 3%–4.5% of all cultivated land.[35] Native grasslands are among the most endangered ecosystems due to their agricultural suitability and widespread conversion to cropland. In the United States, rising corn and soybean prices led to rapid grassland conversion in Minnesota, where, between 2008 and 2012, the native prairie was ploughed at a rate four times higher than in the previous 15 years. Grassland loss also affects groundwater systems. In Nebraska, converting grasslands to forests has resulted in a nearly 17% reduction in groundwater recharge, underscoring the broader hydrological impacts of grassland degradation.[36,37]

Freshwater Degradation

Freshwater ecosystems are impacted by land-use changes, deforestation, drainage, and water abstraction, all of which alter water quantity and degrade water quality. Changes in land and water use within a catchment area alter the morphology of rivers and streams, affecting water flow rates. Faster flows increase runoff and siltation, while reduced flows from water abstraction lower water velocities and shrink fish habitats.

River control structures, such as stopbanks, channel realignment, and streambank planting, help protect floodplains and infrastructure from flooding, and alter the river's width, slope, depth, and ecological connectivity. These modifications disrupt flood pulses, critical for seed dispersal, nutrient cycling, sediment transport, and species diversity in riverine floodplains. Similarly, dams for hydropower and irrigation alter channel structures, impeding fish migration and thereby affecting their reproduction and feeding patterns.[38]

Vegetation cover plays a key role in maintaining suitable habitat conditions for fish. Some species depend on shaded streams for survival, but deforestation and agricultural expansion remove tree canopies, causing higher summer temperatures, reduced oxygen levels, and altered pH, all of which can make waterways unsuitable for fish.[39,40] Urbanization in watersheds further degrades freshwater systems by reducing groundwater recharge, altering river flows, and disrupting natural hydrological cycles.[41]

Freshwater ecosystems are also impacted by pollution from agriculture, households, industry, and manufacturing. These activities introduce a range of organic and inorganic contaminants into streams, rivers, lakes, and aquifers, including heavy metals, fertilizers, and industrial chemicals. Heavy metals bind to sediment, degrading water quality and interfering with the processes of aquatic photosynthesis, respiration, growth, and reproduction. Additionally, surface runoff carrying sediments and discharges of warm water into cooler waterways further contribute to water quality deterioration.[42]

Wetland Degradation

Wetlands play a vital role in protecting water quality, storing floodwater, controlling erosion, and providing habitats for fish and wildlife.[43] However, human activities have led to significant wetland degradation and loss by altering water quality, quantity, and flow rates, increasing pollution, disrupting species composition, and introducing non-native species. Approximately 87% of wetlands have been lost over the past 300 years, with 54% of this loss occurring since 1900.[44] The primary drivers of wetland degradation include urban expansion, agricultural conversion, pollution, infrastructure development, and the introduction of invasive species. A summary of common activities that contribute to wetland degradation is provided in Table 5.1.[45,46,47,48,49,50]

Marine Degradation

Fishing is a major driver of ocean wildlife decline, with the number of overfished stocks worldwide tripling over the past 50 years. Nearly 90% of the world's marine fish stocks are fully exploited, overexploited, or depleted.[51] Overfishing disrupts food webs, as seen in the Baltic Sea, where excessive cod fishing has led to a surge in sprat populations, negatively affecting herring and reducing guillemot seabird reproduction due to changes in prey availability.[52] Bycatch, or the unintended capture of non-target marine life, is another significant consequence of overfishing, affecting marine mammals, seabirds, and other fish species.[53]

Table 5.1 Threats to Wetlands

Threat	*Description*
Land conversion	Urban development degrades wetlands by reducing habitat for wildlife and compromising services like water filtration, flood control, and biodiversity support.
Road building	Roads built through wetlands disrupt surface and groundwater flows, degrade habitat quality, and increase wildlife-vehicle conflict.
Dams	Dams alter river flows and trap sediment, disrupting downstream wetlands by changing water regimes and cutting off critical sediment supply.
Grazing	Uncontrolled livestock access increases nutrient loads, causes erosion, disturbs wildlife, and spreads invasive plants.
Pollution	High levels of sediment, nutrients, sewage, heavy metals, and pesticides from agriculture, industry, and urban runoff exceed wetlands' filtration capacity, harming species and degrading habitat.
Plant and animal pests	Invasive plants outcompete natives, and exotic animals like non-native fish disturb vegetation and water quality, threatening wetland biodiversity.
Fires	Natural fires are part of wetland ecology, but intense or uncontrolled burns due to reduced wetland size can lead to species loss and habitat degradation.
Water extraction	Excessive water withdrawal lowers wetland levels, reducing habitat and increasing pollutant concentration and eutrophication.
Changes to water quality	Shifts in water pH or nutrient levels can release harmful pollutants or limit access to vital minerals, harming plant and animal life.

Aquaculture is rapidly expanding, often in concentrated areas, resulting in localized environmental impacts. Shellfish farms, for instance, reduce phytoplankton levels, which are essential for other marine organisms. Aquaculture operations can also increase nutrient enrichment, depositing feces, uneaten food, shells, and live animals, which can smother seabed species and alter habitats. Additionally, escaped genetically modified fish may interbreed with wild populations, potentially disrupting the genetic diversity of native species and leading to unpredictable consequences.[54,55]

Degradation of Carbon Sinks

Carbon is a fundamental element in plants, animals, the atmosphere, oceans, and geological formations such as limestone and coal. It naturally cycles between these reservoirs through various processes. For example, carbon dioxide dissolves in the ocean, plants absorb it during photosynthesis, and animals consume plant-based carbohydrates, releasing carbon dioxide through respiration. Processes that add carbon dioxide to the atmosphere are referred to as carbon sources, while those that remove and store carbon dioxide are known as carbon sinks.[56]

Soils

The world's soils store between 700 and 3,000 gigatons of carbon, holding more than three times the amount of carbon present in the atmosphere as carbon dioxide. However, due to land clearing and cultivation, agricultural soils have lost 50%–70% of their original soil organic carbon pool, which consists of microbes, fungi, invertebrates, root matter, and decomposing vegetation.[57,58] The total historical loss of soil organic carbon from human activities, primarily due to the conversion of natural ecosystems into agricultural land, is estimated at 176 gigatons. Future projections indicate that from 2010 to 2050, an additional 27 gigatons of soil organic carbon could be released. Of this, 16 gigatons are expected to come from new land conversion for agriculture, while 11 gigatons will result from ongoing land degradation and declining productivity.[59] Land degradation is one of the most significant contributors to climate change, as soils play a crucial role in carbon storage. An increase of just 1% in the carbon stocks within the top one meter of soil could store more carbon than the total annual carbon dioxide emissions from fossil fuel burning.[60]

Forests

Forests store carbon dioxide in several key pools:

- Vegetation: Living plant biomass, including both wood and non-wood materials. While above-ground biomass is visible, the below-ground root system also contains significant carbon stores. The carbon content of biomass ranges from 35% to 65% of its dry weight.
- Deadwood and litter: Decomposing plant material is an essential source of nutrients for plant growth.
- Soil organic matter (humus): Formed from the decomposition of litter, this is a key component in long-term carbon storage in forest ecosystems.[61]

Globally, forests store an estimated 296 gigatons of carbon dioxide in their biomass, both above and below ground. Each year, forests absorb approximately 2.4 billion tons of carbon dioxide, which is roughly one-third of the annual carbon dioxide emissions from the combustion of fossil fuels.[62] The highest carbon densities are found in forests in South America and Western and Central Africa, which store around 120 tons of carbon per hectare in living biomass alone, compared to the global average of 74 tons per hectare. However, over the past 25 years, carbon stocks in forest biomass have declined by nearly 11.1 gigatons, primarily due to deforestation, land conversion for agriculture and settlements, and forest degradation.[63]

According to FAO, the global forest area shrank by 3% between 1990 and 2015, with the most significant losses occurring in tropical regions.[64] These tropical forests release more carbon than they absorb, primarily due to deforestation and land degradation. One study estimates that tropical forests emit 425 million metric tons of carbon annually.[65] Further research in Borneo has shown that drought conditions can temporarily halt the forest carbon sink by increasing tree mortality. In contrast, forest fragmentation has turned many edge-affected forests into carbon sources.[66] Deforestation is the second-largest driver of climate change, after fossil fuel combustion, and accounts for nearly 20% of total greenhouse gas emissions, more than the entire global transport sector.[67]

Grasslands

Between 1998 and 2013, 19% of global grasslands experienced a persistent decline in productivity due to the use of agrochemicals, invasive plant and animal species, and increased crop production.[68] Although grasslands store less carbon than forests, semi-arid Australian grasslands store around five tons of carbon per hectare, compared to mature, dry sclerophyll forests, which contain about 100 tons per hectare. They are expected to become more resilient carbon sinks in a changing climate.[69] In wildfire-prone regions such as California, forests are struggling to adapt to unmitigated climate change, potentially shifting from carbon sinks to carbon sources by mid-century due to the increasing frequency and severity of wildfires. In contrast, grasslands are expected to remain stable carbon sinks throughout the 21st century, as they are more resilient to disturbances such as fire and drought.[70]

Wetlands

Wetland soils store some of the highest levels of carbon in the biosphere, containing 20%–30% of the estimated 1,500 petagrams of global soil carbon despite covering only 5%–8% of the Earth's land surface. The anoxic conditions in wetlands slow decomposition, allowing organic matter to accumulate over time and making wetlands a significant carbon sink for atmospheric carbon dioxide.[71] Coastal wetlands play a crucial role in carbon sequestration. Although coastal wetland vegetation covers just 2% of the world's seabed area, it accounts for around 50% of the carbon transferred to ocean sediments. The carbon sequestration rate in coastal vegetation is 100 times faster than in terrestrial forests, highlighting their importance in mitigating climate change.[72] Despite their ecological significance, coastal wetlands are threatened globally, especially in rapidly urbanizing coastal areas. In China, a study found that nearly 24% of the natural coastal wetland area, originally 18,243 square kilometers in 1990, had disappeared by 2015, primarily due to urban expansion and land reclamation.[73]

Peatlands

Peatlands cover only about 3% of the Earth's land area but store around 30% of global soil organic carbon, making them one of the most effective natural carbon sinks. Near-natural peatlands span over three million square kilometers worldwide and sequester approximately 0.37 gigatons of carbon dioxide annually, storing more carbon than all other vegetation types combined. Despite their crucial role in carbon storage, peatlands are highly vulnerable to human activities, particularly agricultural expansion.[74] In Indonesia, peatlands are extensively drained and deforested to make way for palm oil plantations and croplands, leading to significant carbon loss. Carbon is released directly through carbon dioxide emissions from drained peatland and indirectly through soil leaching, with approximately 50% of the leached carbon decomposing and emitting carbon dioxide into rivers.[75] If peatland drainage continues at current rates, peat soil degradation and associated fires could release an estimated nine gigatons of carbon between 2010 and 2050, converting these critical carbon sinks into primary carbon sources and exacerbating climate change.[76]

Ocean Plants

Blue carbon ecosystems play a crucial role in oceanic carbon sequestration, accounting for up to 33% of total carbon dioxide uptake in marine environments. Unlike terrestrial soils, which typically store carbon for decades, blue carbon ecosystems can retain carbon for millennia or longer, making them highly effective long-term carbon sinks.[77] Seagrasses, flowering marine plants that thrive in shallow coastal waters, are key contributors to the storage of blue carbon. With over 70 species worldwide, seagrasses provide essential habitat for marine life, including invertebrates, crabs, turtles, large fish, and birds. These meadows capture up to 83 million metric tons of carbon annually, and despite covering just 0.1% of the ocean floor, they store up to 11% of the total organic carbon buried in the ocean. It is estimated that one acre of seagrass can sequester 83 grams of carbon per square meter per year, equivalent to the carbon dioxide emissions from a car traveling approximately 6,200 kilometers.[78] However, seagrass meadows are rapidly declining due to coastal development, declining water quality resulting from sewage and agricultural runoff, and the impacts of climate change. Since 1980, seagrass has been disappearing at a rate of 110 square kilometers per year, with accelerated losses recorded over time. Decline rates have surged from 0.9% per year before 1940 to 7% per year since 1990, underscoring the urgent need for conservation efforts to protect these critical carbon sinks.[79,80]

Habitat Loss and Fragmentation

Habitat loss and fragmentation are the primary threats to global biodiversity, primarily driven by human activities such as agriculture, mining, and urbanization.[81] Habitat loss occurs when natural landscapes are cleared or converted for human use. At the same time, fragmentation happens when large, continuous habitats are broken into smaller, isolated patches, disrupting ecosystems and reducing connectivity between species populations. Fragmentation affects both terrestrial and aquatic ecosystems. On land, forests are converted into farmland, leaving fragmented patches that can no longer support the same level of biodiversity. In aquatic systems, fragmentation occurs when dams, roads, and water abstraction disrupt connectivity between upstream and downstream habitats, affecting species that rely on free-flowing water.[82,83] Despite appearing continuous, ocean habitats also experience fragmentation due to physiochemical barriers that limit movement, particularly for smaller organisms. Coral reefs face severe habitat loss from dynamite fishing, coral bleaching, and ocean acidification, while seagrass meadows are degraded by dredging, boating activities, nutrient pollution (eutrophication), and extreme weather events. These disruptions weaken ecosystems, reduce species resilience, and threaten the biodiversity that depends on these critical habitats.[84]

Pollution

Pollution significantly threatens biodiversity, affecting ecosystems through the presence of ocean litter, pesticides, fertilizers, and air pollution. Human activities introduce pollutants into the environment, disrupting ecosystems and harming wildlife.

Ocean litter originates from multiple sources, including items lost at sea during storms and urban waste washed into waterways.[85] Marine debris, particularly plastic waste, degrades slowly and persists in the ocean for years. Plastics pose a serious risk to biodiversity. Larger debris entangle marine animals, while microplastics are ingested by plankton and invertebrates, allowing harmful chemicals and pollutants to accumulate in the food chain.[86] These contaminants reduce the nutritional value of diets, cause physical harm, expose species to pathogens, and facilitate the transport of invasive species. Additionally, plastics contain chemical additives, which can leach into organisms after ingestion, further threatening marine life.

Pesticides and fertilizers from agricultural lands pollute nearby soil, streams, and groundwater when they are carried by rainfall. This type of pollution is a significant contributor to biodiversity loss and ecosystem disruption. Pesticides kill non-target species, including beneficial insects, soil microbes, and fish, while fertilizers overburden freshwater and marine ecosystems with excessive nutrients. This triggers harmful algal blooms, which deplete oxygen in the water, creating dead zones where fish and other aquatic life cannot survive.

Air pollution also impacts biodiversity. Sulfur dioxide and nitrogen oxides contribute to acid rain, which increases soil and water acidity, harming plants, aquatic life, and soil organisms. Additionally, ground-level ozone damages plant cell membranes, disrupting photosynthesis and growth. The cumulative impact of these pollutants weakens ecosystems, threatening the survival of species and reducing overall biodiversity.[87,88]

Environmental Degradation Impacts on Biodiversity

Environmental degradation threatens biodiversity, encompassing the variety of life at the genetic, species, and ecosystem levels. Biodiversity is crucial in regulating the climate, filtering air and water, forming soil, and mitigating the effects of natural disasters. It also supports essential resources and services, including timber, fish, crops, pollination, and ecotourism, while contributing to medical advancements and overall physical and mental well-being.[89,90]

Impacts on Birds

Land-based bird populations are increasingly at risk of extinction due to habitat loss and degradation. A study on Galliformes, a group of ground-dwelling birds including pheasants, grouse, and quail, in Sundaland (covering Thailand's peninsula, Borneo, Malaysia, Sumatra, Java, and Bali) found that 25% of Galliform populations in the region have disappeared outside of nature reserves. In Sumatra, this figure rises to 50%, with habitat loss primarily driven by agricultural expansion and infrastructure development outside protected areas.[91]

Migratory birds, which travel across land and sea in response to seasonal changes, are also experiencing widespread declines. Over 30 years, more than half of migratory bird species across all major flyways have declined. These species face threats at multiple points in their annual migration cycle, meaning a single disruption can affect an entire population. Despite these challenges, only 9% of 1,451 migratory bird species are adequately protected throughout their full migration cycle, compared to 45% of non-migratory species.[92] Migratory birds rely on multiple, often disconnected habitats, which reduces the likelihood that all necessary locations will be sufficiently protected. Less than 3% of threatened migratory birds have adequate protected area coverage across all parts of their migration routes.[93]

Impacts on Insects

Over 40% of insect species are at risk of extinction, with their decline primarily driven by habitat loss, land conversion for intensive agriculture and urbanization, pollution from synthetic pesticides and fertilizers, invasive species, pathogens, and climate change. These losses affect specialized insects occupying specific ecological niches and common and generalist species.[94] The decline of insect populations has significant consequences for ecosystem stability, as insects play crucial roles in pollination and serve as a key food source for birds, mammals, and amphibians. An estimated 80% of wild plants depend on insects for pollination, while 60% of bird species rely on insects as a primary food source. A study in Germany's nature reserves found a 76% decline in flying insect populations between 1989 and 2016.[95] The threat extends to freshwater insects, with approximately 25% of species facing extinction. Biodiversity loss in freshwater ecosystems occurs at a faster rate than in terrestrial or marine environments, disrupting essential ecosystem functions such as nutrient cycling, decomposition, and food sources for indigenous communities and other organisms.[96]

Impacts on Plants

Pollution of air, water, and soil poses serious risks to organisms. For example, particulate matter combined with pollutants like sulfur dioxide can impair plant growth. Coarse particles, such as dust, accumulate on leaf surfaces, disrupting gas exchange, increasing leaf temperature, and reducing photosynthesis. Toxic elements such as arsenic and fluoride can accumulate in agricultural soils and plants, which are then ingested by animals, negatively impacting their health.[97] Ozone pollution also harms vegetation and ecosystems. When ozone enters the leaves of sensitive plants, it can reduce photosynthesis, slow growth, and make plants more vulnerable to disease, insect damage, extreme weather, and other pollutants. These impacts extend to entire ecosystems, resulting in the loss of species diversity, shifts in plant composition, habitat degradation, and disruptions in water and nutrient cycles.[98]

Impacts on Freshwater Fish

Habitat fragmentation in freshwater ecosystems is increasing rapidly, resulting in greater biodiversity declines than in the most severely affected terrestrial ecosystems. Human-made structures such as dams, culverts, fords, and weirs disrupt natural water flows, degrade habitats, reduce connectivity, and block access to critical spawning areas, significantly impacting fish migration. For example, nearly half of New Zealand's native freshwater fish are diadromous, requiring both freshwater and marine environments to complete their life cycles. In locations where dams obstruct fish passage, species richness is significantly lower.[99]

Sediment pollution also poses a significant threat to freshwater fish populations. Sediments originate from natural sources, such as upland erosion, riparian areas, and stream banks. However, human activities like agriculture, mining, logging, and urban development significantly increase sediment loads. Forestry, for instance, contributes to sedimentation through deforestation and soil erosion, resulting in high turbidity levels in waterways.[100,101] Fine sediment pollution affects freshwater fish in multiple ways:

1. Directly harming fish, leading to mortality, reduced growth rates, and increased disease susceptibility.
2. Inhibiting the development of fish eggs and larvae reduces the recruitment success of fish.
3. Altering natural habitats and migration patterns makes it difficult for fish to reach essential spawning and feeding grounds.

4. Reducing food availability, as sedimentation smothers aquatic vegetation and invertebrates.
5. Interfering with feeding efficiency, making it harder for fish to locate and capture prey.[102]

Water pollution is another major driver of freshwater biodiversity loss, driven by economic and demographic pressures. Nutrients, pathogens, heavy metals, organic pollutants, and micro-pollutants from human activities contribute to the decline in water quality. Untreated wastewater is a significant concern, with up to 90% of global wastewater being discharged directly into water bodies without treatment.[103] A study in the Barnoi River, Bangladesh, found that after eight years of untreated wastewater discharge, fish abundance declined by over 47% and species richness dropped by 35% at downstream sites compared to unaffected upstream areas. Water transparency, dissolved oxygen levels, and pH were significantly lower in polluted downstream sites.[104]

Impacts on Marine Life

Marine ecosystems face multiple threats, including ocean acidification, warming, habitat loss, and pollution. Rising ocean temperatures and increasing acidity reduce survival and growth rates, alter species distributions, and modify habitats. Additionally, marine species are affected by habitat destruction, which results in breeding, predation, and competition for food due to commercial fishing. In New Zealand, a Ministry for the Environment study identified excess sedimentation as one of the biggest threats to marine habitats, alongside seabed trawling, dredging for fish and shellfish, marine pests, and nutrient pollution from land runoff. Sediment accumulation is primarily driven by forest harvesting, land clearing for agriculture, and urban development, with rocky reefs, seagrass meadows, and sand and mudflats particularly vulnerable. In a study on the impacts of sediment on reef fish species (Amphiprion melanopus, Amphiprion percula, and Acanthochromis polyacanthus), Amphiprion melanopus was found to be the most sensitive, exhibiting gill damage, reduced oxygen uptake, and impaired oxygen delivery to tissues, ultimately compromising its performance and survival.[105,106]

Toxic chemicals and heavy metals entering the ocean pose severe risks to marine organisms, causing disease, genetic mutations, congenital disabilities, reproductive issues, behavioral changes, and mortality. The impact varies by species, with top predators most affected due to biomagnification.[107] In the Gulf of Maine, methylmercury biomagnifies more than tenfold in species such as swordfish, Atlantic bluefin tuna, harbor porpoises, Atlantic white-sided dolphins, and common thresher sharks.[108] Plastics in the marine environment also pose significant chemical hazards. Many plastics contain endocrine-disrupting chemicals that are also found in pesticides, metals, food additives, and personal care products. These chemicals can cause impaired reproduction, low birth rates, biodiversity loss, thyroid dysfunction, metabolic disorders, and hormone-sensitive cancers in marine organisms.[109,110]

Notes

1 United Nations, "Glossary of Environment Statistics, Studies in Methods," (1997), https://unstats.un.org/unsd/publication/seriesf/seriesf_67e.pdf.
2 D.L. Johnson et al., "Meanings of Environmental Terms," *Journal of Environmental Quality* 26, no. 3 (1997).
3 IPBES, "Media Release: Nature's Dangerous Decline 'Unprecedented'; Species Extinction Rates 'Accelerating'," (n.d.), https://ipbes.net/news/Media-Release-Global-Assessment.
4 IPBES, "Media Release: Nature's Dangerous Decline 'Unprecedented'; Species Extinction Rates 'Accelerating'."
5 S.A. Moges and A.S. Gebregiorgis, "5.21 – Climate Vulnerability on the Water Resources Systems and Potential Adaptation Approaches in East Africa: The Case of Ethiopia," in *Climate Vulnerability*, ed. Roger A. Pielke (Oxford: Academic Press, 2013).
6 WHO, "Land Degradation and Desertification," (n.d.), https://who.int/globalchange/ecosystems/desert/en/.
7 UNEP, "Global Environment Outlook 6," (2019), https://unenvironment.org/resources/global-environment-outlook-6.
8 Pasquale Borrelli et al., "An Assessment of the Global Impact of 21st Century Land Use Change on Soil Erosion," *Nature Communications* 8, no. 1 (2017).
9 Ontario Ministry of Agriculture, Food and Rural Affairs, "Soil Erosion – Causes and Effects," (n.d.), http://omafra.gov.on.ca/english/engineer/facts/12-053.htm.
10 FoodPrint, "How Industrial Agriculture Affects Our Soil," (n.d.), https://foodprint.org/issues/how-industrial-agriculture-affects-our-soil/.
11 FoodPrint, "How Industrial Agriculture Affects Our Soil."
12 United Nations Convention to Combat Desertification, "Global Land Outlook," (2017), https://knowledge.unccd.int/sites/default/files/2018-06/GLO%20English_Full_Report_rev1.pdf.
13 Robert L. Zimdahl, "11 – A Glimpse Ahead," in *Agriculture's Ethical Horizon (Second Edition)*, ed. Robert L. Zimdahl (Oxford: Elsevier, 2012).
14 R.C. Brears, *The Green Economy and the Water-Energy-Food Nexus* (London: Palgrave Macmillan UK, 2017).
15 Queensland Government, "Soil Acidification," (2015), https://publications.qld.gov.au/dataset/science-notes-soils/resource/2a2b6fc7-671d-4031-96ad-377eaabba6fa.
16 Government of South Australia, "Managing Soil Acidity," (2011), https://naturalresources.sa.gov.au/files/003653 65-5cda-4bbd-a55c-a31c00a21cb6/sustainable-agriculture-managing-soil-acidity-fact.pdf.
17 PBR, "What's Behind Desertification?," (n.d.), https://prb.org/whatsbehinddesertification/.
18 UNCCD, "Frequently Asked Questions (Faq)," (n.d.), https://unccd.int/frequently-asked-questions-faq.
19 Moges and Gebregiorgis, "5.21 – Climate Vulnerability on the Water Resources Systems and Potential Adaptation Approaches in East Africa: The Case of Ethiopia."

20 Matt Rovero, "Nutrient Cycling Causes and Impacts of Desertification," in *Reference Module in Earth Systems and Environmental Sciences* (Amsterdam: Elsevier, 2017).
21 UNCCD, "Frequently Asked Questions (Faq)."
22 IUCN, "Deforestation and Forest Degradation," (2017), https://iucn.org/resources/issues-briefs/deforestation-and-forest-degradation.
23 FAO, "State of the World's Forests 2018: Forests and Agriculture: Land Use Challenges and Opportunities," (2018), http://fao.org/3/I9535EN/i9535en.pdf.
24 FAO, "Global Forest Resources Assessment 2020 Key Findings," (2020), https://openknowledge.fao.org/server/api/core/bitstreams/9f24d451-2e56-4ae2-8a4a-1bc511f5e60e/content?utm_source=chatgpt.com.
25 Simon L. Lewis, David P. Edwards, and David Galbraith, "Increasing Human Dominance of Tropical Forests," *Science* 349, no. 6250 (2015).
26 FAO, "Deforestation," (n.d.), http://fao.org/3/j9345e/j9345e07.htm.
27 Noriko Hosonuma et al., "An Assessment of Deforestation and Forest Degradation Drivers in Developing Countries," *Environmental Research Letters* 7, no. 4 (2012).
28 Moges and Gebregiorgis, "5.21 – Climate Vulnerability on the Water Resources Systems and Potential Adaptation Approaches in East Africa: The Case of Ethiopia."
29 FAO, "Reducing Forest Degradation," (2018), http://fao.org/sustainable-forest-management/toolbox/modules/reducing-forest-degradation/cases/en/?type=111.
30 United States Department of Agriculture Forest Service, "Issues and Influences: Forest Fragmentation and Parcelization," (n.d.), https://nrs.fs.fed.us/futures/issues_influences/fragmentation_parcelization/.
31 Hussein M. Sulieman, "Exploring Drivers of Forest Degradation and Fragmentation in Sudan: The Case of Erawashda Forest and Its Surrounding Community," *Science of the Total Environment* 621 (2018).
32 U.S. Forest Service, "Ecosystem Services from National Grasslands," (n.d.), https://fs.fed.us/grasslands/ecoservices/index.shtml.
33 Bianca O. Andrade et al., "Grassland Degradation and Restoration: A Conceptual Framework of Stages and Thresholds Illustrated by Southern Brazilian Grasslands," *Natureza & Conservação* 13, no. 2 (2015).
34 United Nations Convention to Combat Desertification, "Global Land Outlook."
35 United Nations Convention to Combat Desertification, "Global Land Outlook."
36 Tyler J. Lark et al., "Accelerated Conversion of Native Prairie to Cropland in Minnesota," *Environmental Conservation* 46, no. 2 (2018).
37 Zablon A. Adane et al., "Impact of Grassland Conversion to Forest on Groundwater Recharge in the Nebraska Sand Hills," *Journal of Hydrology: Regional Studies* 15 (2018).
38 G.M. Reid, T.C. MacBeath, and K. Csatádi, "Global Challenges in Freshwater-Fish Conservation Related to Public Aquariums and the Aquarium Industry," *International Zoo Yearbook* 47, no. 1 (2013).
39 Robert I. McDonald et al., "Estimating Watershed Degradation over the Last Century and Its Impact on Water-Treatment Costs for the World's Large Cities," *Proceedings of the National Academy of Sciences* 113, no. 32 (2016).
40 NIWA, "Habitat Loss," (n.d.), https://niwa.co.nz/our-science/freshwater/tools/guide-to-restoring-freshwater-native-fish/identifying-the-problem/habitat-loss.
41 R.C. Brears, *Blue and Green Cities: The Role of Blue-Green Infrastructure in Managing Urban Water Resources* (London, UK: Palgrave Macmillan UK, 2018).
42 R.C. Brears, *Urban Water Security* (Chichester and Hoboken, NJ: John Wiley & Sons, 2016).

43 US EPA, "Threats to Wetlands," (2001), http://virginia.edu/blandy/blandy_web/education/Bay/WetlandThreats_EPA.pdf.
44 Ramsar, "75% of Earth's Land Areas Are Degraded; Wetlands Have Been Hit Hardest, with 87% Lost Globally in the Last 300 Years," (2017), https://ramsar.org/news/75-of-earths-land-areas-are-degraded-wetlands-have-been-hit-hardest-with-87-lost-globally-in.
45 National Park Service, "Threats to Wetlands," (n.d.), https://nps.gov/subjects/wetlands/threats-to-wetlands.htm.
46 US EPA, "Threats to Wetlands."
47 Conservation Department of Biodiversity, and Attractions, "Threats to Wetlands," (n.d.), https://dpaw.wa.gov.au/management/wetlands/conserving-and-managing-our-wetlands/threats-to-wetlands.
48 International Rivers, "Environmental Impacts of Dams," (n.d.), https://internationalrivers.org/environmental-impacts-of-dams.
49 University of Montana, "Threats to Wetlands," (n.d.), http://cas.umt.edu/watershedclinic/links/clarkfork/wetlands/threatstowetlands.htm.
50 Qiaoying Lin and Shen Yu, "Losses of Natural Coastal Wetlands by Land Conversion and Ecological Degradation in the Urbanizing Chinese Coast," *Scientific Reports* 8, no. 1 (2018).
51 United Nations Conference on Trade and Development, "90% of Fish Stocks Are Used up – Fisheries Subsidies Must Stop," (2018), https://unctad.org/en/pages/newsdetails.aspx?OriginalVersionID=1812.
52 Christian Möllmann and Rabea Diekmann, "Chapter 4 – Marine Ecosystem Regime Shifts Induced by Climate and Overfishing: A Review for the Northern Hemisphere," in *Advances in Ecological Research*, ed. Guy Woodward, Ute Jacob, and Eoin J. O'Gorman (London, UK: Academic Press, 2012).
53 WWF, "Overfishing," (n.d.), https://worldwildlife.org/threats/overfishing.
54 Ministry for the Environment, "Environment Aotearoa 2015," (2015), https://mfe.govt.nz/publications/environmental-reporting/environment-aotearoa-2015.
55 Cyrus Martin, "Not So Many Fish in the Sea," *Current Biology* 27, no. 11 (2017).
56 NIWA, "What Is a Carbon Sink?," (n.d.), https://niwa.co.nz/atmosphere/faq/what-is-a-carbon-sink.
57 Parliament of Australia, "Carbon Sequestration," (n.d.), https://aph.gov.au/About_Parliament/Parliamentary_Departments/Parliamentary_Library/Browse_by_Topic/ClimateChangeold/responses/mitigation/carbon.
58 IUCN, "Land Degradation and Climate Change," (2015), https://iucn.org/sites/dev/files/import/downloads/land_degradation_issues_brief_cop21_031215.pdf.
59 United Nations Convention to Combat Desertification, "Global Land Outlook."
60 IUCN, "Land Degradation and Climate Change."
61 FAO, "Forests and Climate Change – Instruments Related to the United Nations Framework Convention on Climate Change and Their Potential for Sustainable Forest Management in Africa, Forests and Climate Change Working Paper No. 1," (2003), http://fao.org/forestry/climatechange/53622/en/.
62 IUCN, "Deforestation and Forest Degradation."
63 FAO, "Global Forest Resources Assessment 2015: How Are the World's Forests Changing?," (2016), http://fao.org/3/a-i4793e.pdf.
64 FAO, "Global Forest Resources Assessment 2015: How Are the World's Forests Changing?"

65 A. Baccini et al., "Tropical Forests Are a Net Carbon Source Based on Aboveground Measurements of Gain and Loss," *Science* 358, no. 6360 (2017).
66 Lan Qie et al., "Long-Term Carbon Sink in Borneo's Forests Halted by Drought and Vulnerable to Edge Effects," *Nature Communications* 8, no. 1 (2017).
67 FAO, "The State of the World's Forests 2018 – Forest Pathways to Sustainable Development," (2018), http://fao.org/3/I9535EN/i9535en.pdf.
68 United Nations Convention to Combat Desertification, "Global Land Outlook."
69 Australian Government Chief Scientist, "Which Plants Store More Carbon in Australia: Forests or Grasses?," (2009), https://chiefscientist.gov.au/2009/12/which-plants-store-more-carbon-in-australia-forests-or-grasses/.
70 Pawlok Dass et al., "Grasslands May Be More Reliable Carbon Sinks Than Forests in California," *Environmental Research Letters* 13, no. 7 (2018).
71 A.M. Nahlik and M.S. Fennessy, "Carbon Storage in Us Wetlands," *Nature Communications* 7 (2016).
72 NIWA, "Muddy Sinks," (n.d.), https://niwa.co.nz/news/muddy-sinks.
73 Lin and Yu, "Losses of Natural Coastal Wetlands by Land Conversion and Ecological Degradation in the Urbanizing Chinese Coast."
74 Sirui Wang et al., "Potential Shift from a Carbon Sink to a Source in Amazonian Peatlands under a Changing Climate," *Proceedings of the National Academy of Sciences* 115, no. 49 (2018).
75 Francisca Wit et al., "The Invisible Carbon Footprint as a Hidden Impact of Peatland Degradation Inducing Marine Carbonate Dissolution in Sumatra, Indonesia," *Scientific Reports* 8, no. 1 (2018).
76 United Nations Convention to Combat Desertification, "Global Land Outlook."
77 Maria E. Röhr et al., "Blue Carbon Storage Capacity of Temperate Eelgrass (Zostera Marina) Meadows," *Global Biogeochemical Cycles* 32, no. 10 (2018).
78 Smithsonian Ocean, "Seagrass and Seagrass Beds," (n.d.), https://ocean.si.edu/ocean-life/plants-algae/seagrass-and-seagrass-beds.
79 Michelle Waycott et al., "Accelerating Loss of Seagrasses across the Globe Threatens Coastal Ecosystems," *Proceedings of the National Academy of Sciences* 106, (2009).
80 Benjamin L. Jones, Leanne C. Cullen-Unsworth, and Richard K.F. Unsworth, "Tracking Nitrogen Source Using δ15n Reveals Human and Agricultural Drivers of Seagrass Degradation across the British Isles," *Frontiers in Plant Science* 9, no. 133 (2018).
81 Jordan E. Rogan and Thomas E. Lacher, "Impacts of Habitat Loss and Fragmentation on Terrestrial Biodiversity," in *Reference Module in Earth Systems and Environmental Sciences* (Amsterdam: Elsevier, 2018).
82 Melanie Hagen et al., "2 – Biodiversity, Species Interactions and Ecological Networks in a Fragmented World," in *Advances in Ecological Research*, ed. Ute Jacob and Guy Woodward (London, UK: Academic Press, 2012).
83 Matthew R. Fuller, Martin W. Doyle, and David L. Strayer, "Causes and Consequences of Habitat Fragmentation in River Networks," *Annals of the New York Academy of Sciences* 1355, no. 1 (2015).
84 Hagen et al., "2 – Biodiversity, Species Interactions and Ecological Networks in a Fragmented World."
85 WWF, "Pollution," (n.d.), https://worldwildlife.org/threats/pollution.
86 Carlo G. Avio, Stefania Gorbi, and Francesco Regoli, "Plastics and Microplastics in the Oceans: From Emerging Pollutants to Emerged Threat," *Marine Environmental Research* 128 (2017).

87 WWF, "Pollution."

88 UNECE, "Air Pollution, Ecosystems and Biodiversity," (n.d.), http://unece.org/environmental-policy/conventions/envlrtapwelcome/cross-sectoral-linkages/air-pollution-ecosystems-and-biodiversity.html.

89 UNEP, "Global Environment Outlook 6."

90 Francisco Sánchez-Bayo and Kris A.G. Wyckhuys, "Worldwide Decline of the Entomofauna: A Review of Its Drivers," *Biological Conservation* 232 (2019).

91 Elizabeth H. Boakes, Richard A. Fuller, and Philip J.K. McGowan, "The Extirpation of Species Outside Protected Areas," *Conservation Letters* 12, no. 1 (2019).

92 Claire A. Runge et al., "Protected Areas and Global Conservation of Migratory Birds," *Science* 350, no. 6265 (2015).

93 Runge et al., "Protected Areas and Global Conservation of Migratory Birds."

94 Sánchez-Bayo and Wyckhuys, "Worldwide Decline of the Entomofauna: A Review of Its Drivers."

95 Caspar A. Hallmann et al., "More Than 75 Percent Decline over 27 Years in Total Flying Insect Biomass in Protected Areas," *PLOS ONE* 12, no. 10 (2017).

96 Craig R. Macadam and Jenni A. Stockan, "More Than Just Fish Food: Ecosystem Services Provided by Freshwater Insects," *Ecological Entomology* 40, no. S1 (2015).

97 Bruce G. Miller, "Chapter 3 – The Effect of Coal Usage on Human Health and the Environment," in *Coal Energy Systems*, ed. Bruce G. Miller (Burlington: Academic Press, 2005).

98 US EPA, "Ecosystem Effects of Ozone Pollution," (n.d.). https://epa.gov/ground-level-ozone-pollution/ecosystem-effects-ozone-pollution.

99 Department of Conservation, "Protecting New Zealand's Rivers," (2011), https://doc.govt.nz/about-us/statutory-and-advisory-bodies/nz-conservation-authority/publications/protecting-new-zealands-rivers/.

100 Michael E. Kjelland et al., "A Review of the Potential Effects of Suspended Sediment on Fishes: Potential Dredging-Related Physiological, Behavioral, and Transgenerational Implications," *Environment Systems and Decisions* 35, no. 3 (2015).

101 Minnesota Pollution Control Agency, "Turbidity: Description, Impact on Water Quality, Sources, Measures," (2008), https://pca.state.mn.us/sites/default/files/wq-iw3-21.pdf.

102 Minnesota Pollution Control Agency, "Turbidity: Description, Impact on Water Quality, Sources, Measures."

103 UN-Water, "Towards a Worldwide Assessment of Freshwater Quality: A Un-Water Analytical Brief," (2016), http://unwater.org/publications/towards-worldwide-assessment-freshwater-quality/.

104 Shams M. Galib et al., "Municipal Wastewater Can Result in a Dramatic Decline in Freshwater Fishes: A Lesson from a Developing Country," *Knowledge and Management of Aquatic Ecosystems*, 419 (2018).

105 Ministry for the Environment & Statistics New Zealand, "Our Marine Environment 2016," (2016), https://mfe.govt.nz/sites/default/files/media/Environmental%20reporting/our-marine-environment.pdf.

106 Sybille Hess et al., "Species-Specific Impacts of Suspended Sediments on Gill Structure and Function in Coral Reef Fishes," *Proceedings of the Royal Society B: Biological Sciences* 284, no. 1866 (2017).

107 D.W. Blowes et al., "9.05 – the Geochemistry of Acid Mine Drainage," in *Treatise on Geochemistry*, ed. Heinrich D. Holland and Karl K. Turekian (Oxford: Pergamon, 2003).

108 Gareth Harding, John Dalziel, and Peter Vass, "Bioaccumulation of Methylmercury within the Marine Food Web of the Outer Bay of Fundy, Gulf of Maine," *PLOS ONE* 13, no. 7 (2018).

109 Claude Monneret, "What Is an Endocrine Disruptor?," *Comptes Rendus Biologies* 340, no. 9 (2017).

110 Frederic Gallo et al., "Marine Litter Plastics and Microplastics and Their Toxic Chemicals Components: The Need for Urgent Preventive Measures," *Environmental sciences Europe* 30, no. 1 (2018).

References

Adane, Zablon A., Paolo Nasta, Vitaly Zlotnik, and David Wedin. "Impact of Grassland Conversion to Forest on Groundwater Recharge in the Nebraska Sand Hills." *Journal of Hydrology: Regional Studies* 15 (2018/02/01): 171–83.

Andrade, Bianca O., Christiane Koch, Ilsi I. Boldrini, Eduardo Vélez-Martin, Heinrich Hasenack, Julia-Maria Hermann, Johannes Kollmann, Valério D. Pillar, and Gerhard E. Overbeck. "Grassland Degradation and Restoration: A Conceptual Framework of Stages and Thresholds Illustrated by Southern Brazilian Grasslands." *Natureza & Conservação* 13, no. 2 (2015/07/01): 95–104.

Australian Government Chief Scientist. "Which Plants Store More Carbon in Australia: Forests or Grasses?" (2009). https://chiefscientist.gov.au/2009/12/which-plants-store-more-carbon-in-australia-forests-or-grasses/.

Avio, Carlo G., Stefania Gorbi, and Francesco Regoli. "Plastics and Microplastics in the Oceans: From Emerging Pollutants to Emerged Threat." *Marine Environmental Research* 128 (2017/07/01): 2–11.

Baccini, A., W. Walker, L. Carvalho, M. Farina, D. Sulla-Menashe, and R.A. Houghton. "Tropical Forests Are a Net Carbon Source Based on Aboveground Measurements of Gain and Loss." *Science* 358, no. 6360 (2017): 230.

Blowes, D.W., C.J. Ptacek, J.L. Jambor, and C.G. Weisener. "9.05 – The Geochemistry of Acid Mine Drainage." In *Treatise on Geochemistry*, edited by Heinrich D. Holland, and Karl K. Turekian, 149–204. Oxford: Pergamon, 2003.

Boakes, Elizabeth H., Richard A. Fuller, and Philip J.K. McGowan. "The Extirpation of Species Outside Protected Areas." *Conservation Letters* 12, no. 1 (2019/01/01): e12608.

Borrelli, Pasquale, David A. Robinson, Larissa R. Fleischer, Emanuele Lugato, Cristiano Ballabio, Christine Alewell, Katrin Meusburger, et al. "An Assessment of the Global Impact of 21st Century Land Use Change on Soil Erosion." *Nature Communications* 8, no. 1 (2017/12/08): 2013.

Brears, R.C. *Blue and Green Cities: The Role of Blue-Green Infrastructure in Managing Urban Water Resources*. London, UK: Palgrave Macmillan UK, 2018.

Brears, R.C. *The Green Economy and the Water-Energy-Food Nexus*. London: Palgrave Macmillan UK, 2017.

Brears, R.C. *Urban Water Security*. Chichester and Hoboken, NJ: John Wiley & Sons, 2016.

Dass, Pawlok, Benjamin Z. Houlton, Yingping Wang, and David Warlind. "Grasslands May Be More Reliable Carbon Sinks Than Forests in California." *Environmental Research Letters* 13, no. 7 (2018/07/01): 074027.

Department of Biodiversity, Conservation, and Attractions. "Threats to Wetlands." (n.d.). https://dpaw.wa.gov.au/management/wetlands/conserving-and-managing-our-wetlands/threats-to-wetlands.

Department of Conservation. "Protecting New Zealand's Rivers." (2011). https://doc.govt.nz/about-us/statutory-and-advisory-bodies/nz-conservation-authority/publications/protecting-new-zealands-rivers/.

FAO. "Deforestation." (n.d.). http://fao.org/3/j9345e/j9345e07.htm.

FAO. "Forests and Climate Change – Instruments Related to the United Nations Framework Convention on Climate Change and Their Potential for Sustainable Forest Management in Africa, Forests and Climate Change Working Paper No. 1." (2003). http://fao.org/forestry/climatechange/53622/en/.

FAO. "Global Forest Resources Assessment 2015: How Are the World's Forests Changing?" (2016). http://fao.org/3/a-i4793e.pdf.

FAO. "Global Forest Resources Assessment 2020 Key Findings." (2020). https://openknowledge.fao.org/server/api/core/bitstreams/9f24d451-2e56-4ae2-8a4a-1bc511f5e60e/content?utm_source=chatgpt.com.

FAO. "Reducing Forest Degradation." (2018). http://fao.org/sustainable-forest-management/toolbox/modules/reducing-forest-degradation/cases/en/?type=111.

FAO. "State of the World's Forests 2018: Forests and Agriculture: Land Use Challenges and Opportunities." (2018). http://fao.org/3/I9535EN/i9535en.pdf.

FAO. "The State of the World's Forests 2018 – Forest Pathways to Sustainable Development." (2018). http://fao.org/3/I9535EN/i9535en.pdf.

FoodPrint. "How Industrial Agriculture Affects Our Soil." (n.d.). https://foodprint.org/issues/how-industrial-agriculture-affects-our-soil/.

Fuller, Matthew R., Martin W. Doyle, and David L. Strayer. "Causes and Consequences of Habitat Fragmentation in River Networks." *Annals of the New York Academy of Sciences* 1355, no. 1 (2015/10/01): 31–51.

Galib, Shams M., A.B.M. Mohsin, Md. T. Parvez, Martyn C. Lucas, Nipa Chaki, Sadman S. Arnob, Md. I. Hossain, and M. Nazrul Islam. "Municipal Wastewater Can Result in a Dramatic Decline in Freshwater Fishes: A Lesson from a Developing Country." *Knowledge and Management of Aquatic Ecosystems* 419 (2018): 37.

Gallo, Frederic, Cristina Fossi, Roland Weber, David Santillo, Joao Sousa, Imogen Ingram, Angel Nadal, and Dolores Romano. "Marine Litter Plastics and Microplastics and Their Toxic Chemicals Components: The Need for Urgent Preventive Measures." [In English]. *Environmental sciences Europe* 30, no. 1 (2018): 13–13.

Government of South Australia. "Managing Soil Acidity." (2011). https://naturalresources.sa.gov.au/files/00365365-5cda-4bbd-a55c-a31c00a21cb6/sustainable-agriculture-managing-soil-acidity-fact.pdf.

Hagen, Melanie, W. Daniel Kissling, Claus Rasmussen, Marcus A.M. De Aguiar, Lee E. Brown, Daniel W. Carstensen, Isabel Alves-Dos-Santos, et al. "2 – Biodiversity, Species *Advances in Ecological Research* Interactions and Ecological Networks in a Fragmented World." In, edited by Ute Jacob, and Guy Woodward, 89–210. London, UK: Academic Press, 2012.

Hallmann, Caspar A., Martin Sorg, Eelke Jongejans, Henk Siepel, Nick Hofland, Heinz Schwan, Werner Stenmans, et al. "More Than 75 Percent Decline over 27 Years in Total Flying Insect Biomass in Protected Areas." *PLOS ONE* 12, no. 10 (2017): e0185809.

Harding, Gareth, John Dalziel, and Peter Vass. "Bioaccumulation of Methylmercury within the Marine Food Web of the Outer Bay of Fundy, Gulf of Maine." *PLOS ONE* 13, no. 7 (2018): e0197220.

Hess, Sybille, J.P. Leteisha, S.H. Andrew, A.M. Shannon, S.W. Amelia, and L.R. Jodie. "Species-Specific Impacts of Suspended Sediments on Gill Structure and Function in Coral Reef Fishes." *Proceedings of the Royal Society B: Biological Sciences* 284, no. 1866 (2017/11/15): 20171279.

Hosonuma, Noriko, Martin Herold, Veronique De Sy, Ruth S. De Fries, Maria Brockhaus, Louis Verchot, Arild Angelsen, and Erika Romijn. "An Assessment of Deforestation and Forest Degradation Drivers in Developing Countries." *Environmental Research Letters* 7, no. 4 (2012/10/08): 044009.

International Rivers. "Environmental Impacts of Dams." (n.d.). https://internationalrivers.org/environmental-impacts-of-dams.

IPBES. "Media Release: Nature's Dangerous Decline 'Unprecedented'; Species Extinction Rates 'Accelerating'." (n.d.). https://ipbes.net/news/Media-Release-Global-Assessment.

IUCN. "Deforestation and Forest Degradation." (2017). https://iucn.org/resources/issues-briefs/deforestation-and-forest-degradation.

IUCN. "Land Degradation and Climate Change." (2015). https://iucn.org/sites/dev/files/import/downloads/land_degradation_issues_brief_cop21_031215.pdf

Johnson, D.L., S.H. Ambrose, T.J. Bassett, M.L. Bowen, D.E. Crummey, J.S. Isaacson, D.N. Johnson, et al. "Meanings of Environmental Terms." [In English]. *Journal of Environmental Quality* 26, no. 3 (1997): 581–89.

Jones, Benjamin L., Leanne C. Cullen-Unsworth, and Richard K.F. Unsworth. "Tracking Nitrogen Source Using δ15n Reveals Human and Agricultural Drivers of Seagrass Degradation across the British Isles." [In English]. *Frontiers in Plant Science* 9, no. 133 (2018/02/06).

Kjelland, Michael E., Christa M. Woodley, Todd M. Swannack, and David L. Smith. "A Review of the Potential Effects of Suspended Sediment on Fishes: Potential Dredging-Related Physiological, Behavioral, and Transgenerational Implications." *Environment Systems and Decisions* 35, no. 3 (2015/09/01): 334–50.

Lark, Tyler J., Ben Larson, Ian Schelly, Sapna Batish, and Holly K. Gibbs. "Accelerated Conversion of Native Prairie to Cropland in Minnesota." *Environmental Conservation* 46, no. 2 (2018): 155–62.

Lewis, Simon L., David P. Edwards, and David Galbraith. "Increasing Human Dominance of Tropical Forests." *Science* 349, no. 6250 (2015): 827.

Lin, Qiaoying, and Shen Yu. "Losses of Natural Coastal Wetlands by Land Conversion and Ecological Degradation in the Urbanizing Chinese Coast." *Scientific Reports* 8, no. 1 (2018/10/09): 15046.

Macadam, Craig R., and Jenni A. Stockan. "More Than Just Fish Food: Ecosystem Services Provided by Freshwater Insects." *Ecological Entomology* 40, no. S1 (2015/09/01): 113–23.

Martin, Cyrus. "Not So Many Fish in the Sea." *Current Biology* 27, no. 11 (2017/06/05): R439–R43.

McDonald, Robert I., Katherine F. Weber, Julie Padowski, Tim Boucher, and Daniel Shemie. "Estimating Watershed Degradation over the Last Century and Its Impact on Water-Treatment Costs for the World's Large Cities." *Proceedings of the National Academy of Sciences* 113, no. 32 (2016): 9117–22.

Miller, Bruce G. "Chapter 3 – The Effect of Coal Usage on Human Health and the Environment." In *Coal Energy Systems*, edited by Bruce G. Miller, 77–122. Burlington: Academic Press, 2005.

Ministry for the Environment. "Environment Aotearoa 2015." (2015). https://mfe.govt.nz/publications/environmental-reporting/environment-aotearoa-2015.

Ministry for the Environment & Statistics New Zealand. "Our Marine Environment 2016." (2016). https://mfe.govt.nz/sites/default/files/media/Environmental%20reporting/our-marine-environment.pdf.

Minnesota Pollution Control Agency. "Turbidity: Description, Impact on Water Quality, Sources, Measures." (2008). https://pca.state.mn.us/sites/default/files/wq-iw3-21.pdf.

Moges, S.A., and A.S. Gebregiorgis. "5.21 – Climate Vulnerability on the Water Resources Systems and Potential Adaptation Approaches in East Africa: The Case of Ethiopia." In *Climate Vulnerability*, edited by Roger A. Pielke, 335–45. Oxford: Academic Press, 2013.

Möllmann, Christian, and Rabea Diekmann. "Chapter 4 – Marine Ecosystem Regime Shifts Induced by Climate and Overfishing: A Review for the Northern Hemisphere." In *Advances in Ecological Research*, edited by Guy Woodward, Ute Jacob, and Eoin J. O'Gorman, 303–47. London, UK: Academic Press, 2012.

Monneret, Claude. "What Is an Endocrine Disruptor?" *Comptes Rendus Biologies* 340, no. 9 (2017/09/01): 403–05.

Nahlik, A.M., and M.S. Fennessy. "Carbon Storage in Us Wetlands." *Nature Communications* 7 (2016/12/13): 13835.

National Park Service. "Threats to Wetlands." (n.d.). https://nps.gov/subjects/wetlands/threats-to-wetlands.htm.

NIWA. "Habitat Loss." (n.d.). https://niwa.co.nz/our-science/freshwater/tools/guide-to-restoring-freshwater-native-fish/identifying-the-problem/habitat-loss.

NIWA. "Muddy Sinks." (n.d.). https://niwa.co.nz/news/muddy-sinks.

NIWA. "What Is a Carbon Sink?" (n.d.). https://niwa.co.nz/atmosphere/faq/what-is-a-carbon-sink.

Ontario Ministry of Agriculture, Food and Rural Affairs. "Soil Erosion – Causes and Effects." (n.d.). http://omafra.gov.on.ca/english/engineer/facts/12-053.htm.

Parliament of Australia. "Carbon Sequestration." (n.d.). https://aph.gov.au/About_Parliament/Parliamentary_Departments/Parliamentary_Library/Browse_by_Topic/ClimateChangeold/responses/mitigation/carbon.

PBR. "What's Behind Desertification?" (n.d.). https://prb.org/whatsbehinddesertification/.

Qie, Lan, Simon L. Lewis, Martin J.P. Sullivan, Gabriela Lopez-Gonzalez, Georgia C. Pickavance, Terry Sunderland, Peter Ashton, et al. "Long-Term Carbon Sink in Borneo's Forests Halted by Drought and Vulnerable to Edge Effects." *Nature Communications* 8, no. 1 (2017/12/19): 1966.

Queensland Government. "Soil Acidification." (2015). https://publications.qld.gov.au/dataset/science-notes-soils/resource/2a2b6fc7-671d-4031-96ad-377eaabba6fa.

Ramsar. "75% of Earth's Land Areas Are Degraded; Wetlands Have Been Hit Hardest, with 87% Lost Globally in the Last 300 Years." (2017). https://ramsar.org/news/75-of-earths-land-areas-are-degraded-wetlands-have-been-hit-hardest-with-87-lost-globally-in.

Reid, G.M., T.C. MacBeath, and K. Csatádi. "Global Challenges in Freshwater-Fish Conservation Related to Public Aquariums and the Aquarium Industry." *International Zoo Yearbook* 47, no. 1 (2013/01/01): 6–45.

Rogan, Jordan E., and Thomas E. Lacher. "Impacts of Habitat Loss and Fragmentation on Terrestrial Biodiversity." In *Reference Module in Earth Systems and Environmental Sciences*. Amsterdam: Elsevier, 2018.

Röhr, Maria E., Marianne Holmer, Julia K. Baum, Mats Björk, Katharyn Boyer, Diana Chin, Lia Chalifour, et al. "Blue Carbon Storage Capacity of Temperate Eelgrass (Zostera Marina) Meadows." *Global Biogeochemical Cycles* 32, no. 10 (2018): 1457–75.

Rovero, Matt. "Nutrient Cycling Causes and Impacts of Desertification." In *Reference Module in Earth Systems and Environmental Sciences*. Amsterdam: Elsevier, 2017.

Runge, Claire A., James E.M. Watson, Stuart H.M. Butchart, Jeffrey O. Hanson, Hugh P. Possingham, and Richard A. Fuller. "Protected Areas and Global Conservation of Migratory Birds." *Science* 350, no. 6265 (2015): 1255.

Sánchez-Bayo, Francisco, and Kris A.G. Wyckhuys. "Worldwide Decline of the Entomofauna: A Review of Its Drivers." *Biological Conservation* 232 (2019/04/01): 8–27.

Smithsonian Ocean. "Seagrass and Seagrass Beds." (n.d.). https://ocean.si.edu/ocean-life/plants-algae/seagrass-and-seagrass-beds.

Sulieman, Hussein M. "Exploring Drivers of Forest Degradation and Fragmentation in Sudan: The Case of Erawashda Forest and Its Surrounding Community." *Science of the Total Environment* 621 (2018/04/15): 895–904.

U.S. Forest Service. "Ecosystem Services from National Grasslands." (n.d.). https://fs.fed.us/grasslands/ecoservices/index.shtml.

UN-Water. "Towards a Worldwide Assessment of Freshwater Quality: A Un-Water Analytical Brief." (2016). http://unwater.org/publications/towards-worldwide-assessment-freshwater-quality/.

UNCCD. "Frequently Asked Questions (Faq)." (n.d.). https://unccd.int/frequently-asked-questions-faq.

UNECE. "Air Pollution, Ecosystems and Biodiversity." (n.d.). http://unece.org/environmental-policy/conventions/envlrtapwelcome/cross-sectoral-linkages/air-pollution-ecosystems-and-biodiversity.html.

UNEP. "Global Environment Outlook 6." (2019). https://unenvironment.org/resources/global-environment-outlook-6.

United Nations. "Glossary of Environment Statistics, Studies in Methods." (1997). https://unstats.un.org/unsd/publication/seriesf/seriesf_67e.pdf.

United Nations Conference on Trade and Development. "90% of Fish Stocks Are Used Up – Fisheries Subsidies Must Stop." (2018) https://unctad.org/en/pages/newsdetails.aspx?OriginalVersionID=1812.

United Nations Convention to Combat Desertification. "Global Land Outlook." (2017). https://knowledge.unccd.int/sites/default/files/2018-06/GLO%20English_Full_Report_rev1.pdf.

United States Department of Agriculture Forest Service. "Issues and Influences: Forest Fragmentation and Parcelization." (n.d.). https://nrs.fs.fed.us/futures/issues_influences/fragmentation_parcelization/.

University of Montana. "Threats to Wetlands." (n.d.). http://cas.umt.edu/watershedclinic/links/clarkfork/wetlands/threatstowetlands.htm.

US EPA. "Ecosystem Effects of Ozone Pollution." (n.d.). https://epa.gov/ground-level-ozone-pollution/ecosystem-effects-ozone-pollution.

US EPA. "Threats to Wetlands." (2001). http://virginia.edu/blandy/blandy_web/education/Bay/WetlandThreats_EPA.pdf.

Wang, Sirui, Qianlai Zhuang, Outi Lähteenoja, Frederick C. Draper, and Hinsby Cadillo-Quiroz. "Potential Shift from a Carbon Sink to a Source in Amazonian Peatlands under a Changing Climate." *Proceedings of the National Academy of Sciences* 115, no. 49 (2018): 12407.

Waycott, Michelle, Carlos M. Duarte, Tim J.B. Carruthers, Robert J. Orth, William C. Dennison, Suzanne Olyarnik, Ainsley Calladine, et al. "Accelerating Loss of Seagrasses across the Globe Threatens Coastal Ecosystems." *Proceedings of the National Academy of Sciences* 106, (2009): pnas.0905620106.

WHO. "Land Degradation and Desertification." (n.d.). https://who.int/globalchange/ecosystems/desert/en/.

Wit, Francisca, Tim Rixen, Antje Baum, Widodo S. Pranowo, and Andreas A. Hutahaean. "The Invisible Carbon Footprint as a Hidden Impact of Peatland Degradation Inducing Marine Carbonate Dissolution in Sumatra, Indonesia." *Scientific Reports* 8, no. 1 (2018/11/27): 17403.

WWF. "Overfishing." (n.d.). https://worldwildlife.org/threats/overfishing.

WWF. "Pollution." (n.d.). https://worldwildlife.org/threats/pollution.

Zimdahl, Robert L. "11 – A Glimpse Ahead." In *Agriculture's Ethical Horizon (Second Edition)*, edited by Robert L. Zimdahl, 229–74. Oxford: Elsevier, 2012.

6 Climate Change and Environmental Degradation Impacts on People and the Economy

Introduction

Societies have developed their daily lives around established climate and environmental conditions. People and communities are accustomed to a specific range of environmental stability and may struggle to adapt to extreme variations. Climate change and environmental degradation will disrupt social, cultural, and natural resources, affecting everything from infrastructure and agriculture to public health and community well-being.

Climate Change Impacts on Human Well-Being

Climate change is expected to affect people and the economy in various ways, including the following.

Increased Risk of Disaster

Climate change is increasing the frequency, intensity, and scale of disasters, leading to more deaths, injuries, and damage to property and infrastructure.[1] According to the UN Office for Disaster Risk Reduction, weather-related events such as storms, floods, droughts, and heatwaves have caused 90% of major disasters over the past two decades. Climate change is expected to intensify rainfall patterns and increase the frequency and severity of heavy downpours, leading to more frequent and severe flooding. These changes will impact livelihoods, human health and safety, businesses, communities, and cultural heritage. Additionally, alterations to rivers and wetlands due to climate change and human activities may reduce their ability to regulate flooding, thereby further exacerbating the risks they naturally pose.[2]

Exposure to Floods

Flood risks are expected to rise as global temperatures increase, with significant impacts at 1.5°C of warming and even greater consequences at 2°C of warming. Under a 2°C warming scenario, the overall area exposed to

DOI: 10.4324/9781003517948-6

flooding is expected to expand, affecting countries that represent 73% of the global population. Flood risk is projected to double at 1.5°C and increase by 170% at 2°C compared to the baseline period of 1976–2005, assuming constant population sizes. By 2050, limiting warming to 1.5°C could reduce the number of people exposed to flooding by 26–34 million compared to a 2°C scenario. In Europe, without adaptation measures, flood impacts under a 2°C warming scenario could more than double, with an estimated 525,000 people affected annually and €12.5 billion in expected annual losses.[3]

Human activities have intensified flood hazards by altering land use. Deforestation, soil degradation, the expansion of impermeable surfaces, and the construction of drainage networks have increased runoff into streams, resulting in higher peak discharges, greater flood volumes, and more frequent flooding events.[4,5,6] Urban expansion in flood-prone areas has further heightened exposure to flood risks, including inundation and erosion.[7] The impact of Hurricane Harvey in Houston, where more than a meter of rain fell over a densely populated urban area, illustrates this vulnerability. Researchers found that extreme rainfall and increased runoff resulting from urbanization amplified the flood risk during the hurricane by approximately 21 times.[8]

Exposure to Drought and Heatwaves

At 2°C of warming, the number of people exposed to drought is expected to be significantly higher than at 1.5°C of warming. Under constant socioeconomic conditions, projections indicate that between 2021 and 2040, nearly 115 million people will experience extreme drought each month under a 1.5°C warming scenario. However, if global temperatures rise to 2°C between 2041 and 2060, this number is expected to increase to just over 190 million.[9]

Heatwaves in European cities are also projected to become more frequent and intense. All cities across the continent are expected to experience more frequent and hotter heatwaves, with southern European cities seeing the most significant increase in heatwave days, up to 69% more than present levels. Meanwhile, central European cities, where infrastructure and populations are generally less adapted to extreme heat, are likely to face the most significant rises in maximum temperatures, with heatwaves becoming up to 14°C hotter than historical levels.[10]

Food Supply and Security

While warmer temperatures may extend growing seasons in some regions, climate change is expected to increase food production vulnerabilities, particularly due to droughts.[11] For example, modelling indicates that during an exceptional drought, there is more than an 80% probability that wheat production in the United States and Canada will fall below its long-term average.

Similarly, India faces the highest risk of maize yield reductions under drought conditions, while rice production in Vietnam and Thailand is particularly vulnerable. The relationship between drought severity and crop loss is non-linear, with a shift from moderate to severe drought causing greater yield losses than extreme to exceptional drought.[12]

Insect pests may further exacerbate crop losses. Currently, insects consume between 5% and 20% of major grain crops, and this loss could increase by 10%–25% for every degree Celsius of warming, with the highest impacts expected in temperate regions.[13] Flooding also threatens food security, as heavy storm events wash away topsoil. In the United States, the 2019 Midwest floods resulted in approximately $440 million in crop losses in Nebraska alone, one of the area's most severely affected.[14]

Rising ocean temperatures and ocean acidification are expected to impact marine species and their distribution, affecting commercial fisheries and marine aquaculture.[15] Between 1930 and 2010, ocean warming contributed to an estimated 4.1% decline in sustainable catches for many fish and shellfish species. In some regions, including East China and the North Sea, declines reached 15%–35%. These trends are expected to persist, posing a threat to the livelihoods of communities that rely on marine resources.[16]

Water Supply and Security

Climate change is expected to alter the spatial and temporal distribution of precipitation, leading to drier summers and more frequent and intense droughts. In southern Australia, climate change is projected to reduce average river flows by 10%–20%. Rising global temperatures increase evaporation from land and oceans, while a warmer atmosphere holds more water vapor, thereby increasing the likelihood of heavy rainfall, the primary cause of inland flooding.[17]

Modern land-use practices, such as expanding development on floodplains, increasing impermeable surfaces, and degrading natural landscapes, reduce the land's ability to absorb heavy rainfall, heightening flood risks and intensifying their impacts.[18] Increased stormwater runoff will carry higher loads of pathogens, chemicals, nutrients, and suspended sediment into waterways. Climate-induced changes in precipitation alone could increase riverine total nitrogen loading by nearly 20% across the continental United States by the end of the century, contributing to water quality degradation.[19]

Rising sea levels also threaten freshwater supplies in coastal and island regions through saltwater intrusion, which may require water utilities to enhance treatment, relocate intake points, or develop alternative freshwater sources. Higher sea levels will further increase the salinity of coastal aquifers, particularly in areas where groundwater recharge is expected to decline, compromising their long-term usability.[20,21,22,23]

Human Health and Well-Being

Climate change impacts human health in multiple ways. A healthy human maintains physical, mental, and social well-being, but climate change disrupts this balance through direct effects, such as rising air and sea temperatures, storms, and flooding. Indirect impacts arise from environmental and ecosystem changes, while social and economic shifts, including migration pressures, health inequalities, and economic hardship, further contribute to health risks.

By mid-century, climate change is expected to worsen existing health challenges, particularly in communities with inadequate health protection and limited adaptive capacity. Beyond this, significant health impacts are expected to emerge, including increased injuries, diseases, and mortality from extreme heatwaves and wildfires, higher risks of flooding, undernutrition due to declining food production, and a rise in waterborne, foodborne, and vector-borne diseases. Additionally, declining air quality will exacerbate respiratory illnesses, and reduced labor productivity due to heat stress will further strain economies. Specific examples of climate change's effects on human health and well-being are outlined in Table 6.1.[24,25,26,27,28,29,30,31,32,33,34]

Economic Costs

Without significant climate change mitigation, rising temperatures, sea-level rise, and extreme weather events will increasingly disrupt critical infrastructure, damage property, reduce labor productivity, and weaken community resilience. Industries reliant on stable climatic conditions and natural resources, such as agriculture, tourism, and fisheries, are particularly vulnerable. Rising temperatures will lower power generation efficiency while driving up demand, increasing electricity costs. Businesses with global supply chains and overseas operations will also face disruptions due to climate-related economic instability.[35]

If global temperatures rise by 4°C above pre-industrial levels by 2100, GDP losses could reach up to 10% by the end of the century.[36] Even under a 2°C warming scenario, median global GDP per capita is projected to be around 5% lower than a 1.5°C scenario, with the most significant losses occurring in lower-income countries.[37] In Europe, one study estimates that climate-related damages could triple by the 2020s, increase sixfold by mid-century, and exceed ten times the current annual damage of €3.4 billion by 2100.[38] Meanwhile, the long-term global economic benefit of limiting warming to 2°C instead of 4°C is estimated at approximately $17.6 billion per year by the end of the century.[39]

Migration

An estimated 143 million people in Sub-Saharan Africa, South Asia, and Latin America could be displaced within their own countries due to

Table 6.1 Climate Change Impacts on Human Health and Well-Being

Climate Change Impact	*Effects*
Increased temperatures	Hotter days and prolonged heatwaves will lead to an increase in heat-related illnesses and deaths.
Lower air quality	Worsening air quality due to warmer temperatures and changing weather patterns can lead to asthma, respiratory and cardiovascular issues. Wildfires and higher CO_2 levels increase airborne allergens.
Increases in ozone	Rising temperatures increase the number of days with high ground-level ozone, aggravating asthma and damaging lung tissue.
Changes in particulate matter	Climate change increases exposure to fine particles from sources like wildfire smoke and dust, contributing to respiratory and cardiovascular diseases.
Changes in allergens and asthma triggers	Longer pollen seasons due to climate change can worsen allergies and asthma symptoms.
Extreme weather events	Storms and floods can disrupt access to food, water, healthcare, and cause injuries, trauma, and mental health disorders such as PTSD and depression.
Vector-borne disease	Warming and rainfall shifts expand the geographic range and seasonal activity of disease vectors such as mosquitoes and ticks.
Illnesses	Increased rain and flooding heighten exposure to contaminated water, causing gastrointestinal, nervous, respiratory, liver, and kidney illnesses.
Food safety and nutrition	Extreme weather disrupts food supply chains, increases foodborne illnesses, and reduces nutritional quality of crops due to higher CO_2 levels.
Mental health	Climate-related changes and disasters can result in psychological stress, anxiety, and depression, especially after life or property loss.
Populations of concern	Vulnerable groups such as children, elderly, low-income communities, and people with chronic conditions face greater risks due to limited adaptive capacity and higher sensitivity to climate stressors.

slow-onset climate change. These movements will be driven by declining water availability, reduced crop productivity, rising sea levels, and increasing storm surges, forcing populations to relocate from increasingly uninhabitable areas.[40] In Europe, climate change is expected to lead to a significant increase in asylum applications. By the end of the century, under a moderate emissions scenario (RCP 4.5), asylum applications are expected to increase by 28%, equivalent to approximately 98,000 additional applications per year. Under a high-emissions scenario (RCP 8.5), this increase could reach 188%, resulting in approximately 660,000 additional applications annually.[41]

Environmental Degradation Impacts on Human Well-Being

Environmental degradation, driven by non-climatic factors, has widespread consequences for people and the economy. These impacts include the following.

Health Impacts of Desertification

Desertification, the expansion of desert-like conditions due to human activities and climatic fluctuations, is a significant form of land degradation that primarily affects the world's drylands. It has both direct and indirect health consequences. Indirectly, desertification contributes to mass migrations, poverty, food and water shortages, malnutrition, and conflicts over land and water resources.[42,43] Directly, desertification affects human health by degrading water and air quality. The accumulation of dust particles in the air can lead to respiratory diseases, cardiovascular issues, silicosis, conjunctivitis, meningococcal meningitis, and coccidioidomycosis (Valley Fever).[44] In Seoul, South Korea, for example, dust clouds carried by winds from the Mongolian desert have been linked to increased mortality rates, including a 1.7% rise in overall deaths, a 2.2% increase in deaths among individuals aged 65 and older, and a 4.1% rise in deaths related to cardiovascular and respiratory conditions.[45]

Vector-Borne Disease

Changes to the natural environment, including deforestation, agriculture, livestock farming, water management projects, and urban expansion, are increasing human vulnerability to preventable outbreaks of vector-borne diseases.[46] Deforestation alters disease transmission by modifying vegetation, introducing livestock, expanding human settlements, and reducing biodiversity. It can create ecological conditions that favor the spread of disease vectors and parasites. For example, water puddles in deforested areas often have lower salinity and acidity than those in intact forests, making them more suitable breeding sites for particular Anopheles mosquito species, the primary vectors of malaria. The growing demand for food is driving the conversion of natural ecosystems into agricultural land, which alters local microclimates and increases the risk of malaria in cultivated areas. Similarly, urban expansion into previously undisturbed environments can heighten exposure to disease vectors and nonhuman hosts, as seen with yellow fever. Poorly managed wastewater drainage in urban areas can also create breeding grounds for mosquitoes, contributing to the spread of diseases such as dengue fever.[47,48]

Health Effects of Air Pollution

Air pollution is a leading global cause of death and disease, affecting human health in multiple ways. Different air pollutants impact tissues and organs through various mechanisms, as outlined in Table 6.2.[49,50]

Table 6.2 How Different Types of Air Pollution Damage Tissue

Pollutant	*Injury Determinants*	*Tissues Affected*
Sulphur dioxide	Highly soluble	Upper airway and skin damage
Nitrogen dioxide	Less soluble (nitrogen dioxide and ozone are irritating)	Deeper lung penetration
Ozone		Bronchial and bronchiolar injury
Carbon monoxide		Tissue hypoxia due to impaired oxygen delivery
Particulate matter	Toxicity depends on size, structure, and composition	Large particles: mucous membranes, upper airways Small particles: bronchioles and alveoli Ultrafine particles: systemic tissue reactions

Costs of Land Degradation

Land degradation imposes a significant economic burden, costing the global economy approximately $10.6 trillion annually, equivalent to 17% of global GDP. More than half of the world's arable land is moderately or severely degraded, resulting in economic losses due to reduced agricultural productivity, lower incomes, and the diminished value of essential ecosystem services, including water filtration, air purification, and nutrient recycling.[51] The rural poor are disproportionately affected, as they rely directly on land for their sustenance and livelihoods.[52] Sub-Saharan Africa bears a particularly high burden, accounting for 26% of the total global costs of land degradation caused by land-use and land-cover changes. Of this, only 46% of the costs stem from direct losses in provisioning services, such as food and raw materials, while the remaining costs are due to the degradation of broader ecosystem services, including carbon sequestration, biodiversity, genetic resources, and cultural values – many of which provide benefits beyond the local land users.[53]

Loss of Food Production

Expanding food production to meet rising demand has often come at the cost of biodiversity and essential ecosystem services that sustain global food systems. Agricultural intensification accelerates soil erosion, a critical environmental issue that depletes topsoil and nutrients, ultimately reducing crop yields and productivity. The global economic losses of soil erosion are estimated at $8 billion annually.[54] In the European Union, approximately 12 million hectares of agricultural land experience severe erosion, resulting in a 0.43% annual decline in crop productivity. This productivity loss translates into an estimated annual economic cost of €1.25 billion. Additionally, up to $577 billion in global crop production is at risk each year due to the decline of pollinators, further threatening food security and agricultural sustainability.[55]

Losses from Overfishing

Many of the world's fish stocks are depleted due to overexploitation, which reduces productivity and leads to significant economic losses. A study analyzing 14 FAO fishing areas from 1950 to 2010 estimated that approximately 35% of global marine fish stocks have been or are currently overexploited. The cumulative global catch losses over this period amounted to 332.8 million tons, resulting in approximately $298.9 billion in direct economic losses.[56] The economic impact of overfishing extends beyond direct losses. A study examining the depletion of nine severely overfished species between 2005 and 2009, including black sea bass in the South Atlantic and red snapper in the Gulf of Mexico, found that the most significant indirect losses – nearly $53 million per year on average – were due to fewer fishing trips targeting South Atlantic black sea bass. This decline affected businesses reliant on anglers, such as boat rentals, charter services, and bait and tackle shops. When factoring in the broader economic impact, including reduced spending at hotels, restaurants, and wholesale suppliers, the total estimated loss from fewer trips for black sea bass alone reached $138 million in the region.[57]

Livelihoods Threatened by Forest Degradation

Forest degradation affects livelihoods and income generation by reducing direct economic benefits and diminishing essential ecosystem services.[58] Many of the world's rural poor rely on healthy forests and trees for sustenance and economic stability. Approximately 40% of the extreme rural poor, around 250 million people, live in forest and savannah areas, where access to forest products, goods, and services is crucial. In developing countries, forests and trees contribute approximately 20% of rural household income, underscoring their crucial role in supporting food security and economic resilience.[59]

Contaminated Groundwater Impacts

Groundwater contamination can result from the presence of microorganisms and thousands of synthetic chemicals, posing significant health risks. Drinking water contaminated with bacteria and viruses can cause diseases such as hepatitis, cholera, and giardiasis. Benzene, a component of petrol, is a known human carcinogen. Lead exposure can lead to learning disabilities in children, as well as nerve, kidney, and liver damage and pregnancy complications.[60] High nitrate levels in drinking water can cause methemoglobinemia, also known as "blue-baby syndrome," where nitrates are converted to nitrites in the body, interfering with the ability of red blood cells to carry oxygen. Bottle-fed infants under three months are particularly vulnerable, often showing symptoms such as blue discoloration around the mouth, hands, and feet.[61,62]

Health Effects of Chemical Pollution of Oceans

Chemical pollution in the oceans can impact human health through direct exposure to pollutants from industrial and accidental sources; however, the most significant risk arises from consuming contaminated seafood. Environmental contaminants in seafood have been linked to health issues such as allergies, neurological developmental deficits, and disruptions to the thyroid hormone system. Heavy metals, such as lead, cadmium, and mercury, pose direct toxicity risks in marine environments. Additionally, diethylhexyl phthalate, found in some plastics, is a known carcinogen, while other plastic-derived toxins have been associated with cancers, congenital disabilities, immune disorders, and developmental problems in children.[63,64,65]

Disasters Leading to Disease Outbreaks

Disasters significantly increase the risk of disease outbreaks. Following floods, mortality rates can increase by up to 50% in the first year. Floods often lead to outbreaks of diseases such as hepatitis E, gastrointestinal infections, and leptospirosis, especially in areas with inadequate hygiene and displaced populations.[66] Psychological distress in survivors can further weaken their immune systems and exacerbate physical illnesses. Coastal floods contribute to the highest mortality rates after windstorms, including hurricanes, cyclones, and typhoons. Losing coastal habitats and coral reefs has reduced natural protection, heightening flood and hurricane risks for the 100–300 million people in coastal 100-year flood zones.[67]

Cost of Air Pollution

By 2060, air pollution is expected to cost the global economy approximately 1% of GDP, or $2.6 trillion annually, due to lost workdays, medical expenses, and reduced agricultural productivity. Over this period, annual healthcare costs related to air pollution are projected to rise from $21 billion in 2015 to $176 billion, while the number of workdays lost due to air pollution-related illnesses is expected to increase from 1.2 billion to 3.7 billion. The annual global welfare costs associated with premature deaths from outdoor air pollution are projected to rise from $3 trillion in 2015 to $18 trillion and $25 trillion by 2060. Additionally, the annual welfare costs related to pain and suffering from pollution-induced illnesses are expected to reach $2.2 trillion by 2060, up from approximately $300 billion in 2015.[68]

Loss of Ecosystem Services

Biodiversity and ecosystem services are essential to the global economy, supporting food production, water purification, flood protection, and climate change mitigation at local, regional, and global levels. In 2011, the total estimated value of global ecosystem services ranged between $125 trillion and $140 trillion, which is more than one and a half times the world's GDP that year. However, between 1997 and 2011, the loss of ecosystem services due to human activities resulted in annual economic losses of $4 trillion to $20 trillion.[69] In Morocco, for example, environmental degradation cost the economy approximately $3.9 billion in 2014, equivalent to 3.5% of GDP, with the most significant impacts stemming from water-related issues, air pollution, and agricultural land degradation.[70]

Increased Water Treatment Costs

Water treatment costs are influenced by the quality of the source water, which is directly affected by land use in the surrounding watershed. Agriculture and urban development are significant contributors to water source degradation, leading to increased sediment and nutrient pollution. As a result, water treatment costs have risen in many areas. Nearly one-third of large cities worldwide have experienced higher treatment expenses, with operating and maintenance capital costs increasing by an average of 53% and replacement capital costs rising by approximately 44%.[71]

Notes

1 IUCN, "Nature-Based Solutions to Disasters," (2017), www.iucn.org/sites/dev/files/nbs_to_disasters_issues_brief_final.pdf.

2 Scottish Natural Heritage, "Impacts on People and Our Economy," (n.d.), www.nature.scot/climate-change/climate-change-impacts-scotland/impacts-people-and-our-economy.

3 J.-C.C. Martinez, D.I. Ruiz, A.S. Ramirez, A. Dosio, A. Toreti, A. Ceglar, D. Fumagalli, F. Dentener, R. Lecerf, A. Zucchini, L. Panarello, S. Niemeyer, I. Perez-Dominguez, T. Fellmann, A. Kitous, J. Després, A. Christodoulou, H. Demirel, L. Alfieri, F. Dottori, M. Vousdoukas, L. Mentaschi, E. Voukouvalas, C. Cammalleri, P.M.F. Barbosa, F. Micale, J. Vogt, J.I.B. Cano, G. Caudullo, A. Mauri, D. de Rigo, G. Libertà, T.H. Durrant, T.A. Vivancos, J. San-Miguel-Ayanz, S. Gosling, J. Zaherpour, A. De Roo, B. Bisselink, J. Bernhard, A. Bianchi, M. Rozsai, W. Szewczyk, I. Mongelli, and L. Feyen, "Climate Impacts in Europe: Final Report of the Jrc Peseta Iii Project," (2018), www.researchgate.net/publication/328956833_Climate_impacts_in_Europe_final_report_of_the_JRC_PESETA_III_project.

4 USGS, "Effects of Urban Development on Floods," (n.d.), https://pubs.usgs.gov/fs/fs07603/.

5 James D. Miller and Michael Hutchins, "The Impacts of Urbanisation and Climate Change on Urban Flooding and Urban Water Quality: A Review of the Evidence Concerning the United Kingdom," *Journal of Hydrology: Regional Studies* 12 (2017).

6 R.C. Brears, *Urban Water Security* (Chichester and Hoboken, NJ: John Wiley & Sons, 2016).

7 USGS, "Effects of Urban Development on Floods."

8 Wei Zhang et al., "Urbanization Exacerbated the Rainfall and Flooding Caused by Hurricane Harvey in Houston," *Nature* 563, no. 7731 (2018).

9 IPCC, "Special Report. Global Warming of 1.5 °C," (2018), www.ipcc.ch/sr15/.
10 Selma B. Guerreiro et al., "Future Heat-Waves, Droughts and Floods in 571 European Cities," *Environmental Research Letters* 13, no. 3 (2018).
11 Tsugihiro Watanabe et al., "Management of Climatic Extremes with Focus on Floods and Droughts in Agriculture," *Irrigation and Drainage* 67, no. 1 (2018).
12 Guoyong Leng and Jim Hall, "Crop Yield Sensitivity of Global Major Agricultural Countries to Droughts and the Projected Changes in the Future," *Science of the Total Environment* 654 (2019).
13 Curtis A. Deutsch et al., "Increase in Crop Losses to Insect Pests in a Warming Climate," *Science* 361, no. 6405 (2018).
14 National Public Radio, "Nebraska Faces over $1.3 Billion in Flood Losses," (2019), www.npr.org/2019/03/21/705408364/nebraska-faces-over-1-3-billion-in-flood-losses.
15 Scottish Natural Heritage, "Impacts on People and Our Economy."
16 Christopher M. Free et al., "Impacts of Historical Warming on Marine Fisheries Production," *Science* 363, no. 6430 (2019).
17 Francis Chiew and Ian Prosser, "Water and Climate," in *Water: Science and Solutions for Australia*, ed. Ian Prosser (Clayton: CSIRO, 2011).
18 Union of Concerned Scientists, "Climate Change, Extreme Precipitation and Flooding: The Latest Science," (2018), www.ucsusa.org/global-warming/global-warming-impacts/floods#.XBB5fWgzY2w.
19 E. Sinha, A.M. Michalak, and V. Balaji, "Eutrophication Will Increase during the 21st Century as a Result of Precipitation Changes," *Science* 357, no. 6349 (2017).
20 Brears, *Urban Water Security*.
21 Richard G. Taylor et al., "Ground Water and Climate Change," *Nature Climate Change* 3 (2012).
22 US EPA, "Climate Impacts on Water Resources," (n.d.), https://archive.epa.gov/epa/climate-impacts/climate-impacts-water-resources.html.
23 US EPA, "Climate Adaptation and Saltwater Intrusion," (n.d.), www.epa.gov/arc-x/climate-adaptation-and-saltwater-intrusion.
24 US EPA, "Climate Impacts on Human Health," (2017), https://19january2017snapshot.epa.gov/climate-impacts/climate-impacts-human-health_.html.
25 Ruth M. Doherty, Mathew R. Heal, and Fiona M. O'Connor, "Climate Change Impacts on Human Health over Europe through Its Effect on Air Quality," *Environmental Health* 16, no. 1 (2017).
26 Massimo Franchini and Pier M. Mannucci, "Impact on Human Health of Climate Changes," *European Journal of Internal Medicine* 26, no. 1 (2015).
27 U.S. Global Change Research Program, "The Impacts of Climate Change on Human Health in the United States: A Scientific Assessment," (n.d.), https://health2016.globalchange.gov/.
28 Patrick L. Kinney, "Interactions of Climate Change, Air Pollution, and Human Health," *Current Environmental Health Reports* 5, no. 1 (2018).
29 Simon Hales, "Climate Change, Extreme Rainfall Events, Drinking Water and Enteric Disease," *Reviews on Environmental Health* 34 (2019).
30 Summer Allen and Alan de Brauw, "Nutrition-Sensitive Value Chain Development in a Changing Climate," in *The Climate-Smart Agriculture Papers: Investigating the Business of a Productive, Resilient and Low Emission Future*, ed. Todd S. Rosenstock, Andreea Nowak, and Evan Girvetz (Cham: Springer International Publishing, 2019).

31 Gary W. Evans, "Projected Behavioral Impacts of Global Climate Change," *Annual Review of Psychology* 70, no. 1 (2019).

32 Kimberley Thomas et al., "Explaining Differential Vulnerability to Climate Change: A Social Science Review," *Wiley Interdisciplinary Reviews: Climate Change* 10, no. 2 (2019).

33 Patricia Schwerdtle, Kathryn Bowen, and Celia McMichael, "The Health Impacts of Climate-Related Migration," *BMC Medicine* 16, no. 1 (2018).

34 Jordan L. Schnell and Michael J. Prather, "Co-Occurrence of Extremes in Surface Ozone, Particulate Matter, and Temperature over Eastern North America," *Proceedings of the National Academy of Sciences* 114, no. 11 (2017).

35 U.S. Global Change Research Program, "Fourth National Climate Assessment," (2018), www.globalchange.gov/nca4.

36 OECD, "The Economic Consequences of Climate Change," (2015), https://espas.secure.europarl.europa.eu/orbis/sites/default/files/generated/document/en/OECD%20Climate%20Change.pdf.

37 Felix Pretis et al., "Uncertain Impacts on Economic Growth When Stabilizing Global Temperatures at 1.5°C or 2°C Warming," *Philosophical Transactions of the Royal Society A: Mathematical, Physical and Engineering Sciences* 376, no. 2119 (2018).

38 Giovanni Forzieri et al., "Escalating Impacts of Climate Extremes on Critical Infrastructures in Europe," *Global Environmental Change* 48 (2018).

39 Tom Kompas, Van H. Pham, and Tuong N. Che, "The Effects of Climate Change on Gdp by Country and the Global Economic Gains from Complying with the Paris Climate Accord," *Earth's Future* 6, no. 8 (2018).

40 World Bank, "Groundswell: Preparing for Internal Climate Migration," (2018), https://openknowledge.worldbank.org/handle/10986/29461.

41 Anouch Missirian and Wolfram Schlenker, "Asylum Applications Respond to Temperature Fluctuations," *Science* 358, no. 6370 (2017).

42 WHO, "Land Degradation and Desertification," (n.d.), www.who.int/globalchange/ecosystems/desert/en/.

43 A.S. Goudie, "Desertification," in *Encyclopedia of Environmental Health*, ed. J.O. Nriagu (Burlington: Elsevier, 2011).

44 Goudie, "Desertification."

45 Ho-Jang Kwon et al., "Effects of the Asian Dust Events on Daily Mortality in Seoul, Korea," *Environmental Research* 90, no. 1 (2002).

46 Neil Vora, "Impact of Anthropogenic Environmental Alterations on Vector-Borne Diseases," *Medscape Journal of Medicine* 10, no. 10 (2008).

47 Nathan D. Burkett-Cadena and Amy Y. Vittor, "Deforestation and Vector-Borne Disease: Forest Conversion Favors Important Mosquito Vectors of Human Pathogens," *Basic and Applied Ecology* 26 (2018).

48 Vora, "Impact of Anthropogenic Environmental Alterations on Vector-Borne Diseases."

49 WHO, "Ambient Air Pollution: Health Impacts," (n.d.), www.who.int/airpollution/ambient/health-impacts/en/.

50 Dean E. Schraufnagel et al., "Air Pollution and Noncommunicable Diseases: A Review by the Forum of International Respiratory Societies' Environmental Committee, Part 1: The Damaging Effects of Air Pollution," *CHEST* 155, no. 2 (2019).

51 ELD Initiative, "The Value of Land: Prosperous Lands and Positive Rewards through Sustainable Land Management," (2015), www.eld-initiative.org/fileadmin/pdf/ELD-main-report_05_web_72dpi.pdf.

52 ELD Initiative, "The Value of Land: Prosperous Lands and Positive Rewards through Sustainable Land Management."
53 Ephraim Nkonya et al., "Global Cost of Land Degradation," in *Economics of Land Degradation and Improvement – A Global Assessment for Sustainable Development*, ed. Ephraim Nkonya, Alisher Mirzabaev, and Joachim von Braun (Cham: Springer International Publishing, 2016).
54 Martina Sartori et al., "A Linkage between the Biophysical and the Economic: Assessing the Global Market Impacts of Soil Erosion," *Land Use Policy* 86 (2019).
55 IPBES, "Media Release: Nature's Dangerous Decline 'Unprecedented'; Species Extinction Rates 'Accelerating'," (2017), www.ipbes.net/news/Media-Release-Global-Assessment.
56 Qi Ding et al., "Estimation of Catch Losses Resulting from Overexploitation in the Global Marine Fisheries," *Acta Oceanologica Sinica* 36, no. 12 (2017).
57 Pew, "Southeast Suffered Millions in Economic Losses from Overfishing," (2012), www.pewtrusts.org/en/about/news-room/press-releases-and-statements/2012/09/04/southeast-suffered-millions-in-economic-losses-from-overfishing.
58 ODI, "Understanding the Effects of Resource Degradation on Socio-Economic Outcomes in Developing Countries," (2014), www.odi.org/sites/odi.org.uk/files/odi-assets/publications-opinion-files/8830.pdf.
59 FAO, "The State of the World's Forests 2018 – Forest Pathways to Sustainable Development," (2018), www.fao.org/3/I9535EN/i9535en.pdf.
60 US EPA, "Getting up to Speed on Groundwater Contamination," (2015), www.epa.gov/sites/production/files/2015-08/documents/mgwc-gwc1.pdf.
61 Michael A. Mallin, "Chapter 4 – Effect of Human Land Development on Water Quality," in *Handbook of Water Purity and Quality*, ed. Satinder Ahuja (Amsterdam: Academic Press, 2009).
62 WHO, "Water-Related Diseases," (n.d.), www.who.int/water_sanitation_health/diseases-risks/diseases/methaemoglob/en/.
63 Dan Wilhelmsson et al., "Chapter 6 – Marine Pollution," in *Managing Ocean Environments in a Changing Climate*, ed. Kevin J. Noone, Ussif R. Sumaila, and Robert J. Diaz (Boston, MA: Elsevier, 2013).
64 Hui Gao et al., "Bisphenol a and Hormone-Associated Cancers: Current Progress and Perspectives," *Medicine* 94, no. 1 (2015).
65 Neeti Rustagi, S.K. Pradhan, and Ritesh Singh, "Public Health Impact of Plastics: An Overview," *Indian Journal of Occupational and Environmental Medicine* 15, no. 3 (2011).
66 Katarzyna Alderman, Lyle R. Turner, and Shilu Tong, "Floods and Human Health: A Systematic Review," *Environment International* 47 (2012).
67 IPBES, "Media Release: Nature's Dangerous Decline 'Unprecedented'; Species Extinction Rates 'Accelerating'."
68 OECD, "The Economic Consequences of Outdoor Air Pollution," (2016), www.oecd.org/environment/the-economic-consequences-of-outdoor-air-pollution-9789264257474-en.htm.
69 Robert Costanza et al., "Changes in the Global Value of Ecosystem Services," *Global Environmental Change* 26 (2014).
70 L. Croitoru and M. Sarraf, "How Much Does Environmental Degradation Cost? The Case of Morocco," *Journal of Environmental Protection* 9 (2018).
71 Robert I. McDonald et al., "Estimating Watershed Degradation over the Last Century and Its Impact on Water-Treatment Costs for the World's Large Cities," *Proceedings of the National Academy of Sciences* 113, no. 32 (2016).

References

Alderman, Katarzyna, Lyle R. Turner, and Shilu Tong. "Floods and Human Health: A Systematic Review." *Environment International* 47 (2012/10/15): 37–47.

Allen, Summer, and Alan de Brauw. "Nutrition-Sensitive Value Chain Development in a Changing Climate." In *The Climate-Smart Agriculture Papers: Investigating the Business of a Productive, Resilient and Low Emission Future*, edited by Todd S. Rosenstock, Andreea Nowak, and Evan Girvetz, 247–56. Cham: Springer International Publishing, 2019.

Brears, R.C. *Urban Water Security*. Chichester and Hoboken, NJ: John Wiley & Sons, 2016.

Burkett-Cadena, Nathan D., and Amy Y. Vittor. "Deforestation and Vector-Borne Disease: Forest Conversion Favors Important Mosquito Vectors of Human Pathogens." *Basic and Applied Ecology* 26 (2018/02/01): 101–10.

Chiew, Francis, and Ian Prosser. "Water and Climate." Chap. 3 In *Water: Science and Solutions for Australia*, edited by Ian Prosser. Clayton: CSIRO, 2011.

Ciscar Martinez, J.-C., D. Ibarreta Ruiz, A. Soria Ramirez, A. Dosio, A. Toreti, A. Ceglar, D. Fumagalli, F. Dentener, R. Lecerf, A. Zucchini, L. Panarello, S. Niemeyer, I. Perez-Dominguez, T. Fellmann, A. Kitous, J. Després, A. Christodoulou, H. Demirel, L. Alfieri, F. Dottori, M. Vousdoukas, L. Mentaschi, E. Voukouvalas, C. Cammalleri, P. Marinho Ferreira Barbosa, F. Micale, J. Vogt, J.I. Barredo Cano, G. Caudullo, A. Mauri, D. de Rigo, G. Libertà, T. Houston Durrant, T. Artés Vivancos, J. San-Miguel-Ayanz, S. Gosling, J. Zaherpour, A. De Roo, B. Bisselink, J. Bernhard, A. Bianchi, M. Rozsai, W. Szewczyk, I. Mongelli and L. Feyen. "Climate Impacts in Europe: Final Report of the Jrc Peseta Iii Project." (2018). www.researchgate.net/publication/328956833_Climate_impacts_in_Europe_final_report_of_the_JRC_PESETA_III_project.

Costanza, Robert, Rudolf de Groot, Paul Sutton, Sander van der Ploeg, Sharolyn J. Anderson, Ida Kubiszewski, Stephen Farber, and R.K. Turner. "Changes in the Global Value of Ecosystem Services." *Global Environmental Change* 26 (2014/05/01): 152–58.

Croitoru, L., and M. Sarraf. "How Much Does Environmental Degradation Cost? The Case of Morocco." *Journal of Environmental Protection* 9 (2018): 254–65.

Deutsch, Curtis A., Joshua J. Tewksbury, Michelle Tigchelaar, David S. Battisti, Scott C. Merrill, Raymond B. Huey, and Rosamond L. Naylor. "Increase in Crop Losses to Insect Pests in a Warming Climate." *Science* 361, no. 6405 (2018): 916.

Ding, Qi, Xinjun Chen, Yong Chen, and Siquan Tian. "Estimation of Catch Losses Resulting from Overexploitation in the Global Marine Fisheries." *Acta Oceanologica Sinica* 36, no. 12 (2017/12/01): 37–44.

Doherty, Ruth M., Mathew R. Heal, and Fiona M. O'Connor. "Climate Change Impacts on Human Health over Europe through Its Effect on Air Quality." *Environmental Health* 16, no. 1 (2017/12/05): 118.

ELD Initiative. "The Value of Land: Prosperous Lands and Positive Rewards through Sustainable Land Management." (2015). www.eld-initiative.org/fileadmin/pdf/ELD-main-report_05_web_72dpi.pdf.

Evans, Gary W. "Projected Behavioral Impacts of Global Climate Change." *Annual Review of Psychology* 70, no. 1 (2019/01/04): 449–74.

FAO. "The State of the World's Forests 2018 – Forest Pathways to Sustainable Development." (2018). www.fao.org/3/I9535EN/i9535en.pdf.

Forzieri, Giovanni, Alessandra Bianchi, Filipe Batista e Silva, Mario A. Marin Herrera, Antoine Leblois, Carlo Lavalle, Jeroen C.J.H. Aerts, and Luc Feyen. "Escalating

Impacts of Climate Extremes on Critical Infrastructures in Europe." *Global Environmental Change* 48 (2018/01/01): 97–107.

Franchini, Massimo, and Pier M. Mannucci. "Impact on Human Health of Climate Changes." *European Journal of Internal Medicine* 26, no. 1 (2015): 1–5.

Free, Christopher M., James T. Thorson, Malin L. Pinsky, Kiva L. Oken, John Wiedenmann, and Olaf P. Jensen. "Impacts of Historical Warming on Marine Fisheries Production." *Science* 363, no. 6430 (2019): 979.

Gao, Hui, Bao-Jun Yang, Nan Li, Li-Min Feng, Xiao-Yu Shi, Wei-Hong Zhao, and Si-Jin Liu. "Bisphenol A and Hormone-Associated Cancers: Current Progress and Perspectives." [In eng]. *Medicine* 94, no. 1 (2015): e211–e11.

Goudie, A.S. "Desertification." In *Encyclopedia of Environmental Health*, edited by J.O. Nriagu, 30–35. Burlington: Elsevier, 2011.

Guerreiro, Selma B., Richard J. Dawson, Chris Kilsby, Elizabeth Lewis, and Alistair Ford. "Future Heat-Waves, Droughts and Floods in 571 European Cities." *Environmental Research Letters* 13, no. 3 (2018/02/20): 034009.

Hales, Simon. "Climate Change, Extreme Rainfall Events, Drinking Water and Enteric Disease." *Reviews on Environmental Health* 34, no.1 (2019).

IPBES. "Media Release: Nature's Dangerous Decline 'Unprecedented'; Species Extinction Rates 'Accelerating'." (2017). www.ipbes.net/news/Media-Release-Global-Assessment.

IPCC. "Special Report. Global Warming of 1.5 °C." (2018). www.ipcc.ch/sr15/.

IUCN. "Nature-Based Solutions to Disasters." (2017). www.iucn.org/sites/dev/files/nbs_to_disasters_issues_brief_final.pdf.

Kinney, Patrick L. "Interactions of Climate Change, Air Pollution, and Human Health." *Current Environmental Health Reports* 5, no. 1 (2018/03/01): 179–86.

Kompas, Tom, Van H. Pham, and Tuong N. Che. "The Effects of Climate Change on Gdp by Country and the Global Economic Gains from Complying with the Paris Climate Accord." *Earth's Future* 6, no. 8 (2018/08/01): 1153–73.

Kwon, Ho-Jang, Soo-Hun Cho, Youngsin Chun, Frederic Lagarde, and Göran Pershagen. "Effects of the Asian Dust Events on Daily Mortality in Seoul, Korea." *Environmental Research* 90, no. 1 (2002/09/01): 1–5.

Leng, Guoyong, and Jim Hall. "Crop Yield Sensitivity of Global Major Agricultural Countries to Droughts and the Projected Changes in the Future." *Science of the Total Environment* 654 (2019/03/01): 811–21.

Mallin, Michael A. "Chapter 4 – Effect of Human Land Development on Water Quality." In *Handbook of Water Purity and Quality*, edited by Satinder Ahuja, 67–94. Amsterdam: Academic Press, 2009.

McDonald, Robert I., Katherine F. Weber, Julie Padowski, Tim Boucher, and Daniel Shemie. "Estimating Watershed Degradation over the Last Century and Its Impact on Water-Treatment Costs for the World's Large Cities." *Proceedings of the National Academy of Sciences* 113, no. 32 (2016): 9117–22.

Miller, James D., and Michael Hutchins. "The Impacts of Urbanisation and Climate Change on Urban Flooding and Urban Water Quality: A Review of the Evidence Concerning the United Kingdom." *Journal of Hydrology: Regional Studies* 12 (2017/08/01): 345–62.

Missirian, Anouch, and Wolfram Schlenker. "Asylum Applications Respond to Temperature Fluctuations." *Science* 358, no. 6370 (2017): 1610.

National Public Radio. "Nebraska Faces over $1.3 Billion in Flood Losses." (2019). www.npr.org/2019/03/21/705408364/nebraska-faces-over-1-3-billion-in-flood-losses.

Nkonya, Ephraim, Weston Anderson, Edward Kato, Jawoo Koo, Alisher Mirzabaev, Joachim von Braun, and Stefan Meyer. "Global Cost of Land Degradation." In *Economics of Land Degradation and Improvement – A Global Assessment for Sustainable Development*, edited by Ephraim Nkonya, Alisher Mirzabaev, and Joachim von Braun, 117–65. Cham: Springer International Publishing, 2016.

ODI. "Understanding the Effects of Resource Degradation on Socio-Economic Outcomes in Developing Countries." (2014). www.odi.org/sites/odi.org.uk/files/odi-assets/publications-opinion-files/8830.pdf.

OECD. "The Economic Consequences of Climate Change." (2015). https://espas.secure.europarl.europa.eu/orbis/sites/default/files/generated/document/en/OECD%20Climate%20Change.pdf.

OECD. "The Economic Consequences of Outdoor Air Pollution." (2016). www.oecd.org/environment/the-economic-consequences-of-outdoor-air-pollution-9789264257474-en.htm.

Pew. "Southeast Suffered Millions in Economic Losses from Overfishing." (2012). www.pewtrusts.org/en/about/news-room/press-releases-and-statements/2012/09/04/southeast-suffered-millions-in-economic-losses-from-overfishing.

Pretis, Felix, Moritz Schwarz, Kevin Tang, Karsten Haustein, and R. Allen Myles. "Uncertain Impacts on Economic Growth When Stabilizing Global Temperatures at 1.5°C or 2°C Warming." *Philosophical Transactions of the Royal Society A: Mathematical, Physical and Engineering Sciences* 376, no. 2119 (2018/05/13): 20160460.

Rustagi, Neeti, S.K. Pradhan, and Ritesh Singh. "Public Health Impact of Plastics: An Overview." [In eng]. *Indian Journal of Occupational and Environmental Medicine* 15, no. 3 (2011): 100–03.

Sartori, Martina, George Philippidis, Emanuele Ferrari, Pasquale Borrelli, Emanuele Lugato, Luca Montanarella, and Panos Panagos. "A Linkage between the Biophysical and the Economic: Assessing the Global Market Impacts of Soil Erosion." *Land Use Policy* 86 (2019/07/01): 299–312.

Schnell, Jordan L., and Michael J. Prather. "Co-Occurrence of Extremes in Surface Ozone, Particulate Matter, and Temperature over Eastern North America." *Proceedings of the National Academy of Sciences* 114, no. 11 (2017): 2854.

Schraufnagel, Dean E., John R. Balmes, Clayton T. Cowl, Sara De Matteis, Soon-Hee Jung, Kevin Mortimer, Rogelio Perez-Padilla, et al. "Air Pollution and Noncommunicable Diseases: A Review by the Forum of International Respiratory Societies' Environmental Committee, Part 1: The Damaging Effects of Air Pollution." *CHEST* 155, no. 2 (2019): 409–16.

Schwerdtle, Patricia, Kathryn Bowen, and Celia McMichael. "The Health Impacts of Climate-Related Migration." *BMC Medicine* 16, no. 1 (2018/01/04): 1.

Scottish Natural Heritage. "Impacts on People and Our Economy." (n.d.). www.nature.scot/climate-change/climate-change-impacts-scotland/impacts-people-and-our-economy.

Sinha, E., A.M. Michalak, and V. Balaji. "Eutrophication Will Increase during the 21st Century as a Result of Precipitation Changes." *Science* 357, no. 6349 (2017): 405.

Taylor, Richard G., Bridget Scanlon, Petra Döll, Matt Rodell, Rens van Beek, Yoshihide Wada, Laurent Longuevergne, et al. "Ground Water and Climate Change." *Nature Climate Change* 3 (2012/11/25): 322.

Thomas, Kimberley, R.D. Hardy, Heather Lazrus, Michael Mendez, Ben Orlove, Isabel Rivera-Collazo, J.T. Roberts, et al. "Explaining Differential Vulnerability to Climate

Change: A Social Science Review." *Wiley Interdisciplinary Reviews: Climate Change* 10, no. 2 (2019/03/01): e565.

Union of Concerned Scientists. "Climate Change, Extreme Precipitation and Flooding: The Latest Science." (2018). www.ucsusa.org/global-warming/global-warming-impacts/floods#.XBB5fWgzY2w.

US EPA. "Climate Adaptation and Saltwater Intrusion." (n.d.). www.epa.gov/arc-x/climate-adaptation-and-saltwater-intrusion.

US EPA. "Climate Impacts on Human Health." (2017). https://19january2017snapshot.epa.gov/climate-impacts/climate-impacts-human-health_.html.

US EPA. "Climate Impacts on Water Resources." (n.d.). https://archive.epa.gov/epa/climate-impacts/climate-impacts-water-resources.html.

US EPA. "Getting Up to Speed on Groundwater Contamination." (2015). www.epa.gov/sites/production/files/2015-08/documents/mgwc-gwc1.pdf.

U.S. Global Change Research Program. "Fourth National Climate Assessment." (2018). www.globalchange.gov/nca4.

U.S. Global Change Research Program. "The Impacts of Climate Change on Human Health in the United States: A Scientific Assessment." (n.d.). https://health2016.globalchange.gov/.

USGS. "Effects of Urban Development on Floods." (n.d.). https://pubs.usgs.gov/fs/fs07603/.

Vora, Neil. "Impact of Anthropogenic Environmental Alterations on Vector-Borne Diseases." [In eng]. *Medscape Journal of Medicine* 10, no. 10 (2008): 238–38.

Watanabe, Tsugihiro, Johannes Cullmann, Chandra S. Pathak, Mika Turunen, Kamran Emami, Graziano Ghinassi, and Yasmin Siddiqi. "Management of Climatic Extremes with Focus on Floods and Droughts in Agriculture." *Irrigation and Drainage* 67, no. 1 (2018/02/01): 29–42.

WHO. "Ambient Air Pollution: Health Impacts." (n.d.). www.who.int/airpollution/ambient/health-impacts/en/.

WHO. "Land Degradation and Desertification." (n.d.). www.who.int/globalchange/ecosystems/desert/en/.

WHO. "Water-Related Diseases." (n.d.). www.who.int/water_sanitation_health/diseases-risks/diseases/methaemoglob/en/.

Wilhelmsson, Dan, Richard C. Thompson, Katrin Holmström, Olof Lindén, and Hanna Eriksson-Hägg. "Chapter 6 – Marine Pollution." In *Managing Ocean Environments in a Changing Climate*, edited by Kevin J. Noone, Ussif R. Sumaila, and Robert J. Diaz, 127–69. Boston, MA: Elsevier, 2013.

World Bank. "Groundswell: Preparing for Internal Climate Migration." (2018). https://openknowledge.worldbank.org/handle/10986/29461.

Zhang, Wei, Gabriele Villarini, Gabriel A. Vecchi, and James A. Smith. "Urbanization Exacerbated the Rainfall and Flooding Caused by Hurricane Harvey in Houston." *Nature* 563, no. 7731 (2018/11/01): 384–88.

7 Nature-Based Solutions in Climate Change Mitigation

Introduction

Climate change mitigation involves reducing greenhouse gas emissions and enhancing natural carbon storage. This includes cutting emissions from the energy sector through renewable energy policies, energy-efficient practices, and biological carbon sequestration – capturing and storing carbon in natural systems. Carbon sequestration refers to the removal of carbon dioxide from the atmosphere by soils and plants in terrestrial and aquatic environments, such as wetlands. It also includes preventing carbon dioxide emissions from terrestrial ecosystems, helping maintain and expand natural carbon storage. These actions contribute to conserving existing carbon pools and increasing carbon storage capacity, making them essential strategies for mitigating climate change.[1,2]

Ecosystem-Based Mitigation

In climate change mitigation, nature-based solutions (NBS) are also known as ecosystem-based mitigation (EbM), which encompasses a range of strategies that utilize natural ecosystems to reduce greenhouse gas emissions and enhance carbon storage. These approaches include sustainable forest management, reforestation with native species, conservation and restoration of peatlands and wetlands, protection of ocean carbon sinks, improved grassland management, and environmentally sustainable agricultural practices.[3]

Co-Benefits of Ecosystem-Based Mitigation

In addition to reducing greenhouse gas emissions, EbM provides multiple co-benefits, including:

- Improved air quality leads to better public health and a reduction in respiratory illnesses.
- Higher labor productivity with lower exposure to air pollution reducing health-related work absences.

DOI: 10.4324/9781003517948-7

- Enhanced environmental quality, characterized by cleaner air and water, yields improved recreational and aesthetic benefits.
- Stronger ecosystem services can be achieved by conserving and expanding forest cover while reducing emissions from deforestation and degradation.
- Reduced crop damage, as lower ground-level ozone levels improve agricultural yields.
- Increased agricultural productivity through soil carbon conservation and enrichment.
- Less coastal damage, as EbM helps limit sea-level rise and strengthens natural coastal defenses.
- Fewer heat-related deaths can be mitigated by mitigating extreme heat events.
- Slower biodiversity loss, as stabilizing the climate reduces ecosystem disruptions.[4,5]

Ecosystem-Based Mitigation Policy Measures

The Convention on Biological Diversity (CBD) has provided key recommendations for policymakers to maximize EbM's benefits while preventing negative impacts on biodiversity. These guidelines help design, implement, and monitor EbM strategies. Table 7.1 summarizes these recommendations.[6]

Soils

Soils are the largest terrestrial carbon sink, storing 74% of all terrestrial carbon within the top one meter.[7] When managed sustainably, soils effectively store

Table 7.1 Enhancing the Biodiversity Benefits of Ecosystem-Based Mitigation

Ecosystem-Based Mitigation Focus	*Description*
Soil	Support biodiversity conservation by conserving and restoring organic carbon in soils and biomass, including in peatlands, wetlands, grasslands, and drylands.
Agriculture	Promote sustainable agricultural practices that help maintain or increase carbon stocks while also conserving and sustainably using biodiversity. Activities should be assessed, implemented, and monitored to ensure long-term impact.
Forest	Manage forests sustainably and enhance forest carbon stocks, integrating biodiversity conservation and sustainable land management. Ensure the full and effective participation of indigenous peoples and local communities in policy-making and implementation, with attention to land ownership.
Marine and coastal	Strengthen the conservation, sustainable use, and restoration of climate-vulnerable marine and coastal habitats – such as mangroves, peatlands, tidal salt marshes, and seagrass beds – that also contribute to climate change mitigation.

carbon; however, erosion, poor management, and unsustainable agricultural practices can release this stored carbon as carbon dioxide (CO_2) into the atmosphere, thereby contributing to climate change. Various soil conservation and management practices can be implemented to prevent carbon loss and enhance soil health. These strategies improve crop yields, boost economic returns, and enhance soil fertility, ensuring long-term sustainability.[8]

Soil Erosion Control and Vegetative Measures

Soil erosion control involves strategies to prevent or mitigate wind and water erosion, thereby safeguarding soil health and minimizing environmental damage. Wind erosion can be managed by decreasing wind velocity using windbreaks and barriers, increasing soil surface roughness by maintaining stable soil aggregates, or reshaping the land to limit erosion. Water erosion is controlled by reducing surface runoff using structural and vegetative barriers, as well as increasing soil cover with cover crops and mulching. These methods help prevent soil loss, water pollution, and property damage. Additionally, vegetative measures such as grass strips, tree and shrub buffers, riparian vegetation, and grassed waterways slow runoff, reduce sediment transport, and enhance sediment deposition, further protecting soil integrity.[9]

Case 7.1: Enhancing Soil Health and Carbon Sequestration through California's Healthy Soils Program

The Healthy Soils Program (HSP) is an initiative in California aimed at enhancing soil health and reducing greenhouse gas (GHG) emissions through conservation management practices. The program operates under the California Healthy Soils Initiative, a multi-agency collaboration aimed at promoting sustainable agricultural practices on the state's farmlands and ranchlands. HSP consists of two primary components: the HSP Incentives Program and the HSP Demonstration Projects. The Incentives Program provides financial support to California growers and ranchers to implement conservation practices, including cover cropping, no-till and reduced-till farming, mulching, compost application, and conservation plantings. These practices contribute to soil health improvement, carbon sequestration, and GHG reduction. The program is funded by California Climate Investments, with an allocation of $40.5 million between 2016 and 2019, supplemented by a $10 million grant from the California Drought, Water, Parks, Climate, Coastal Protection, and Outdoor Access for All Act of 2018. The Demonstration Projects facilitate knowledge-sharing by showcasing real-world applications of HSP practices on farms and ranches. These projects are divided into two categories. Type A projects focus on the implementation of conservation management practices while collecting field data on GHG emissions, soil health metrics, and cost-benefit analyses. Type B projects emphasize demonstrating conservation practices and analyzing adoption barriers, without requiring the measurement of GHG emissions. By providing financial support and facilitating knowledge exchange, the HSP encourages the widespread adoption of conservation management practices, fostering long-term sustainability in California's agricultural sector.[10]

CDFA. "Healthy Soils Program." (n.d.). www.cdfa.ca.gov/oars/healthysoils/.

Sustainable Agricultural Techniques

In agricultural systems, the amount and duration of carbon storage in soils depend largely on soil management practices. Enhancing soil carbon storage helps remove carbon dioxide from the atmosphere. It provides multiple benefits, including improved soil and water quality, reduced nutrient loss, lower soil erosion, better water conservation, and increased crop yields. Various management techniques can effectively create a net carbon sink in soils, including the following:

Conservation Tillage

Conservation tillage aims to minimize soil disturbance while maintaining soil health and productivity. It includes various techniques, with zero tillage (no-till) and reduced or minimum tillage being the most common. In no-till farming, soil disturbance is limited to planting and nutrient application, thereby preserving the soil's structure. Reduced or minimum tillage balances the need for crop production and weed control while limiting soil disruption by reducing the intensity, depth, or frequency of tillage operations. These practices help improve soil structure, moisture retention, and carbon storage, making them essential for sustainable agriculture.[11]

Cover Cropping

Cover crops are planted after a main crop is harvested and terminated before the next crop is sown. Typically, annual plants establish quickly, adding organic matter to the soil, reducing erosion, and preventing nutrient loss from leaching. Different cover crop species offer unique benefits, so farmers often plant multiple species together as a cover crop mixture or intercrop, thereby enhancing soil health and improving the overall resilience of their cropping system.[12,13]

Crop Rotation

Crop rotation is the practice of growing different crops in a recurring sequence on the same land, thereby promoting soil health and biodiversity, as opposed to intensive monocropping. Using a variety of crop types helps increase soil organic matter.[14] For example, rotations that include cash crops and meadow crops, such as legumes or perennial grasses, contribute significantly to the soil's organic matter by decomposing plant matter, particularly root systems. These roots help maintain or enhance soil humus, a long-lasting organic matter that improves soil fertility and structure.[15]

Compost

Composting is a highly effective long-term strategy for enhancing soil fertility in organic farming. It supports soil restoration, carbon sequestration, improved water retention, and reduced reliance on chemical inputs such as fertilizers, pesticides, and fuel.[16] Composting is particularly beneficial for recovering degraded soils and boosting soil health.

There are three standard composting methods used on farms:

- *Passive pile method*: Compost materials are stacked without turning, ventilation, or temperature monitoring.
- *Windrow method*: Compost feedstock is arranged in long, narrow piles that are regularly turned to maintain aeration and decomposition.
- *Aerated static pile*: Compost materials are piled, but instead of turning, oxygen is supplied through PVC pipes placed at the bottom of the pile to enhance decomposition.[17]

Zero Pesticide Use

Initially, fertilizers and pesticides enhance crop growth and biomass production. However, certain fertilizers, particularly nitrogen (N) fertilizers and pesticides, can increase microbial activity, thereby accelerating the decomposition of organic matter. These chemicals provide microorganisms with readily available nitrogen, which speeds up decomposition, especially in soils with a high carbon-to-nitrogen (C:N) ratio, where decomposition is typically slow due to nitrogen limitations. In organic farming, the absence of synthetic fertilizers and pesticides typically results in a higher organic matter content, which enhances soil fertility, stability, and moisture retention. This reduces the risk of erosion and desertification, contributing to more sustainable and resilient agricultural systems.[18,19]

Integrated Pest Management

Effective pest and disease management promotes healthy crops, resulting in optimal biomass production and contributing to the accumulation of organic matter in the soil. Diversified cropping systems and mixed crop-livestock farming enhance natural pest control by encouraging beneficial species interactions. Integrated pest management (IPM) helps farmers maintain a balanced ecosystem by regularly monitoring their crops to assess pest-to-predator ratios and identify the causes of damage. Based on these observations, farmers decide whether to apply natural or chemical treatments and determine the appropriate application levels. While IPM does not eliminate pesticide use, it ensures pesticides are used more safely and selectively, reducing environmental and health risks.[20,21]

Agroforestry

Agroforestry is the practice of integrating trees, shrubs, and other woody perennials into farming systems. It combines food crops, forestry, and pasture species to enhance crop production and land conservation. There are two main approaches to agroforestry. The first uses crops or pastures as a temporary land use until forest plantations become fully established. The second is to permanently integrate trees and shrubs into crop or livestock systems, providing long-term benefits such as soil protection, biodiversity enhancement, and resource sustainability. Perennial trees and shrubs in agroforestry systems help recycle nutrients by drawing them from deeper soil layers through their root systems, improving soil fertility and overall ecosystem health.[22]

Mixed Crop-Livestock Farming System

Traditional farms utilize livestock to maintain soil fertility and plant material that would otherwise go to waste. Mixed crop-livestock farming systems integrate pasture, livestock, and crop production within the same farm, offering multiple benefits. These include higher soil organic matter, improved soil fertility, and greater soil biodiversity, all of which contribute to a more sustainable and resilient agricultural system.[23,24]

Benefits of Sustainable Agricultural Techniques

The sustainable agriculture techniques discussed above contribute to healthy soil, providing several key benefits:

- *Enhanced carbon sequestration*: Regenerative practices, such as cover cropping, help build soil organic matter and store carbon, thereby reducing atmospheric CO_2.
- Improved water retention: Soils rich in organic matter retain more water, improving drought resilience and reducing irrigation needs.
- *Reduced pesticide dependence*: Healthy soils with diverse microbial life support stronger plants, making them less susceptible to pests and better able to defend against attacks.
- *Eliminates the need for synthetic fertilizers*: Sustainable soil management reduces reliance on synthetic, fossil fuel-based fertilizers, promoting natural nutrient cycling.

- *Higher crop yields and economic returns*: Improved soil health leads to greater productivity, better harvests and increased farm profitability.[25,26,27]

Case 7.2: The MAIS Program in the Jacuípe Basin, Brazil

The Modulo Agroclimático Inteligente e Sustentável (MAIS) Program is supporting family farmers in Brazil's semi-arid Jacuípe Basin to adapt to climate change. It is one of the first climate-smart agricultural programs in the country to mainstream climate-disruptive technologies among smallholder farmers. Over 12 years, the program has engaged more than 650 farmers, promoting best practices in animal nutrition, farm management, food and water security, and soil restoration. The program has led to a 63% increase in dairy production, a 204% rise in farmers' income, and a 30% improvement in pastureland conditions. Additionally, it has reduced the water footprint of local farms by 50%, decreased production oscillations by 30%, and offsets over 3 tons of CO_2 per hectare of restored pastureland. Each USD 1 invested has generated approximately USD 7 in the region. The MAIS Program integrates climate-smart solutions, such as using Opuntia ficus (cactus) as a substitute for corn, implementing silvopastoral systems, and restoring degraded pasturelands with drought-resistant trees. It operates through Adapta Group and Instituto Adapta Sertão, training cooperative-linked field technicians to assist farmers. A dedicated financial mechanism facilitates access to credit for implementing these practices. The program fosters community-driven adaptation, where local organizations and farmer leaders adapt technologies to regional conditions, ensuring sustainability. If expanded to 15 major dairy corporations, the MAIS approach could generate a USD 5 billion market opportunity and reduce Brazil's agricultural greenhouse gas emissions by 4%.[28]

UNFCCC. "Mais Program: Climate-Smart Agriculture| Brazil." (n.d.). https://unfccc.int/climate-action/momentum-for-change/financing-for-climate-friendly-investment/mais-program-climate-smart-agriculture.

Forests

Forests store approximately 45% of the world's carbon stock, making them a crucial component of land-based carbon sequestration strategies. Preserving and enhancing this natural carbon sink is a key focus of land management efforts aimed at reducing carbon dioxide emissions. Several forestry-based strategies can help mitigate carbon emissions, including reforestation, which restores forests in areas where they recently existed, and afforestation, which establishes forests in previously unforested areas. Increasing the carbon density of existing forests, reducing emissions from deforestation and degradation, and using biomass as a renewable energy source to replace fossil fuels are also practical approaches.

Forests provide a cost-effective means of addressing climate change, as carbon sequestration through forestry is relatively inexpensive compared to other mitigation measures. However, deforestation and forest degradation, whether driven by human activities or natural disturbances, result in carbon loss, greenhouse gas emissions, and a reduction in ecosystem services. Sustainable forest management practices must be adopted to maximize carbon sequestration and protect existing carbon stocks, thereby ensuring that forests remain essential for mitigating climate change.[29,30,31]

Afforestation, Reforestation, and Forest Restoration

Expanding forest area and density through afforestation, reforestation, and forest restoration enhances the absorption of atmospheric carbon dioxide. Forest carbon sequestration rates depend on management practices, tree species, and geographic location. When agricultural land is converted into forests through afforestation, sequestration rates vary by region and species. On average, a planted forest in a temperate zone can sequester approximately four tons of carbon dioxide per hectare per year.[32]

A study on forestry-based carbon mitigation strategies in Oregon, the Pacific Northwest, found that reforesting areas affected by fire and beetle infestations could increase carbon stocks by 315 billion metric tons by 2100, resulting in a 5% reduction in forest sector net emissions compared to a business-as-usual scenario.[33] Expanding forest area and density can be achieved through various approaches, including the following.

Increase of Tree Cover Outside Forests

Outside forests, trees are found in agroforestry systems, farms, rural landscapes, along roads and rivers, near human settlements, and within urban areas. Expanding tree cover in these areas helps mitigate climate change by storing carbon, preventing land degradation, providing renewable fuel as an alternative to fossil fuels, and fixing nitrogen to reduce fertilizer use. These trees enhance environmental sustainability and ecosystem resilience beyond traditional forested areas.[34]

Enhancement of Forest Carbon Stocks

Implementing effective forest management practices allows forests to provide multiple ecosystem services while enhancing carbon storage. Key strategies include silvicultural treatments such as tending operations and gap enrichment, selecting suitable tree species, and adjusting rotation cycles to optimize carbon sequestration. These practices enhance the climate change mitigation potential of forests while maintaining their ecological and economic benefits.[35]

Sustainable Forest Management Practices

Sustainable forest management practices can ensure that productive or multi-purpose forests continue to store carbon and maintain their capacity to provide other ecosystem services. Management practices should be actively planned and tailored to each specific ecosystem and evolving situation to avoid overexploitation and forest degradation.[36]

Integrated Fire Management

Wildfires are a significant source of greenhouse gas emissions and are expected to become more frequent and intense due to climate change. Strengthening wildfire prevention and risk reduction measures can help protect carbon storage and sequestration in natural forests, planted forests, and other ecosystems such as floodplains.[37] Overly dense managed forests are more susceptible to large-scale wildfires, threatening lives, property, and biodiversity. These forests are also less resilient to invasive species, increasing the risk of widespread tree mortality. Removing excess biomass can improve forest health, reduce fire risks, and restore ecological balance. The harvested biomass can also be repurposed for electricity generation, heat and steam power production, and home heating, offering a sustainable energy alternative.[38]

Management of Protected Areas and Wildlife

Protected areas play a vital role in conserving forests and wildlife. Strengthening the management of these areas and expanding their coverage is recognized as one of the key mitigation strategies for Reducing Emissions from Deforestation and Forest Degradation (REDD+). The wise management of protected areas is critical for the successful conservation and sustainable use of wild animals in a changing climate. Priority actions include:

- Developing adapted strategies, policies, and legislation
- Setting up long-term monitoring and reporting schemes for indicator plant and animal species
- Facilitating transboundary cooperation
- Better involvement of local people
- Implementing measures that help reduce non-climate change-related, human-induced pressures, for example, habitat loss or degradation and illegal hunting and trade, and therefore overall vulnerability to climate change
- Expanding or shifting protected area boundaries for the conservation of forest ecosystems and species at risk from climate change
- Improving habitat connectivity and providing migration corridors and seasonal feeding areas for migratory species
- Integrating carbon sequestration and storage, along with other ecosystem services, into management and finance plans[39]

Case 7.3: Ghana's Cocoa Forest REDD+ Programme

Ghana's Cocoa Forest REDD+ Programme (GCFRP) is a landmark national initiative designed to reduce carbon emissions stemming from deforestation and cocoa-driven land conversion, while simultaneously promoting sustainable cocoa farming practices. Coordinated by the Ghana Cocoa Board and the Forestry Commission, the program spans 5.9 million hectares within Ghana's High Forest Zone. It is supported by the Forest Carbon Partnership Facility Carbon Fund, through which Ghana stands to receive up to $50 million in payments for verified emission reductions. The GCFRP is structured around six key pillars that collectively aim to align cocoa production with environmental sustainability. First, the climate-smart cocoa pillar focuses on enhancing cocoa yields through sustainable intensification methods and agroforestry practices. Second, landscape-level governance is promoted by establishing Hotspot Intervention Areas, enabling communities to co-manage land use. Third, forest rehabilitation efforts, such as assisted natural regeneration and improved deforestation monitoring, are central to curbing illegal land-use practices. Fourth, institutional coordination fosters partnerships among government agencies, cocoa companies, non-governmental organizations, and farmers. Fifth, the program enhances financial inclusion by providing farmers with access to credit, yield insurance, and climate-resilient investment opportunities. Finally, policy and legislative reforms are being advanced to improve forest governance, clarify tree tenure rights, and establish fair benefit-sharing mechanisms. Together, these integrated strategies aim to reduce emissions from deforestation, support Ghana's climate commitments, ensure a resilient cocoa sector, and enhance rural livelihoods through sustainable income generation. The GCFRP also aspires to serve as a replicable model for other cocoa-producing nations across West Africa.[40]

Forest Carbon Partnership Facility. "Ghana." (n.d.). www.forestcarbonpartnership.org/country/ghana.

Grasslands

Mitigation strategies for grassland ecosystems include reducing grazing intensity, regulating fire frequency, preventing the conversion of grasslands to cropland, and restoring degraded grasslands. Proper grazing management can help maintain carbon storage, as lower grazing intensities typically result in reduced carbon loss. However, the optimal grazing intensity for carbon sequestration depends on climate, soil type, and vegetation. For example, intermediate grazing levels in tropical grasslands promote the highest carbon sequestration, while in other systems, even moderate grazing may result in soil carbon loss. Grazing by wild or domesticated animals also helps reduce

Table 7.2 Grassland Management Practices That Enhance Carbon Stocks

Grassland Management Practice	*Description*
Sowing improved species	Introduce plant species that are better suited to local climates and more resilient to drought and grazing. These species enhance soil fertility and increase plant productivity, which results in higher carbon inputs to the soil and greater long-term carbon storage.
Applying water, fertilizer, or organic matter	Improve water and nutrient availability by applying irrigation, fertilizers, or organic matter (e.g. compost or manure). These inputs stimulate plant growth, enhance photosynthesis, and contribute to the accumulation of soil organic carbon.
Restoring degraded lands	Rehabilitate overgrazed or eroded grasslands by reseeding, soil amendments, or controlled grazing. Restored areas show increased vegetation cover and root biomass, improving soil structure and boosting carbon sequestration.
Including grass in crop rotations	Incorporate grasses into crop rotation cycles on arable land to enhance organic matter inputs through root systems and residue. Grazed forage grasses reduce tillage needs, thereby minimizing soil disturbance and limiting carbon losses through decomposition.

fire risk by lowering fuel loads, significantly decreasing carbon dioxide emissions from wildfires.[41] Additional grassland management practices that support carbon storage are outlined in Table 7.2.[42]

Case 7.4: Boomitra Grassland Restoration and Soil Enrichment Project

The Boomitra Grassland Restoration through Soil Enrichment project is working to restore degraded grasslands across 1.2 million acres in the Pampas region of Argentina and Paraguay. The initiative addresses the decline of Soil Organic Carbon (SOC) caused by overgrazing, unsustainable land management, and land-use changes that threaten both local ecosystems and the livelihoods of ranchers. The project promotes sustainable grazing practices, including rotational grazing and improved water management, in collaboration with local organizations. These techniques help rebuild SOC, enhance soil fertility, and increase grassland productivity. By improving land conditions, ranchers gain better economic stability, reducing the need for land-use conversion. The initiative also provides ranchers with carbon credit payments, offering an alternative income stream while incentivizing continued adoption of climate-positive practices. Beyond carbon sequestration, the project delivers ecological and social benefits. Healthier grasslands improve biodiversity, providing critical habitats for local flora and fauna. Improved soil structure and water retention enhance drought resilience, reducing erosion and supporting regional water cycles. The project also integrates digital agronomic tools, equipping ranchers with data-driven insights to optimize land use and maximize carbon sequestration. Boomitra's approach contributes to multiple UN Sustainable Development Goals by addressing climate change, supporting local economies, and improving agricultural sustainability. By restoring degraded landscapes and ensuring long-term viability for ranchers, the initiative strengthens both environmental and community resilience. Compliance with California Assembly Bill 1305 (2023) further aligns the project with evolving climate policies, ensuring transparency in carbon credit markets.[43]

Boomitra. "Climate, Community, and Biodiversity." (n.d.). https://boomitra.com/projects/boomitra-grassland-restoration-in-the-temperate-grasslands-of-south-america-through-soil-enrichment/#:~:text=Boomitra's%20grassland%20restoration%20project%20not,security%2C%20and%20biodiversity%20also%20increase.

Peatlands

Restoring degraded peatlands is a key mitigation strategy to prevent further peat decomposition and reduce fire risk. One practical approach is re-wetting drained peatlands, either partially or entirely, to restore natural water levels. Full restoration, which involves re-establishing peat-forming vegetation, provides the most significant emission reductions and helps reinstate carbon sequestration processes. When full restoration is not feasible, shifting to less drainage-intensive land uses – such as transitioning from crop cultivation to pasture or planting tree species that tolerate high water levels – can be a viable alternative. EbM measures in peatland areas offer multiple co-benefits beyond carbon sequestration. These include regulating the water cycle, controlling floods, purifying water, and removing excess nutrients and pollutants. Additionally, peatland restoration reduces air pollution risks by lowering the likelihood of peat fires. It helps counteract subsidence, where peat decomposition causes soil shrinkage, leading to worsened drainage, more frequent flooding, and increased saltwater intrusion.[44,45]

Case 7.5: Peatland ACTION – Restoring Scotland's Peatlands for Climate Resilience

The Peatland ACTION program is a national initiative in Scotland aimed at restoring degraded peatlands to mitigate climate change, enhance biodiversity, and support rural economies. Funded by the Scottish Government, the program has played a key role in reversing peatland degradation since its launch in 2012. With an investment commitment of £250 million up to 2030, the initiative supports land managers in undertaking restoration activities across Scotland's peatlands. Peatlands cover nearly two million hectares of Scotland, but approximately 75% of these are degraded, making them a significant source of carbon emissions. When intact, peatlands function as carbon sinks; however, degradation causes them to release the stored carbon into the atmosphere. Recognizing their role in climate mitigation and water regulation, Peatland ACTION prioritizes hydrological restoration, including blocking drainage ditches with peat dams, rewetting landscapes, and stabilizing eroded peat surfaces. The program has restored over 51,000 hectares of peatland to date, with a record 10,360 hectares restored in the 2023–24 fiscal year alone. Restoration efforts focus on reducing greenhouse gas emissions, improving water quality, mitigating flood risks, and enhancing biodiversity. The initiative also contributes to Scotland's net-zero targets by helping reduce the nation's 15% share of emissions linked to degraded peatlands. Delivered through NatureScot, Forestry and Land Scotland, Cairngorms National Park Authority, Loch Lomond & the Trossachs National Park Authority, and Scottish Water, Peatland ACTION provides technical guidance, funding support, and expertise to landowners. The initiative has also played a role in developing skills and creating rural employment opportunities, supporting Scotland's transition to a low-carbon economy while ensuring the long-term health of its peatlands.[46]

NatureScot. "Peatland Action." (n.d.). www.nature.scot/climate-change/nature-based-solutions/nature-based-solutions-practice/peatland-action.

Coastal and Marine

The coastal and marine environment is a significant carbon sink, playing a crucial role in carbon storage and climate regulation. Protecting and restoring these ecosystems is essential to maintaining this vital natural service. Several measures, including the following, can be implemented to safeguard and enhance coastal and marine carbon sinks.

Coastal Wetlands

Coastal wetlands are long-term carbon sinks, absorbing atmospheric carbon dioxide through the production of biomass and the burial of organic matter, primarily from roots and rhizomes, in the sediment. The amount of carbon stored in these wetlands varies depending on the type and location of the wetland. Generally, vegetation productivity increases from the saline areas of estuaries and deltas to the freshwater regions at the head of these systems, whether temperate or tropical. The preservation of soil carbon in wetlands is facilitated by regular tidal flooding, which creates saturated soil conditions. These conditions reduce oxygen levels, slowing the decomposition of organic matter and minimizing carbon dioxide release.[47] Restoring degraded coastal wetlands and near-shore marine ecosystems involves several actions, such as adding sediment to raise the land above water, allowing wetland plants to re-colonize, modifying erosion processes that degrade the area, rewetting drained wetlands by blocking drains and reducing groundwater extraction, and relocating flood defenses inland or on higher ground to restore intertidal habitats. These measures help attenuate storm surges and restore wetlands as natural buffers.[48] Restoring coastal wetlands offers multiple benefits, including reversing greenhouse gas emissions, enhancing carbon stocks, and improving various ecosystem services. These include reducing shoreline erosion, absorbing pollutants to maintain water quality, preventing eutrophication, providing feeding habitats for commercial fish, and offering recreational opportunities.[49]

Case 7.6: Blue Carbon Ecosystem Restoration Grants

The Blue Carbon Ecosystem Restoration Grants program provides funding from 2021–22 to 2024–25 for projects that restore degraded coastal wetland ecosystems in Australia. These ecosystems, which include mangroves, seagrass, and tidal marshes, play a vital role in carbon sequestration, biodiversity conservation, shoreline protection, and cultural preservation.

Five projects were awarded funding (AUD) under this initiative:

- Sunshine Coast Regional Council and Partners *($2,036,000)*: The Blue Heart – Blue Carbon Wetland Restoration Project in Sunshine Coast, Queensland, focuses on restoring former farmland to coastal wetlands. It provides benefits in carbon sequestration, biodiversity, flood mitigation, recreation, and First Nations engagement.
- Greening Australia *($1,779,988)*: The Mungalla Blue Carbon Project in Ingham, Queensland, restores tidal flows to former grazing land, enhancing Indigenous heritage, ecotourism, the Great Barrier Reef's water quality, and the biodiversity of birds and marine life.
- The University of Adelaide *($1,972,500)*: The Gulf St Vincent Seagrass Restoration Project at Port Gawler, South Australia, enhances marine diversity, sediment stabilization, shoreline protection, and nutrient processing.

- Southern Regional Natural Resource Management *($793,947)*: The Temperate Saltmarsh Restoration Project in PittWater-Orielton Lagoon, Tasmania, strengthens coastal resilience, biodiversity, recreational fisheries, and tourism.
- The Nature Conservancy Australia and Partners *($2,896,526)*: The South Australian Blue Carbon Ecosystem Restoration Project in Upper Gulf St Vincent focuses on biodiversity, social, and cultural benefits.

The grants are part of the Blue Carbon Conservation, Restoration, and Accounting Program, supporting climate action, investment scaling, and sustainable management. Environmental economic accounting is being conducted to assess the long-term benefits of these restoration projects. A best practice guideline for hydrologic assessments has also been developed to support future coastal ecosystem projects.[50]

Department of Climate Change, Energy, the Environment and Water. "Blue Carbon Ecosystem Restoration Grants." (n.d.). www.dcceew.gov.au/environment/marine/coastal-blue-carbon-ecosystems/conservation/restoration-grants.

Mangroves

Mangroves are among the most effective ecosystems for carbon capture and storage, with their soils holding over six billion tons of carbon and sequestering three to four times more carbon than terrestrial forests.[51] These ecosystems absorb carbon dioxide from the atmosphere, storing it in their biomass and organic-rich soils, which remain stable over long periods. Mangroves create thick, waterlogged soils, where dead plant material accumulates rather than decomposing due to the low oxygen levels within their complex root systems. This results in the formation of carbon-rich peat, which releases minimal methane emissions, making mangroves one of the most efficient carbon sinks. Establishing protected areas is a crucial strategy for securing the long-term conservation of mangroves. Protected status provides a legal and management framework to safeguard mangroves, ensuring continued carbon sequestration, biodiversity protection, and resilience against environmental threats.[52]

Case 7.7: Mangrove Restoration and Coastal Greenbelt Protection in Sumatra

The Livelihoods-Yagasu Mangrove Restoration Project is a large-scale initiative in Aceh and North Sumatra, Indonesia, aimed at restoring degraded mangrove ecosystems while enhancing local livelihoods.

Implemented by Yagasu since 2011, the project is funded by the Livelihoods Carbon Fund (LCF) I and II, as well as other investors. The project addresses the extensive mangrove deforestation driven by aquaculture, rice production, and palm oil production. To counteract this degradation, Yagasu has restored 10,000 hectares of mangroves, safeguarding 23 mangrove species and numerous bird, fish, reptile, and invertebrate species. The project also enhances coastal defenses against erosion and extreme weather, thereby improving the resilience of local communities. Over 20 years, it is expected to sequester 4.5 million tons of CO_2, making a significant contribution to climate mitigation. Beyond restoration, the project integrates sustainable production and income diversification. The silvofishery approach, which blends 70% mangrove cover with 30% fishponds, provides a fertile habitat for shrimp, crabs, and fish, ensuring stable incomes for fisher communities. Women-led businesses have emerged, producing organic batik, shrimp paste, and mangrove-based food products, leading to a 57% increase in household incomes. Farmers have also adopted fruit tree cultivation, honey production, and sustainable aquaculture, further strengthening economic resilience. Financing is secured through carbon pre-financing and credit offtake agreements with 12 carbon investors, including Danone and Hermes. Additional grants from the Green Climate Fund and Global Environment Facility support adaptation initiatives. The project is certified under the Climate, Community, and Biodiversity Standards and utilizes digital monitoring tools to track progress. By restoring mangroves and establishing a coastal greenbelt corridor, the Livelihoods-Yagasu project is advancing biodiversity conservation, economic security, and climate resilience for vulnerable communities in Sumatra.[53]

South Pole. "Mangrove Restoration and Coastal Greenbelt Protection." 2022. www.wwf.org.uk/sites/default/files/2022-07/case_study_livelihoods_yagasu_mangrove_restoration.pdf.

Seagrass Conservation and Restoration

Seagrass meadows are among the most productive coastal ecosystems, providing essential services for carbon sequestration and oxygen production.[54] Through photosynthesis, a single square meter of seagrass can produce approximately 10 liters of oxygen per day, thereby supporting marine life and enhancing water quality.[55] To conserve and restore seagrass beds, several approaches can be implemented:

- *Natural recolonization*: This method enhances water quality, allowing seagrass to regenerate in favorable conditions naturally. It involves upgrading municipal sewage systems and reducing pollution from industrial, residential, and agricultural sources in coastal zones.

- *Transplantation*: A proactive restoration approach that involves relocating viable seagrass seedlings grown in aquaria or mature plants from healthy donor beds to restoration sites once conditions support their survival.
- *Seeding*: Seagrass restoration can also be achieved by dispersing seeds in estuarine waters, allowing natural propagation and expansion.[56]

Case 7.8: Seagrass Conservation and Restoration in Fiji

Fiji is home to six species of seagrass, including Halodule pinifolia and Halophila ovalis, which play a crucial role in maintaining marine biodiversity and supporting coastal communities. These meadows provide essential ecosystem services, including breeding grounds for fish and turtles, water filtration, and carbon sequestration. Despite their importance, seagrasses in Fiji face increasing threats from coastal development, pollution, and destructive fishing practices, contributing to their global decline at an estimated rate of 7% per year. To address this challenge, a pilot seagrass restoration project was implemented as part of the Management and Conservation of Blue Carbon Ecosystems project. In collaboration with Fiji National University's College of Agriculture, Fisheries & Forestry, the Locally Managed Marine Area network, the Tailevu Provincial Council, and four coastal villages in the Dawasamu district (Driti, Silana, Nataleira, and Nasinu), the project tested internationally recognized seagrass restoration techniques. The trial identified the Sods/Turf method – which involves transplanting healthy seagrass patches into degraded areas – as the most effective in terms of survival rates and planting efficiency. Community members, particularly youth groups, actively participated in the restoration and monitoring of the restored sites. Key recommendations from the project included prioritizing materials that do not trap sediments, handling seagrass sods with care, conducting restoration during low tide, and integrating traditional knowledge into restoration activities. Fiji has prioritized seagrass conservation through its commitments under the Ramsar Convention on Wetlands, the UN Convention on Biological Diversity, and the Paris Agreement. Locally, biodiversity assessments are being conducted in Rewa and Lomaiviti provinces, as well as parts of Vanua Levu, to evaluate the health of seagrass meadows and their carbon storage potential. These findings will contribute to Fiji's national greenhouse gas inventories and reporting under its Nationally Determined Contributions.[57]

The Fiji Times. "Conservation | Hidden Treasures of Our Ocean." (n.d.). www.fijitimes.com.fj/conservation-hidden-treasures-of-our-ocean/.

Seaweed

Seaweed production plays a significant role in carbon dioxide removal, making it an effective tool for mitigating climate change. Seaweed communities capture more carbon through photosynthesis than they release through respiration, contributing significantly to carbon sequestration in marine ecosystems.

Expanding seaweed aquaculture helps mitigate climate change and provides economic and environmental benefits. Seaweed can be harvested for high-value compounds used in the food, pharmaceutical, and biotechnology industries. Additionally, after extracting carbohydrates for hydrocolloid production or biofuel development, the remaining low-value biomass can be converted into protein concentrates for use in the animal feed industry, maximizing its value across multiple sectors.[58]

Case 7.9: CH4 Global's Scalable Solution for Methane Reduction in Livestock

Methane emissions from ruminant animals pose a significant sustainability challenge, accounting for approximately 30% of global methane emissions from human activities. As methane is a potent greenhouse gas, reducing its emissions is considered one of the most effective ways to slow climate change in the near term. CH4 Global, a company specializing in methane reduction solutions, has developed a scalable technology platform designed to deliver gigaton-scale CO_2-equivalent reductions over the next decade. One of CH4 Global's key innovations is Methane Tamer™ Beef Feedlot, a feed supplement containing Asparagopsis seaweed, which has been scientifically proven to reduce enteric methane emissions by up to 90%. The product is guaranteed to achieve a minimum 70% reduction in emissions throughout its shelf life. In addition to its climate benefits, Methane Tamer™ enhances feed efficiency, supports digestion, and improves weight gain, leading to the production of lower-carbon, export-friendly beef products. CH4 Global is collaborating with CirPro, which will be the first commercial supplier of Asparagopsis-based supplements in Australia. This partnership marks a milestone in achieving a carbon-neutral beef industry, demonstrating a shift from merely offsetting emissions to actively reducing them at the source. The company's end-to-end approach encompasses growing, processing, and distributing Asparagopsis-based solutions, optimized explicitly for beef feedlot cattle. As CH4 Global continues to commercialize this technology at scale, its goal remains clear: to drive gigaton-level CO2-equivalent reductions by 2030, thereby contributing to urgent global climate mitigation efforts.[59]

CH4 Global. "The Surest, Fastest Route to Bend the Climate Curve Is to Radically Reduce Methane Emissions in the Next 10 Years." (n.d.). https://ch4global.com.

Marine Protected Areas

Marine Protected Areas (MPAs) were initially established as an NBS to restore overexploited fish stocks, protect marine habitats, and conserve biodiversity. Within MPAs, marine reserves are categorized as fully protected areas, which prohibit all extractive activities, and strongly protected areas, which allow only low-impact fishing methods. Marine reserves contribute to climate change mitigation in the following ways.

Protecting Coastal Habitats

Marine reserves protect coastal wetlands, which store large amounts of inorganic carbon in their underlying sediments. Even minor disturbances, such as vegetation clearance, can trigger significant carbon losses by exposing stored carbon to the atmosphere. Additionally, ecosystems like mangroves, seagrasses, salt marshes, mudflats, and sandbars form interconnected intertidal habitats that sequester and store substantial amounts of organic carbon. Marine reserves help preserve these ecosystems, maintaining their natural carbon storage capacity while preventing the release of carbon caused by human disturbances.[60]

Reducing Trophic Downgrading

Trophic downgrading, or the removal of predators, has accelerated in many areas outside marine reserves, disrupting ecosystem balance. The loss of predators increases herbivore populations, leading to overgrazing of algal communities and, in some cases, complete habitat loss. This degradation impacts carbon uptake, storage, and release in vegetated marine ecosystems. Restoring predator populations helps regulate herbivore numbers, thereby reducing overgrazing and allowing primary producers, such as algae and seagrasses, to thrive. This supports carbon sequestration and storage, strengthening the ecosystem's ability to absorb and retain carbon while preserving biodiversity.[61]

Reducing Trawling and Other Fishing Impacts

Trawling and other mobile fishing methods disturb the seafloor, resuspending buried carbon and altering benthic ecosystems. These activities reduce the abundance of suspension feeders, vital in nutrient cycling and carbon storage. By maintaining healthy populations, suspension feeders enrich sediments with nutrients, improve light penetration, and promote plant growth and productivity. They also act as a carbon pathway, transferring organic carbon from the water column to the seafloor. Protecting marine areas from mobile fishing gear allows suspension feeder populations to recover, enhancing carbon sequestration and preventing further carbon loss from disturbed sediments. In addition, marine reserves help safeguard seabed ecosystems from the impacts of seabed mining, which can further disrupt stored carbon and damage marine habitats.[62]

Case 7.10: Advancing Marine Protected Areas in Ireland

Ireland is working to expand and enhance its Marine Protected Area (MPA) network to meet international commitments for ocean conservation. Currently, only 8.1% of Ireland's marine waters are designated as MPAs, primarily through Special Areas of Conservation and Special Protection Areas under European Union directives. To align with global targets, Ireland aims to protect 30% of its marine waters by 2030. The Irish government is developing new MPA legislation to strengthen marine conservation efforts. This legislation will establish legal protections for vulnerable marine ecosystems and ensure a more structured approach to managing human activities, including fishing, aquaculture, and pollution, within protected zones. The expansion of MPAs will focus on critical habitats such as saltmarshes, seagrass meadows, kelp forests, and oyster reefs, which provide essential ecosystem services, including carbon sequestration, biodiversity support, and coastal protection. Despite existing designations, many of Ireland's MPAs currently permit harmful human activities with little regulation or enforcement, resulting in the degradation of marine habitats. Strengthening protection measures will help restore biodiversity, support sustainable fisheries, and enhance climate resilience. Scientists emphasize that scaling up protection, improving fisheries management, and addressing pollution are crucial to restoring Ireland's marine life by 2050. With the forthcoming MPA legislation and an emphasis on expanding and effectively managing marine conservation areas, Ireland is taking steps to align with international biodiversity commitments and enhance the long-term health of its ocean ecosystems.[63]

Irish Wildlife Trust. "Marine Protected Area Network for Ireland." (n.d.). https://iwt.ie/what-we-do/campaigns/marine-protected-areas/.

Case Study Summary

Various EbM strategies have been applied in terrestrial and marine environments to help reduce the impacts of climate change. A summary of these approaches is provided in Table 7.3.

Table 7.3 Ecosystem-Based Mitigation Approaches

Habitat	*Project Name*	*Activities*	*Co-Benefits*
Soils	California Healthy Soils Program	Supports conservation management practices such as cover cropping, no-till and reduced-till farming, mulching, compost application, and conservation plantings to improve soil health and sequester carbon. Includes incentives and demonstration projects to encourage adoption.	Improves soil fertility, reduces GHG emissions, enhances resilience in agriculture, and shares knowledge through real-world examples.
Soils/ agriculture	MAIS Program in the Jacuípe Basin, Brazil	Implements climate-smart agriculture by restoring degraded pastures, using drought-resistant species like Opuntia-ficus, promoting silvopastoral systems, and training field technicians to support family farmers.	Increases dairy production and farmer income, improves pasture conditions, reduces water footprint, and lowers GHG emissions.
Forests/ agriculture	Ghana's Cocoa Forest REDD+ Programme	Combines climate-smart cocoa farming with landscape-level governance, forest rehabilitation, and institutional coordination. Supports financial access, yield insurance, and reforms in tree tenure.	Reduces deforestation, supports resilient cocoa farming, protects forests, and provides sustainable rural incomes.
Grasslands	Boomitra Grassland Restoration and Soil Enrichment Project	Restores degraded grasslands across Argentina and Paraguay using rotational grazing, improved water management, and digital tools for monitoring soil carbon and optimizing land use.	Increases soil organic carbon, enhances biodiversity, improves drought resilience, and provides carbon credit payments to ranchers.
Peatlands	Peatland ACTION – Scotland	Restores degraded peatlands by rewetting, blocking drainage ditches, and stabilizing eroded areas to stop carbon emissions and revive ecosystem functions. Provides funding and technical support to landowners.	Reduces GHG emissions, improves water quality, mitigates flooding, enhances biodiversity, and supports rural jobs.

(Continued)

Table 7.3 (Continued)

Habitat	*Project Name*	*Activities*	*Co-Benefits*
Coastal wetlands	Blue Carbon Ecosystem Restoration Grants – Australia	Funds restoration of mangroves, seagrass, and saltmarshes through five projects across Australia, focusing on biodiversity, shoreline protection, and carbon sequestration.	Delivers carbon storage, Indigenous and community engagement, habitat restoration, and improved coastal resilience.
Mangroves	Mangrove Restoration and Coastal Greenbelt – Sumatra	Restores 10,000 hectares of mangroves through silvofisheries and income diversification (shrimp paste, organic batik, honey, etc.), with support from carbon investors and digital monitoring tools.	Sequesters carbon, boosts household incomes, strengthens coastal defenses, and supports biodiversity conservation.
Seagrass	Seagrass Conservation and Restoration – Fiji	Uses Sods/Turf method to transplant healthy seagrass into degraded sites. Involves youth groups and integrates traditional knowledge with scientific restoration techniques.	Supports biodiversity, water filtration, carbon sequestration, and strengthens community involvement and marine ecosystem health.
Livestock	CH4 Global's Methane Reduction in Livestock	Introduces Asparagopsis-based Methane Tamer™ feed supplement to reduce enteric methane emissions by up to 90% in ruminants while enhancing feed efficiency.	Reduces methane emissions, improves weight gain, and supports the production of climate-resilient, export-ready beef.
Marine	Marine Protected Areas – Ireland	Expands marine protected areas with new legislation targeting critical habitats like saltmarshes and kelp forests, enhancing marine conservation and restricting harmful activities.	Protects marine biodiversity, promotes carbon sinks, supports fisheries, and enhances climate resilience.

Notes

1 Bruce G. Miller, "13 – Carbon Dioxide Emissions Reduction and Storage," in *Clean Coal Engineering Technology (Second Edition)*, ed. Bruce G. Miller (Oxford, UK: Butterworth-Heinemann, 2017).

2 C. Epple, S.G. Rangel, M. Jenkins, and M. Guth, "Managing Ecosystems in the Context of Climate Change Mitigation: A Review of Current Knowledge and Recommendations to Support Ecosystem-Based Mitigation Actions That Look Beyond Terrestrial Forests," (2016), https://cbd.int/doc/publications/cbd-ts-86-en.pdf.

3 N. Doswald and M. Osti, "Ecosystem-Based Approaches to Adaptation and Mitigation – Good Practice Examples and Lessons Learned in Europe," (2011), https://unep-wcmc.org/system/dataset_file_fields/files/000/000/092/original/BfN_EbAreport.pdf?1398440662.

4 K. Hamilton, M. Brahmbhatt, and J. Liu, "Multiple Benefits from Climate Change Mitigation: Assessing the Evidence," (2017), http://lse.ac.uk/GranthamInstitute/publication/multiple-benefits-from-climate-change-mitigation-assessing-the-evidence/.

5 PBL Netherlands Environmental Assessment Agency, "Costs and Benefits of Climate Change Adaptation and Mitigation: An Assessment on Different Regional Scales," (2014), https://preventionweb.net/publications/view/36768.

6 CBD, "X/33 Biodiversity and Climate Change, Decision Adopted by the Conference of the Parties to the Convention on Biological Diversity at Its Tenth Meeting; Unep/Cbd/Cop/Dec/X/33," (2010), https://cbd.int/climate/doc/cop-10-dec-33-en.pdf.

7 Lucas E. Nave et al., "Reforestation Can Sequester Two Petagrams of Carbon in Us Topsoils in a Century," *Proceedings of the National Academy of Sciences* 115, no. 11 (2018).

8 FAO, "Soils Help to Combat and Adapt to Climate Change by Playing a Key Role in the Carbon Cycle," (2015), http://fao.org/3/a-i4737e.pdf.

9 UNCCD, "Sustainable Land Management Contribution to Successful Landbased Climate Change Adaptation and Mitigation," (2017), https://unccd.int/publications/sustainable-land-management-contribution-successful-land-based-climate-change.

10 CDFA, "Healthy Soils Program," (n.d.), https://cdfa.ca.gov/oars/healthysoils/.

11 Noelia Garcia-Franco et al., "Chapter 23 – Climate-Smart Soil Management in Semiarid Regions," in *Soil Management and Climate Change*, ed. María Á. Muñoz and Raúl Zornoza (Cambridge, Massachusetts: Academic Press, 2018).

12 Sustainable Agriculture Research and Education, "Cover Crops," (n.d.), https://sare.org/Learning-Center/Books/Building-Soils-for-Better-Crops-3rd-Edition/Text-Version/Cover-Crops.

13 Carleton College Science Education Resource Center, "Intercrops and Cover Crops," (n.d.), https://serc.carleton.edu/integrate/teaching_materials/food_supply/student_materials/1192.

14 Ecological Society of America, "Carbon Sequestration in Soils," (2002), https://esa.org/wp-content/uploads/2012/12/carbonsequestrationinsoils.pdf.

15 Soil Quality, "Crop Rotations for Row Crops," (2019), http://soilquality.org/practices/row_crop_rotations.html.

16 Maria Pergola et al., "Composting: The Way for a Sustainable Agriculture," *Applied Soil Ecology* 123 (2018).

17 National Center for Appropriate Technology, "Tipsheet: Compost," (2015), https://ams.usda.gov/sites/default/files/media/Compost_FINAL.pdf.

18 FAO, "The Importance of Soil Organic Matter: Key to Drought-Resistant Soil and Sustained Food Production," (2005), http://fao.org/3/a0100e/a0100e00.htm#Contents.

19 FAO, "Organic Agriculture, Environment and Food Security," (2002), http://fao.org/3/y4137e/y4137e00.htm#TopOfPage.

20 FAO, "The Importance of Soil Organic Matter: Key to Drought-Resistant Soil and Sustained Food Production."

21 P. Karuppuchamy and Sheela Venugopal, "Chapter 21 – Integrated Pest Management," in *Ecofriendly Pest Management for Food Security*, ed. Omkar (San Diego, CA: Academic Press, 2016).
22 FAO, "The Importance of Soil Organic Matter: Key to Drought-Resistant Soil and Sustained Food Production."
23 Bhupinder P. Singh et al., "Chapter 7 – Agricultural Management Practices and Soil Organic Carbon Storage," in *Soil Carbon Storage*, ed. Brajesh K. Singh (Cambridge, Massachusetts: Academic Press, 2018).
24 National Center for Appropriate Technology, "Integrating Livestock and Crops: Improving Soil, Solving Problems, Increasing Income," (2014), https://attra.ncat.org/attra-pub/download.php?id=481.
25 FoodPrint, "How Industrial Agriculture Affects Our Soil," (n.d.), https://foodprint.org/issues/how-industrial-agriculture-affects-our-soil/.
26 Epple, "Managing Ecosystems in the Context of Climate Change Mitigation: A Review of Current Knowledge and Recommendations to Support Ecosystem-Based Mitigation Actions That Look Beyond Terrestrial Forests."
27 Jesse B. Naab et al., "Conservation Agriculture Improves Soil Quality, Crop Yield, and Incomes of Smallholder Farmers in North Western Ghana," *Frontiers in Plant Science* 8, no. 996 (2017).
28 UNFCCC, "Mais Program: Climate-Smart Agriculture| Brazil," (n.d.), https://unfccc.int/climate-action/momentum-for-change/financing-for-climate-friendly-investment/mais-program-climate-smart-agriculture.
29 C. Calfapietra et al., "Carbon Mitigation Potential of Different Forest Ecosystems under Climate Change and Various Managements in Italy," *Ecosystem Health and Sustainability* 1, no. 8 (2015).
30 Beverly E. Law et al., "Land Use Strategies to Mitigate Climate Change in Carbon Dense Temperate Forests," *Proceedings of the National Academy of Sciences* 115, no. 14 (2018).
31 Royal Society, "The Role of Land Carbon Sinks in Mitigating Global Climate Change," (2001), https://royalsociety.org/-/media/Royal_Society_Content/policy/publications/2001/9996.pdf.
32 FAO, "Managing Forests for Climate Change," (2010), http://fao.org/3/i1960e/i1960e00.pdf.
33 Law et al., "Land Use Strategies to Mitigate Climate Change in Carbon Dense Temperate Forests."
34 FAO, "Managing Forests for Climate Change."
35 FAO, "Managing Forests for Climate Change."
36 FAO, "Managing Forests for Climate Change."
37 FAO, "Managing Forests for Climate Change."
38 USDA Forest Service, "Why Biomass Is Important – The Role of the Usda Forest Service in Managing and Using Biomass for Energy and Other Uses," (2006), https://fs.fed.us/research/pdf/biomass_importance.pdf.
39 FAO, "Managing Forests for Climate Change."
40 Forest Carbon Partnership Facility, "Ghana," (n.d.), https://forestcarbonpartnership.org/country/ghana.
41 Epple, "Managing Ecosystems in the Context of Climate Change Mitigation: A Review of Current Knowledge and Recommendations to Support Ecosystem-Based Mitigation Actions That Look Beyond Terrestrial Forests."

42 FAO, "Challenges and Opportunities for Carbon Sequestration in Grassland Systems. A Technical Report on Grassland Management and Climate Change Mitigation," (2010), http://fao.org/fileadmin/templates/agphome/documents/climate/AGPC_grassland_webversion_19.pdf.
43 Boomitra, "Climate, Community, and Biodiversity," (n.d.), https://boomitra.com/projects/boomitra-grassland-restoration-in-the-temperate-grasslands-of-south-america-through-soil-enrichment/#:~:text=Boomitra's%20grassland%20restoration%20project%20not,security%2C%20and%20biodiversity%20also%20increase.
44 Epple, "Managing Ecosystems in the Context of Climate Change Mitigation: A Review of Current Knowledge and Recommendations to Support Ecosystem-Based Mitigation Actions That Look Beyond Terrestrial Forests."
45 FAO, "Peatlands – Guidance for Climate Change Mitigation through Conservation, Rehabilitation and Sustainable Use," (2012), http://fao.org/3/an762e/an762e00.htm.
46 NatureScot, "Peatland Action," (n.d.), https://nature.scot/climate-change/nature-based-solutions/nature-based-solutions-practice/peatland-action.
47 S. Crooks, D. Herr, J. Tamelander, D. Laffoley, and J. Vandever, "Mitigating Climate Change through Restoration and Management of Coastal Wetlands and Near-Shore Marine Ecosystems: Challenges and Opportunities," *Environment Department Paper* (2011).
48 Climate-ADAPT, "Restoration and Management of Coastal Wetlands," (2015), https://climate-adapt.eea.europa.eu/metadata/adaptation-options/restoration-and-management-of-coastal-wetlands.
49 Crooks, "Mitigating Climate Change through Restoration and Management of Coastal Wetlands and near-Shore Marine Ecosystems: Challenges and Opportunities."
50 Energy Department of Climate Change, the Environment and Water, "Blue Carbon Ecosystem Restoration Grants," (n.d.), https://dcceew.gov.au/environment/marine/coastal-blue-carbon-ecosystems/conservation/restoration-grants.
51 UN-Water, "Un-Water Policy Brief on Climate Change and Water," (2019), https://reliefweb.int/report/world/un-water-policy-brief-climate-change-and-water.
52 IUCN, "Mangrove Restoration Potential: A Global Map Highlighting a Critical Opportunity," (2018), https://iucn.org/sites/dev/files/content/documents/mangrove-tnc-report-final.31.10.lowspreads.pdf.
53 South Pole, "Mangrove Restoration and Coastal Greenbelt Protection," (2022), wwf.org.uk/sites/default/files/2022-07/case_study_livelihoods_yagasu_mangrove_restoration.pdf.
54 R. Ramesh et al., "Chapter 14 – Importance of Seagrass Management for Effective Mitigation of Climate Change," in *Coastal Management*, ed. R.R. Krishnamurthy et al. (Cambridge, Massachusetts: Academic Press, 2019).
55 Smithsonian Ocean, "Seagrass and Seagrass Beds," (n.d.), https://ocean.si.edu/ocean-life/plants-algae/seagrass-and-seagrass-beds.
56 Smithsonian Ocean, "Seagrass and Seagrass Beds."
57 The Fiji Times, "Conservation | Hidden Treasures of Our Ocean," (n.d.), https://fijitimes.com.fj/conservation-hidden-treasures-of-our-ocean/.
58 Carlos M. Duarte et al., "Can Seaweed Farming Play a Role in Climate Change Mitigation and Adaptation?," *Frontiers in Marine Science* 4, no. 100 (2017).

59 CH4 Global, "The Surest, Fastest Route to Bend the Climate Curve Is to Radically Reduce Methane Emissions in the Next 10 Years," (n.d.), https://ch4global.com.
60 Callum M. Roberts et al., "Marine Reserves Can Mitigate and Promote Adaptation to Climate Change," *Proceedings of the National Academy of Sciences* 114, no. 24 (2017).
61 Roberts et al., "Marine Reserves Can Mitigate and Promote Adaptation to Climate Change."
62 Roberts et al., "Marine Reserves Can Mitigate and Promote Adaptation to Climate Change."
63 Irish Wildlife Trust, "Marine Protected Area Network for Ireland," (n.d.), https://iwt.ie/what-we-do/campaigns/marine-protected-areas/.

References

Calfapietra, C., A. Barbati, L. Perugini, B. Ferrari, G. Guidolotti, A. Quatrini, and P. Corona. "Carbon Mitigation Potential of Different Forest Ecosystems under Climate Change and Various Managements in Italy." *Ecosystem Health and Sustainability* 1, no. 8 (2015/10/01): 1–9.

CBD. "X/33 Biodiversity and Climate Change, Decision Adopted by the Conference of the Parties to the Convention on Biological Diversity at Its Tenth Meeting; Unep/Cbd/Cop/Dec/X/33." (2010). www.cbd.int/climate/doc/cop-10-dec-33-en.pdf.

Climate-ADAPT. "Restoration and Management of Coastal Wetlands." (2015). https://climate-adapt.eea.europa.eu/metadata/adaptation-options/restoration-and-management-of-coastal-wetlands.

Crooks, S., D. Herr, J. Tamelander, D. Laffoley, and J. Vandever. "Mitigating Climate Change through Restoration and Management of Coastal Wetlands and near-Shore Marine Ecosystems: Challenges and Opportunities." *Environment Department Paper* (2011).

Doswald, N., and M. Osti. "Ecosystem-Based Approaches to Adaptation and Mitigation – Good Practice Examples and Lessons Learned in Europe." (2011). www.unep-wcmc.org/system/dataset_file_fields/files/000/000/092/original/BfN_EbAreport.pdf?1398440662.

Duarte, Carlos M., Jiaping Wu, Xi Xiao, Annette Bruhn, and Dorte Krause-Jensen. "Can Seaweed Farming Play a Role in Climate Change Mitigation and Adaptation?" [In English]. *Frontiers in Marine Science* 4, no. 100 (2017/04/12).

Ecological Society of America. "Carbon Sequestration in Soils." (2002). www.esa.org/wp-content/uploads/2012/12/carbonsequestrationinsoils.pdf.

Epple, C., S.G. Rangel, M. Jenkins, and M. Guth. "Managing Ecosystems in the Context of Climate Change Mitigation: A Review of Current Knowledge and Recommendations to Support Ecosystem-Based Mitigation Actions That Look Beyond Terrestrial Forests." (2016). www.cbd.int/doc/publications/cbd-ts-86-en.pdf.

FAO. "Challenges and Opportunities for Carbon Sequestration in Grassland Systems. A Technical Report on Grassland Management and Climate Change Mitigation." (2010). www.fao.org/fileadmin/templates/agphome/documents/climate/AGPC_grassland_webversion_19.pdf.

FAO. "The Importance of Soil Organic Matter: Key to Drought-Resistant Soil and Sustained Food Production." (2005). www.fao.org/3/a0100e/a0100e00.htm#Contents.

FAO. "Managing Forests for Climate Change." (2010). www.fao.org/3/i1960e/i1960e00.pdf.

FAO. "Organic Agriculture, Environment and Food Security." (2002). www.fao.org/3/y4137e/y4137e00.htm#TopOfPage.

FAO. "Peatlands – Guidance for Climate Change Mitigation through Conservation, Rehabilitation and Sustainable Use." (2012). www.fao.org/3/an762e/an762e00.htm.

FAO. "Soils Help to Combat and Adapt to Climate Change by Playing a Key Role in the Carbon Cycle." (2015). www.fao.org/3/a-i4737e.pdf.

FoodPrint. "How Industrial Agriculture Affects Our Soil." (n.d.). https://foodprint.org/issues/how-industrial-agriculture-affects-our-soil/.

Garcia-Franco, Noelia, Eleanor Hobley, Rico Hübner, and Martin Wiesmeier. "Chapter 23 – Climate-Smart Soil Management in Semiarid Regions." In *Soil Management and Climate Change*, edited by María Á. Muñoz, and Raúl Zornoza, 349–68. Cambridge, Massachusetts: Academic Press, 2018.

Hamilton, K., M. Brahmbhatt, and J. Liu. "Multiple Benefits from Climate Change Mitigation: Assessing the Evidence." (2017). www.lse.ac.uk/GranthamInstitute/publication/multiple-benefits-from-climate-change-mitigation-assessing-the-evidence/.

IUCN. "Mangrove Restoration Potential: A Global Map Highlighting a Critical Opportunity." (2018). www.iucn.org/sites/dev/files/content/documents/mangrove-tnc-report-final.31.10.lowspreads.pdf.

Karuppuchamy, P., and Sheela Venugopal. "Chapter 21 – Integrated Pest Management." In *Ecofriendly Pest Management for Food Security*, edited by Omkar, 651–84. San Diego, CA: Academic Press, 2016.

Law, Beverly E., Tara W. Hudiburg, Logan T. Berner, Jeffrey J. Kent, Polly C. Buotte, and Mark E. Harmon. "Land Use Strategies to Mitigate Climate Change in Carbon Dense Temperate Forests." *Proceedings of the National Academy of Sciences* 115, no. 14 (2018): 3663.

Miller, Bruce G. "13 – Carbon Dioxide Emissions Reduction and Storage." In *Clean Coal Engineering Technology (Second Edition)*, edited by Bruce G. Miller, 609–68. Oxford, UK: Butterworth-Heinemann, 2017.

Naab, Jesse B., George Y. Mahama, Iddrisu Yahaya, and P.V.V. Prasad. "Conservation Agriculture Improves Soil Quality, Crop Yield, and Incomes of Smallholder Farmers in North Western Ghana." [In English]. *Frontiers in Plant Science* 8, no. 996 (2017/06/21).

National Center for Appropriate Technology. "Integrating Livestock and Crops: Improving Soil, Solving Problems, Increasing Income." (2014). https://attra.ncat.org/attra-pub/download.php?id=481.

National Center for Appropriate Technology. "Tipsheet: Compost." (2015). www.ams.usda.gov/sites/default/files/media/Compost_FINAL.pdf.

Nave, Lucas E., Grant M. Domke, Kathryn L. Hofmeister, Umakant Mishra, Charles H. Perry, Brian F. Walters, and Christopher W. Swanston. "Reforestation Can Sequester Two Petagrams of Carbon in Us Topsoils in a Century." *Proceedings of the National Academy of Sciences* 115, no. 11 (2018): 2776.

PBL Netherlands Environmental Assessment Agency. "Costs and Benefits of Climate Change Adaptation and Mitigation: An Assessment on Different Regional Scales." (2014). www.preventionweb.net/publications/view/36768.

Pergola, Maria, Alessandro Persiani, Assunta M. Palese, Vincenzo Di Meo, Vittoria Pastore, Carmine D'Adamo, and Giuseppe Celano. "Composting: The Way for a Sustainable Agriculture." *Applied Soil Ecology* 123 (2018/02/01): 744–50.

Ramesh, R., K. Banerjee, A. Paneerselvam, R. Raghuraman, R. Purvaja, and Ahana Lakshmi. "Chapter 14 – Importance of Seagrass Management for Effective Mitigation of Climate Change." In *Coastal Management*, edited by R.R. Krishnamurthy, M.P. Jonathan, Seshachalam Srinivasalu, and Bernhard Glaeser, 283–99. Cambridge, Massachusetts: Academic Press, 2019.

Roberts, Callum M., Bethan C. O'Leary, Douglas J. McCauley, Philippe M. Cury, Carlos M. Duarte, Jane Lubchenco, Daniel Pauly, et al. "Marine Reserves Can Mitigate and Promote Adaptation to Climate Change." *Proceedings of the National Academy of Sciences* 114, no. 24 (2017): 6167–75.

Royal Society. "The Role of Land Carbon Sinks in Mitigating Global Climate Change." (2001). https://royalsociety.org/-/media/Royal_Society_Content/policy/publications/2001/9996.pdf.

Science Education Resource Center, Carleton College. "Intercrops and Cover Crops." (n.d.). https://serc.carleton.edu/integrate/teaching_materials/food_supply/student_materials/1192.

Singh, Bhupinder P., Raj Setia, Martin Wiesmeier, and Anitha Kunhikrishnan. "Chapter 7 – Agricultural Management Practices and Soil Organic Carbon Storage." In *Soil Carbon Storage*, edited by Brajesh K. Singh, 207–44. Cambridge, Massachusetts: Academic Press, 2018.

Smithsonian Ocean. "Seagrass and Seagrass Beds." (n.d.). https://ocean.si.edu/ocean-life/plants-algae/seagrass-and-seagrass-beds.

Soil Quality. "Crop Rotations for Row Crops." (2019). http://soilquality.org/practices/row_crop_rotations.html.

Sustainable Agriculture Research and Education. "Cover Crops." (n.d.). www.sare.org/Learning-Center/Books/Building-Soils-for-Better-Crops-3rd-Edition/Text-Version/Cover-Crops.

UN-Water. "Un-Water Policy Brief on Climate Change and Water." (2019). https://reliefweb.int/report/world/un-water-policy-brief-climate-change-and-water.

UNCCD. "Sustainable Land Management Contribution to Successful Landbased Climate Change Adaptation and Mitigation." (2017). www.unccd.int/publications/sustainable-land-management-contribution-successful-land-based-climate-change.

USDA Forest Service. "Why Biomass Is Important – The Role of the Usda Forest Service in Managing and Using Biomass for Energy and Other Uses." (2006). www.fs.fed.us/research/pdf/biomass_importance.pdf.

8 Nature-Based Solutions in Climate Adaptation

Introduction

Planners and decision-makers have multiple climate change adaptation pathways, typically categorized into hard (grey) adaptation, soft adaptation, and ecosystem-based adaptation (EbA). Hard adaptation comprises capital-intensive, engineered solutions such as seawalls, flood barriers, and infrastructure modifications. Soft adaptation focuses on institutional, behavioral, and policy measures, including regulatory frameworks, information sharing, and capacity-building initiatives.[1] A third approach, EbA, integrates hard and soft adaptation elements by leveraging nature-based solutions (NBS) to enhance resilience. EbA utilizes natural ecosystems to buffer human communities from the impacts of climate change while ensuring the sustainable delivery of ecosystem services. It involves ecosystem conservation, restoration, and sustainable management to help communities adapt to climate change by mitigating extreme weather events, climate variability, and environmental dis ruptions.[2,3,4]

Ecosystem-Based Adaptation

EbA is typically implemented through targeted management, conservation, and restoration activities focusing on specific ecosystem services to reduce exposure to climate extremes. For example, restoring mangrove forests helps protect coastal communities from the increasing impact of storm surges and rising sea levels. EbA can be applied in three key ways, specifically concerning hard engineering solutions. First, it can complement hard infrastructure by enhancing resilience to climate extremes. Second, it can be a more cost-effective alternative to traditional engineered solutions. Third, it may be the only viable adaptation option in some cases. Unlike hard interventions, which often provide limited benefits beyond their primary adaptation function, EbA offers additional economic, social, and environmental co-benefits. By enhancing climate resilience while promoting biodiversity, food security, and sustainable livelihoods, EbA enables governments to achieve broader development objectives.

DOI: 10.4324/9781003517948-8

EbA also leverages conservation practices to enhance both human and ecosystem resilience. However, while conservation traditionally focuses on restoring or maintaining landscapes under a static climate, EbA is specifically designed to adapt to a changing climate and reduce the vulnerability of people and ecosystems. The key distinctions between EbA and conventional conservation are that EbA measures explicitly address current and future climate impacts and aim to reduce the vulnerability of social-ecological systems, integrating both human and environmental resilience.[5,6,7]

Co-Benefits of Ecosystem-Based Adaptation

EbA uses natural ecosystems to enhance resilience against climate extremes and variability while providing multiple environmental, economic, and social benefits. Restoring or enhancing vegetation, forests, grasslands, wetlands, and water bodies improves natural hazard protection, supports food and fiber production, regulates pest and disease populations, and preserves genetic diversity. A summary of these co-benefits is provided in Table 8.1.[8,9,10]

Ecosystem-Based Adaptation Policy Measures

EbA requires a holistic approach, as policies across different sectors can support or hinder its implementation. Integrating EbA into existing policy frameworks can enhance efficiency by leveraging established legal, institutional, and regulatory systems and available resources. Examples of EbA-relevant existing policy

Table 8.1 Co-Benefits of Ecosystem-Based Adaptation

Ecosystem-Based Adaptation Intervention	*Description*
Restoring fragmented or degraded natural areas	Improves essential ecosystem services such as clean water supply, food production, fishery sustainability, and climate regulation, enhancing livelihoods and local resilience.
Protecting groundwater recharge zones and restoring floodplains	Maintains stable water supplies by increasing groundwater infiltration and moderating flood peaks, helping communities better withstand droughts and floods.
Connecting expanses of forests, grasslands, reefs, and other habitats	Facilitates climate-driven migration and dispersal of species, ecosystems, and even people, ensuring biodiversity continuity and resilience under changing conditions.
Protecting and restoring natural infrastructure (e.g. barrier islands, dunes, mangroves, coral reefs, wetlands, and forests)	Shields communities and assets from coastal hazards – including sea level rise, erosion, and extreme weather – while preserving biodiversity and supporting ecosystem services.

measures are summarized in Table 8.2.[11] When suitable policy frameworks are absent or inadequate, EbA can catalyze the development of new policy measures. These can be EbA-specific, where EbA is the sole focus, allowing policies to be tailored regarding framing, enforcement, and institutional arrangements. However, this approach may not fully address the multi-sectoral nature of EbA. An alternative approach is to incorporate EbA into broader, integrated frameworks, such as Integrated Coastal Zone Management (ICZM) or Marine Spatial Planning (MSP), to enhance coordination across sectors.

To ensure effective EbA implementation, policy measures should:

- Align EbA policies with the broader policy landscape to ensure coordinated and long-term support.
- Enhance awareness, capacity, and engagement across stakeholder sectors to build widespread support and encourage replication.
- Clearly define the roles and responsibilities of institutions to improve coordination, clarify mandates, and strengthen implementation efforts.
- Leverage existing governance mechanisms and promote good governance across all levels, increasing the likelihood of successful EbA delivery.
- Encourage sustainable and innovative long-term financing options to ensure adequate resources for ongoing EbA efforts.
- Strengthen long-term capacity for implementation and enforcement, ensuring the availability of technical expertise and institutional support to maintain EbA measures over time.[12]

Table 8.2 Examples of Existing Policy Measures that Facilitate Ecosystem-Based Adaptation

Existing Policy Measures	*Examples and Applications*
Development policies	Include national and regional strategies such as sustainable development plans and green economy frameworks that integrate climate resilience and environmental sustainability into development goals.
Infrastructure policies	Cover policies guiding water supply, sanitation, solid waste, energy, and transport systems that can incorporate nature-based solutions to reduce climate vulnerabilities.
Environmental policies	Include biodiversity action plans, protected area legislation, and endangered species frameworks that contribute to ecosystem conservation and adaptive capacity.
Area-based planning framework tools	Such as Integrated Coastal Zone Management (ICZM), Marine Spatial Planning (MSP), and Marine Protected Area (MPA) regulations, which promote adaptive, spatially-informed management of natural resources.
Other specific sectoral policies	Examples include sector-specific strategies for tourism, fisheries, forestry, agriculture, aquaculture, health, and energy, which can be aligned with ecosystem-based adaptation objectives.

Soils

Inappropriate agricultural practices on vulnerable soils and marginal lands are a leading cause of soil degradation, resulting in compaction, nutrient depletion, and loss of organic matter. Various soil conservation and management strategies can be implemented to adapt to climate change and improve soil resilience.

Slow-Forming Terraces

Slow-forming terraces are built using infiltration ditches, hedgerows, and earth or stone walls to reduce surface water runoff and enhance water filtration. These terraces help improve agricultural productivity by reducing erosion and increasing soil moisture. They provide multiple ecosystem services, including erosion control, soil retention, nutrient cycling, water regulation, and sustainable production of food, fiber, and fuel.[13]

Integrated Nutrient Management

Integrated Nutrient Management combines natural and synthetic soil nutrients to enhance crop productivity while maintaining long-term soil health. This approach supports key ecosystem services, including food, fiber, fuel production, nutrient cycling, soil formation, and water retention. Adapting plant nutrition and soil fertility management to specific site conditions improves sustainability and resilience in farming systems.[14]

Sustainable Crop Management and Agro-Biodiversity Conservation

Both abiotic and biotic stresses influence crop productivity. Abiotic factors include rising temperatures, decreasing water availability, increased soil salinity, and flooding, while biotic stresses involve increased pest infestations and plant diseases. Additionally, atmospheric carbon dioxide levels, acid rain, and changes in ground-level ozone can further impact crop yields. Adapting to these climate-related challenges will require implementing new technologies and resource conservation strategies to maintain agricultural resilience and productivity.

Crop Diversification

Introducing newly cultivated species and improved crop varieties enhances plant productivity, quality, health, and nutritional value while strengthening resilience to pests, diseases, and climate change. This approach supports key ecosystem services, including food, fiber, and fuel production, pest and disease control, seed dispersal, pollination, and resistance to invasive species. The benefits include greater drought resilience, higher yields, increased resistance to pests and diseases, and improved food security.[15,16]

Ecological Pest Management

Ecological Pest Management (EPM) strengthens natural pest regulation processes to enhance agricultural productivity while reducing reliance on synthetic pesticides. EPM utilizes natural and cultural methods, including host resilience and biological control, to develop self-regulating systems that promote long-term pest management. This approach contributes to key ecosystem services, including food, fiber, and fuel production, as well as pest and disease control. The benefits of EPM include lower pesticide costs for farmers, improved resilience to environmental stresses such as drought and soil compaction, and reduced vulnerability to pest invasions.[17,18]

Sustainable Farming Systems

Farming systems involve more than just a single crop or livestock species. Their ecological and productive resilience to climate change largely depends on managing diverse, integrated systems that combine crops, livestock, and trees.[19]

Mixed Farming

Mixed farming is an agricultural system that combines crop cultivation and livestock production to maximize productivity and resource efficiency. By integrating various agricultural activities, farmers can balance land and labor demands throughout the year under diverse conditions. This system supports key ecosystem services, including food, fiber, and fuel production, genetic diversity, nutrient cycling, and seed dispersal. The benefits of mixed farming include enhanced food security, improved soil fertility and biodiversity, better water conservation, and increased nutrient cycling, all of which contribute to a more resilient and sustainable agricultural system.[20]

Agroforestry

Agroforestry is a land-use system that integrates trees, crops, and livestock on the same land to enhance productivity, sustainability, and climate resilience. By incorporating trees, agroforestry supports the intensification, diversification, and buffering of farming systems, helping to adapt to current climate variability and long-term climate change. This approach provides key ecosystem services, including food, fiber, and fuel production, genetic diversity, soil formation and retention, oxygen production, water cycling, and climate regulation. The benefits of agroforestry include higher land productivity, improved soil protection and fertility, and diversified livelihoods, making agricultural systems more resilient and sustainable.[21]

Case 8.1: Kenya Climate Smart Agriculture Project

The Kenya Climate Smart Agriculture Project (KCSAP) is a government-led initiative, supported by the World Bank, aimed at increasing agricultural productivity and building resilience to climate change in smallholder farming and pastoral communities. Implemented from 2017 to 2022, the project aligns with Kenya's Agriculture Sector Development Strategy and National Climate Change Response Strategy to strengthen food security and promote sustainable land and water management. KCSAP operates across 24 counties, categorized into arid, semi-arid, and medium-to-high rainfall regions, tailoring interventions to local climate conditions. A key focus is Upscaling Climate-Smart Agricultural Practices, which promotes soil and water conservation, agroforestry, crop diversification, and small-scale irrigation. In pastoral communities, the project enhances rangeland management, livestock productivity, and market access, contributing to both climate adaptation and economic stability. The project also strengthens climate-smart agricultural research and seed systems, funding the development of drought-tolerant crop varieties and climate-adaptive livestock breeds. It supports private sector and community-led seed production, ensuring long-term access to high-quality agricultural inputs. To enhance climate resilience, KCSAP improves agro-weather, market, and advisory services by utilizing meteorological forecasting, ICT platforms, and data analytics to provide farmers with real-time climate and market information. The Contingency Emergency Response component offers rapid assistance during droughts, floods, and locust invasions, with KES 12.5 million allocated to locust control efforts. KCSAP integrates monitoring, gender inclusion, and environmental safeguards to ensure equitable and sustainable interventions. By combining scientific research, policy-driven action, and local participation, the project serves as a model for climate-smart agriculture and resilience-building in Africa.[22]

Kenya Climate Smart Agriculture Project. "Kenya Climate Smart Agriculture Project (Kcsap)." (n.d.). https://kcsap.go.ke/home.

Forests

Effective forest management ensures the survival of forest ecosystems while enhancing their environmental, social, and economic benefits. Forest EbA involves monitoring the impacts of climate change, anticipating risks, and implementing measures to mitigate negative consequences. It requires long-term planning across multiple spatial and temporal scales, as well as the adoption of adaptive and collaborative management approaches, to address future climate challenges.[23,24]

Management of Fragile Forest Ecosystems

Certain forests are particularly vulnerable to climate change, including mountain, dryland, and coastal forests, due to their sensitivity to shifts in temperature, changes in precipitation patterns, and extreme weather events.

Mountain Forests

Mountain forests are facing climate-induced hazards, such as rockfalls and landslides, due to the melting of glaciers and the movement of permafrost. Climate-adapted forest management stabilizes slopes, reduces erosion, and regulates water flow, protecting communities. Reducing timber extraction and planting resilient tree species helps forests increase carbon storage, support biodiversity, and improve water conservation. These strategies enhance soil retention, promote watershed health, and provide habitats for wildlife. Reforestation with fruit, nut, and medicinal plants supports local livelihoods, while improved groundwater recharge enhances water quality. Integrating NBS ensures mountain forests remain resilient to climate change while providing essential ecosystem services.[25,26]

Dryland Forests

As climatic conditions become more unpredictable, dryland forests can be managed to support and promote land uses that are more resilient to climate variability. Pastoralism is a more climate-resilient land-use strategy than crop agriculture, livestock ranching, or tourism, making it a more sustainable option in dryland regions. Pastoralism is recognized as a low-input, adaptable land-use system that performs well in variable climatic conditions. By supporting pastoralism, economic development and poverty alleviation can be enhanced, as it provides a less risky and more resilient investment opportunity in areas where other livelihood options may struggle or fail.[27]

Coastal Forests

Coastal forests play a crucial role in protecting shorelines from erosion and storm surges, while providing habitats for wildlife, stabilizing the land, trapping sediments, and improving water quality. They contribute to nutrient cycling in inshore waters, support biodiversity, and serve as nursery grounds for commercially valuable fish and shellfish species, benefiting local economies and marine ecosystems. In addition to their protective functions, coastal forests provide recreational opportunities and generate income through activities such as fisheries and ecotourism. Effective coastal forest management necessitates an interdisciplinary approach, known as Integrated Coastal Area Management, to strike a balance between conservation, economic development, and community resilience.[28]

Case 8.2: Protecting and Restoring Coastal Temperate Rainforests in Chile

The Valdivian Coastal Reserve in Chile's Los Ríos region spans 50,000 hectares of temperate coastal rainforest, forming part of one of the world's last intact temperate forest ecosystems. Established through a collaboration between Conservation International, The Nature Conservancy (TNC), and BHP Billiton, the reserve protects a high-biodiversity landscape along 35 kilometers of Chile's Pacific coastline. It borders the Alerce Costero National Park, which was created following TNC's donation of 9,200 hectares of protected land. The reserve contains Olivillo trees (*Aextoxicon punctatum*), which can live for 400 years, and Alerce trees (*Fitzroya cupressoides*), which can survive for up to 4,000 years. The area supports endemic plants, amphibians, and key species, including the Pudú deer (*Pudu puda*) and Darwin fox (*Lycalopex fulvipes*), classified as critically endangered. Conservation activities include native reforestation, with over 2.7 million native trees planted, and a pilot project to remove non-native species across 1,500 hectares. The reserve generates forest carbon credits under the Verified Carbon Standard, supporting long-term conservation financing. It also provides clean water for neighboring communities through donated water rights. Infrastructure improvements, including new trails, signposts, and research collaborations, enhance sustainable tourism and scientific research while maintaining the ecological integrity of the forest.[29]

Conservation International. "Valdivian Coastal Reserve." (2025). https://conservation.org/docs/default-source/publication-pdfs/valdivia-coastal-reserve_factsheet.pdf?Status=Master&sfvrsn=40b2c330_3#:~:text=The%20Reserve%20is%20an%20important,back%20to%20native%20temperate%20rainforest.

Water Sensitive Forest Management

Forests regulate the water cycle by influencing rainfall interception, evapotranspiration, water infiltration, and groundwater recharge. They help maintain stable base flows during dry seasons and control peak flows during heavy rainfall, which are crucial for adapting to climate variability and change. Additionally, forests stabilize soil, preventing erosion and landslides, which helps protect water users, infrastructure, and communities from environmental hazards.[30,31] Various EbA practices can be implemented to increase water yield, regulate water flow, and reduce drought stress in forests. A summary of these practices is provided in Table 8.3.[32]

Table 8.3 Water Sensitive Forest Management

Ecosystem-Based Adaptation Practice	*Description*
Forest riparian buffers	Wooded buffer zones alongside rivers and streams that absorb nutrients, slow surface runoff, and promote water infiltration, thereby protecting water quality and reducing erosion.
Maintenance of forest cover in headwater areas	Forests in the upper reaches of catchments enhance water infiltration and regulate flow. They stabilize slopes and reduce landslide risks, acting like sponges that gradually release rainfall.
Afforestation of reservoir catchments	Planting trees on degraded or bare lands around reservoirs helps control erosion, reduce sedimentation, and maintain water quality, extending reservoir lifespan.
Land-use conversion (to forestry)	Transforming land uses such as marginal agriculture into forested areas either intentionally or through natural regeneration, enhancing soil stability and water retention.
Continuous cover forestry	Forest management practice that avoids large clear-cuts, maintaining an unbroken canopy to protect soil, reduce erosion, and support consistent water regulation.

Freshwater

EbA enhances water security and supports climate change adaptation by strengthening natural ecosystems, conserving biodiversity, and sustaining essential ecosystem services that benefit human well-being.[33] By preserving and restoring natural landscapes, EbA helps increase water availability, improve water quality, and manage excess water through various strategies.

Basins and Ponds

Detention basins are vegetated depressions that temporarily hold runoff from impermeable surfaces, allowing sediments and pollutants to settle before the water is gradually drained into nearby waterways. Retention ponds are designed to store and regulate surface runoff during rainfall events. They consist of a permanent water body with surrounding banks that provide additional storage capacity, helping to manage excess water and reduce flood risk.[34]

Rehabilitation and Restoration of Rivers

River rehabilitation and restoration focus on restoring the natural functions of rivers that have been lost or degraded due to human activities. Many rivers have historically been modified for specific purposes, such as transportation, often at the expense of ecological balance. With climate change increasing the frequency of storms and floods, there is now a shift toward integrated approaches that enhance climate resilience while improving river health. Restoring rivers helps mitigate the negative impacts of past modifications by improving habitat quality and biodiversity, increasing groundwater recharge, and reducing flood risks. Key rehabilitation and restoration measures include re-meandering rivers, re-naturalizing riverbeds, removing dams and barriers, restoring riverbanks, and eliminating artificial riverbank protections to promote more natural water flow and ecosystem recovery.[35,36]

Case 8.3: Restoration of the Emscher River from Industrial Wastewater Canal to Ecological Resilience

The restoration of the Emscher River is a large-scale ecological and socio-economic transformation project within Germany's Ruhr region, which has historically been dominated by heavy industry. For over a century, the Emscher served as an open wastewater canal, resulting in severe environmental degradation. Since the early 2000s, efforts have focused on rehabilitating the river as part of the broader Emscher Park project, which aims to restore natural ecosystems, upgrade infrastructure, and enhance biodiversity. A key component of the project involved constructing an underground sewer system to divert wastewater away from the river, enabling ecological recovery. The river's course was restructured, replacing its former straight, concrete-lined channel with a meandering, naturalized flow. This design reduces water velocity, improves flood resilience, and enhances stormwater management. Alongside these hydrological improvements, native vegetation has been planted along the riverbanks, restoring habitats and increasing biodiversity. The project integrates blue and green infrastructure, including urban forests, parks, and sustainable drainage systems, which contribute to improved water quality and climate adaptation. Former industrial sites have been repurposed into green spaces, providing recreational opportunities that enhance public well-being and social cohesion. These developments have also contributed to employment opportunities and the redevelopment of industrial areas, helping to prevent abandonment and dereliction. Municipal governments, national institutions, public sector organizations, NGOs, and local communities have collaborated on this initiative, ensuring its long-term success. While the project is ongoing, significant progress has been made in transforming the Emscher into a resilient, ecologically functional river system. The restoration serves as a model for balancing urban development with environmental sustainability in former industrial regions.[37]

Urban Nature Atlas. "Restoration of the Emscher River." (n.d.). https://una.city/nbs/essen/restoration-emscher-river.

Floodplain Restoration and Management

Floodplain loss has resulted from land drainage, urban expansion, and river channelization, reducing their natural retention capacity and ecosystem functions. Restoring floodplains aims to reconnect them to rivers, improving water retention, flood regulation, and biodiversity. Key restoration measures include modifying river channels, removing legacy sediment, creating lakes or ponds, adopting sustainable agricultural practices, afforesting degraded areas, planting native grasses, shrubs, and trees, restoring wetlands, removing invasive species, and establishing riparian buffers. These efforts help enhance flood resilience, improve water quality, and support diverse ecosystems.[38]

Case 8.4: Victorian Murray Floodplain Restoration Project

The Victorian Murray Floodplain Restoration Project (VMFRP) is a large-scale initiative aimed at revitalizing nine high-value floodplains along the Murray River. As part of Victoria's commitments under the Murray-Darling Basin Plan, the project seeks to restore ecological health by ensuring that vital water reaches these floodplains, which have been increasingly disconnected due to river regulation and population growth. To achieve this, environmental infrastructure such as flow regulators, channels, and pumps is being installed to manage water delivery efficiently. These structures enable controlled watering of floodplains, replicating natural flooding cycles without requiring large-scale inundation that could impact private land and urban areas. This approach supports floodplain-dependent native plants and animals, enhancing biodiversity and ecosystem resilience. The VMFRP builds on previous successes, such as those observed at Gunbower Forest and Hattah Lakes, where controlled water flows have resulted in enhanced wetland vegetation and healthier floodplain ecosystems. At Hattah Lakes, for example, regulators have permitted water delivery even during dry periods, leading to the regeneration of native trees and the restoration of key waterbird habitats. The project is a collaborative effort involving state and federal government agencies, water management authorities, and local communities. It is funded by the Australian Government's Department of Climate Change, Energy, the Environment and Water. With climate change increasing drought risks, the VMFRP plays a crucial role in safeguarding these ecologically significant landscapes for future generations.[39]

State Government of Victoria. "Victorian Murray Floodplain Restoration Project." (n.d.). https://water.vic.gov.au/our-programs/murray-darling-basin/victorian-murray-floodplain-restoration-project.

Groundwater Replenishment Using Managed Aquifer Recharge

Managed aquifer recharge is a strategy that enhances natural water storage and increases water availability during periods of low flow and high seasonal variability. This approach involves intentionally recharging aquifers during wet periods to replenish groundwater supplies. The environmental benefits of managed aquifer recharge include maintaining healthy river flows, preventing over-extraction and degradation of water sources, sustaining groundwater-dependent ecosystems, reducing saltwater intrusion and land subsidence, and improving water quality through natural filtration. The socioeconomic benefits include strengthening water supply resilience, minimizing water losses due to evaporation, enabling alternative water sources, and increasing freshwater availability for households, agriculture, and industry.[40]

Case 8.5: Central Arizona Project Recharge Program

The Central Arizona Project (CAP) Recharge Program is a vital water management initiative aimed at enhancing long-term water security in Arizona. Since its inception in 1996, the program has focused on storing renewable surface water underground to ensure a reliable water supply during periods of drought and reduced Colorado River water availability. CAP employs artificial recharge techniques, where water is systematically delivered to recharge basins and allowed to percolate through the soil, replenishing underground aquifers. This process not only stores excess surface water but also replenishes groundwater supplies, mitigates land subsidence, prevents saltwater intrusion, and improves water quality through natural filtration. Currently, six recharge sites operate under CAP's management, collectively capable of storing over 300,000 acre-feet of water annually. Notable projects include:

- Tonopah Desert Recharge Project, the largest site, with an annual storage capacity of 150,000 acre-feet.
- The Hieroglyphic Mountains Recharge Project, which began operations in 2003, helps stabilize groundwater levels.
- Agua Fria Recharge Project, supporting regional water sustainability since 2002.
- Superstition Mountains, Lower Santa Cruz, and Pima Mine Road Recharge Projects, all contributing to enhanced groundwater management.

By strategically integrating recharge infrastructure, CAP ensures greater water resilience for Arizona's growing population. The program remains a cornerstone of sustainable water management in the region, securing future water availability while supporting environmental and economic stability.[41]

Central Arizona Project. "Central Arizona Project (Cap) Recharge Program." (n.d.). https://storymaps.arcgis.com/stories/70f626809fd84e228c0f32aef4222dda.

Lake Restoration

Lakes serve as natural water reservoirs, helping to control floods while supporting water supply, irrigation, fisheries, and tourism. However, many lakes have been drained for agricultural purposes or have silted up due to a lack of maintenance, thereby reducing their capacity and ecological function. Lake restoration focuses on rebuilding their structure and functionality, particularly in areas where they were previously drained. This process helps restore water storage capacity, improve ecosystem health, and enhance their ability to support biodiversity and human activities.[42]

Wetland Restoration

Wetland restoration and management can include large-scale technical interventions, such as installing ditches for rewetting or removing dykes to allow natural flooding. It can also involve smaller-scale measures, such as clearing trees to restore wetland function or adjusting land use, including modifying agricultural practices to better support wetland ecosystems. These actions help improve water retention, enhance biodiversity, and restore ecosystem services.[43]

Case 8.6: Coastal Wetland Restoration on the Hunter River, New South Wales, Australia

The Hunter River floodplain has undergone extensive modification since the mid-19th century, resulting in significant loss and degradation of its wetlands. Flood mitigation structures, dredging, and land conversion for grazing and industrial use altered the hydrology and ecology of the region. By 1984, extensive levees, floodgates, and drainage canals had disrupted tidal exchange, leading to declining water quality, subsidence, and habitat loss. Saltmarsh areas, crucial for migratory birds, have declined by 41% since their designation as Ramsar sites, while mangrove encroachment has reshaped the estuarine landscape. To address these challenges, resource managers initiated large-scale tidal restoration projects, including the Hexham Swamp Rehabilitation Project and the Tomago Wetland Rehabilitation Project. These efforts focused on reinstating tidal inundation by modifying floodgates, removing barriers, and restoring natural hydrological processes. The Hexham Swamp project, completed in 2013, reintroduced tidal flows to 650 hectares of previously isolated wetland. The restoration facilitated the recovery of saltmarsh, improved fish and prawn habitats, and enhanced water quality. The Tomago Wetland project, initiated in 2008 and completed by 2015, restored tidal flows to land previously dominated by pasture grasses. Hydrological modelling guided the staged opening of floodgates, allowing the wetland to transition back to saltmarsh and open-water habitats. The project improved fish passage, enhanced habitat for migratory birds, and contributed to the long-term resilience of the wetland system. These projects demonstrate how strategic land acquisitions, adaptive management, and tidal reinstatement can enhance the capacity of coastal wetlands to respond to sea-level rise while maintaining critical ecosystem services.[44]

CoastAdapt. "A Case Study of Good Coastal Adaptation on the Hunter River, Nsw." (2016). https://coastadapt.com.au/sites/default/files/case_studies/CSO4_Coastal_adaptation_Hunter_River.pdf.

Re-naturalization of Polder Areas

A polder is a low-lying area enclosed by dykes, creating an artificial hydrological system that is isolated from external water bodies except through manually controlled drainage mechanisms. Re-naturalizing polders involves enhancing them with semi-natural features, improving water storage capacity within internal watercourses, and boosting biodiversity while maintaining their flood management functions.[45]

Rainwater Collection from Ground Surfaces

Rainwater collection involves capturing rainfall from ground surfaces using micro-catchments to slow or divert runoff, allowing it to be stored before evaporation. Water can also be collected from rivers, storms, or natural watercourses to enhance soil moisture for agricultural purposes. This practice supports ecosystem services, including freshwater availability, water cycling, and soil formation and retention. The benefits include enhanced groundwater recharge, improved soil conditions, and reduced erosion caused by surface runoff.[46]

Catchment Thinning

Catchment thinning is the strategic removal of trees in densely forested areas experiencing drought to increase surface water runoff and streamflow. This method enhances ecosystem services, including food, fiber, fuel production, and water cycling. The benefits include increased water availability for human use, the provision of fuel and fiber resources, and help in managing forest density in water-stressed regions.[47]

Coastal and Marine

EbA strategies can be tailored to specific ecosystems or sectors, addressing environmental and community resilience. Ecosystem-based approaches focus on conserving and restoring natural systems, such as dune and beach systems, mangroves, seagrasses, and coral reefs, which provide essential ecosystem services for coastal communities. Sector-specific EbA approaches include measures such as marine protected areas (MPAs) and sustainable fisheries management, which ensure the long-term health of marine ecosystems while supporting livelihoods dependent on these resources.[48]

Beach Nourishment

Beach nourishment is a coastal management strategy that counteracts shoreline erosion by artificially adding high-quality sediment to areas with a sediment deficit. This helps maintain beach width and enhances storm protection. It supports multiple ecosystem services, including storm buffering, erosion control, and improved recreational and aesthetic value. The benefits of beach nourishment include reducing erosion impacts, providing a natural storm barrier, enhancing beach usability, creating potential ecological benefits such as improved nesting sites, and preserving the visual and aesthetic appeal of the coastline.[49]

Case 8.7: The Sand Motor Nature-Based Coastal Protection, Netherlands

The Sand Motor is an innovative coastal management project developed off the coast of Kijkduin, near The Hague, in the Netherlands. Constructed in 2011, this artificial peninsula was created using 21.5 million cubic meters of sand, adhering to the "Building with Nature" principle. Instead of traditional dredging and beach nourishment methods, the Sand Motor leverages natural forces – wind, waves, and ocean currents – to gradually distribute sand along the coastline, reinforcing dunes and beaches over the long term. By depositing a large volume of sand in a single operation, the Sand Motor minimizes repeated disturbances to the seabed, reducing environmental impacts while enhancing coastal resilience. The project has resulted in a broader coastline and accelerated dune formation, contributing to natural flood defense and long-term coastal protection. Over time, this evolving landscape has attracted a diverse array of plant and animal species, transforming the site into a dynamic ecosystem that supports rich biodiversity. In addition to its ecological benefits, the Sand Motor provides recreational value, offering a natural area for outdoor activities such as hiking, kite surfing, and birdwatching. The design creates a peaceful retreat for both wildlife and visitors, blending coastal protection with public enjoyment.[50]

De Zandmotor. "About the Sand Motor." (n.d.). https://dezandmotor.nl/en/about-the-sand-motor/#:~:text=nl%20en-,About%20the%20Sand%20Motor,coast%20and%20into%20the%20dunes.

Artificial Sand Dunes and Rehabilitation

Dune rehabilitation restores natural or artificial dunes to improve their overall function and maximize coastal protection. This process, along with artificial dune construction, helps reduce coastal erosion and mitigate flooding in low-lying areas. It supports ecosystem services, including storm protection, erosion control, and recreational and aesthetic value. The benefits include creating a natural buffer against coastal flooding, minimizing erosion impacts, providing habitat for coastal plants and wildlife, and maintaining the integrity of the beach system using natural elements.[51]

Brush Mattressing

A brush mattress is a layer of branches placed on a dune to protect against small-scale erosion from waves and wind. The ecosystem services it addresses include erosion control and nutrient cycling. The benefits include erosion control in low-energy environments, providing potential habitat for various species, and recycling organic matter (such as dead branches).[52]

Revegetation

Revegetation in coastal zones helps prevent and reduce erosion, restore the function of riverbank ecosystems, and improve water quality. It supports ecosystem services such as erosion control, recreational and aesthetic value, and oxygen production. The benefits of revegetation include restoring native plant species, stabilizing dunes by reducing wind-blown erosion, decreasing water turbidity, and creating habitats for wildlife, contributing to a healthier and more resilient coastal environment.[53]

Mangrove Conservation and Restoration

Mangrove conservation and restoration serve as a natural buffer against coastal erosion and flooding, while providing essential habitat and ecosystem services. Mangroves play a crucial role in climate adaptation by mitigating coastal flooding, dissipating wave energy from tropical storms, preventing land erosion, and capturing sediment to stabilize shorelines. Beyond coastal protection, mangroves support livelihoods by providing resources such as honey, medicinal plants, prawns, crabs, and fish while enhancing nutrient cycling, rainfall regulation, and other ecological functions.[54,55] Additionally, mangroves contribute to higher fishery and aquaculture yields, as many commercial fish species rely on mangrove habitats during their juvenile stages, making them vital for sustainable fisheries.[56]

Case 8.8: Community-Led Mangrove Restoration in Umm Al Quwain Lagoon, United Arab Emirates

The mangrove restoration initiative in Umm Al Quwain Lagoon, United Arab Emirates, was launched in 2021 as a collaborative effort to regenerate mangrove ecosystems while fostering community engagement. Implemented by Emirates Nature-WWF in partnership with government agencies and stakeholders, the project follows a structured restoration framework integrating scientific methodologies with local ecological knowledge. The site selection process prioritized degraded areas with less than 10% canopy cover, thereby ensuring long-term security of land tenure and ecological connectivity. Areas with high biodiversity value and minimal human disturbance were identified through a two-stage evaluation process. Khor Al Yeefrah was selected as the primary restoration site, with careful consideration given to local environmental conditions, including tidal hydrology and soil salinity. The restoration phase commenced with the planting of 14,000 mangrove seedlings in March 2023. Seedlings were sourced from a local nursery, averaging 30 centimeters in height, and carefully transported to prevent stem damage. Planting was conducted during low tide to maximize root establishment and minimize competition from existing vegetation. A zig-zag planting pattern was adopted, ensuring optimal spacing for seedling growth while considering factors such as pneumatophores and crab burrows. Post-planting monitoring was implemented following Conservation International's vegetation monitoring framework. Initial survival rates reached 96% after five months, with regular assessments tracking growth metrics, canopy expansion, and soil conditions. The project also integrated community participation through educational initiatives, emphasizing the ecological and socio-economic benefits of mangrove conservation.[57]

WWF, United Arab Emirates. "Restoring Mangrove Ecosystems in the Uae." (n.d.). https://emiratesnaturewwf.ae/en/page/restoring-mangrove-ecosystems-in-the-uae.

Coastal Wetland Conservation and Restoration

Coastal wetlands play a crucial role in climate adaptation by reducing wave and tidal energy, thereby minimizing erosion and coastal flood risks. Their surface roughness slows incoming waves, dissipating energy and lowering the height of storm surges. Additionally, wetlands naturally adapt to sea-level rise by accumulating sediments, gradually increasing their elevation.[58] While coastal wetlands often regenerate naturally, conservation efforts can enhance this process. Fencing wetland areas prevents livestock trampling and browsing, allowing vegetation to regenerate from natural seed sources. Restoring natural water levels and tidal flows may also be necessary, which can involve removing or modifying bunds and reopening floodgates to allow controlled tidal inundation. Major modifications, such as removing artificial drainage or obstructions, should be completed before planting to ensure that new vegetation can establish and adapt to changes in water levels and salinity.[59]

Seagrass Conservation and Restoration

Seagrass beds comprise approximately 60 species of underwater flowering plants that grow in shallow coastal zones on sandy substrates across most continents. The primary method for conserving seagrass ecosystems is to mitigate common threats, such as pollution and boat damage, through regulatory measures. Seagrass restoration typically involves harvesting and transplanting seagrass to degraded areas, followed by active management and monitoring to ensure long-term success. Seagrasses play a vital role in coastal protection by reducing current velocity, dissipating wave energy, and stabilizing sediment, particularly in shallow, low-energy environments where biomass remains stable. By reducing wave energy, seagrass meadows help mitigate coastal flooding and erosion, thereby protecting nearby communities and infrastructure. Additionally, seagrass traps sediment, gradually raising its surface elevation, which may allow it to maintain its coastal protection functions as sea levels rise.[60]

Coral Reef Conservation and Restoration

Coral reef conservation typically involves establishing protected areas or no-take zones to minimize human disturbances and preserve reef ecosystems. Coral reef restoration involves rearing, transplanting, and monitoring coral fragments to aid in the recovery of degraded reefs. As an EbA strategy, coral reef conservation and restoration provide cost-effective coastal protection, reducing wave energy and preventing shoreline erosion, often at a lower cost than artificial structures. Beyond coastal defense, coral reefs offer additional benefits, including supporting fisheries and livelihoods, as they contribute 10%–12% of fish catches in tropical countries. Coral reefs also play a vital role in medical research, producing compounds used to develop treatments for cancer, arthritis, asthma, and heart disease.[61]

Case 8.9: Coral Nurture Program Enhancing Resilience on the Great Barrier Reef

The Coral Nurture Program is a collaborative initiative between researchers and tourism operators aimed at enhancing the resilience of high-value sites on the Great Barrier Reef. Established in 2018, the program integrates scientific research with community-driven coral restoration to mitigate the impacts of environmental change. One of the program's key innovations is the development of the Coralclip®, a stainless-steel attachment device that enables rapid and cost-effective coral out-planting. Unlike traditional chemical adhesives, Coralclip® allows corals to securely attach to the reef substrate, promoting natural growth while maintaining the aesthetic integrity of tourism sites. The program has successfully planted over 105,000 corals across 30 sites on the reef, enhancing biodiversity and coral cover. The initiative prioritizes adaptive science, using research to identify thermally tolerant coral species that can withstand rising ocean temperatures. Through collaboration with the University of Technology Sydney and other institutions, the program integrates genetic and ecological studies to improve coral selection for propagation. Tourism operators play a crucial role in the implementation by using their vessel infrastructure and local knowledge to conduct restoration activities. This partnership reduces operational costs while providing real-time data on reef conditions. Additionally, operators educate thousands of visitors annually, raising awareness about coral conservation and restoration efforts. The Coral Nurture Program is supported by diverse funding sources, including the Australian Government's Reef Trust, Great Barrier Reef Foundation, Rolex Perpetual Planet, and philanthropic donations. The program continues to refine restoration techniques, contributing valuable knowledge to global coral conservation strategies.[62]

Coral Nurture Program. "Boosting Resilience at High-Value Reef Sites for the Great Barrier Reef." (n.d.). https://coralnurtureprogram.org.

Oyster Reefs

Oysters start as free-floating larvae before settling on solid surfaces, where they remain and grow for life. Over time, oyster reefs form on rocks, old shells, shipwrecks, and piers, creating dense clusters that protect underwater vegetation and coastal areas from waves, floods, and tides.[63] Research has shown that oyster reefs can grow quickly enough to keep pace with rising sea levels, making them a valuable natural defense against coastal erosion.[64] Oyster restoration involves building reefs using oyster shells and hatchery-raised larval oysters. In some cases, alternative materials, such as concrete rubble or repurposed bridge decking, create stable reef structures. Restoring oyster reefs provides multiple benefits, including enhancing seafood production, creating habitats for marine species, and improving water quality. A single oyster can filter up to 50 gallons of water per day, helping to maintain healthier coastal ecosystems.[65]

Living Breakwaters

Breakwaters are offshore, often submerged structures that act as a barrier between the sea and the land, reducing wave energy and protecting coastlines. Living breakwaters are designed to restore coastal reef ecosystems and enhance marine habitats while providing coastal protection. These artificial reefs can be made from purpose-built concrete structures or repurposed materials such as old ships and gas platforms. As an EbA measure, living breakwaters help dissipate wave energy, reducing flooding and erosion along the coast. In addition to coastal resilience, they provide habitats and nursery grounds for fish, supporting fisheries and local livelihoods. They also promote tourism and recreation, generating income from scuba diving while helping to alleviate pressure on natural coral reefs.[66]

Integrated Coastal Zone Management and Planning

Integrated Coastal Zone Management (ICZM) is a long-term, adaptive planning process coordinating coastal activities across sectors to ensure sustainable resource use. It integrates environmental, economic, and social objectives, linking land and sea management while involving multiple governance levels. ICZM addresses coastal erosion, pollution, habitat destruction, and climate change impacts through resource allocation, stakeholder participation, and coordinated governance. It helps balance conservation with economic activities, such as fisheries and tourism, thereby reducing conflicts and mitigating disaster risks. Implemented globally in varying forms, ICZM promotes resilience, ensuring coastal ecosystems remain productive and sustainable for future generations while supporting community livelihoods and development.[67,68]

Marine Reserves

Marine reserves support both marine ecosystems and coastal communities in adapting to climate change through various mechanisms, including:

Ocean Acidification

Marine reserves are crucial in rebuilding teleost fish populations, which are essential to the marine inorganic carbon cycle. These fish regulate their salt balance by drinking seawater, then process excess calcium and magnesium, excreting high-magnesium calcite crystals. The dissolution of these carbonates near the ocean surface increases alkalinity, which helps stabilize the surface pH and mitigate the effects of ocean acidification.[69]

Protection from Sea Level Rise

Healthy coastal wetlands, mudflats, and biogenic reefs provide natural protection against sea-level rise, helping to shield people, infrastructure, and property from coastal hazards. Marine reserves safeguard these ecosystems by preventing overharvesting, dredging, and coastal development, ensuring they function as effective natural coastal defenses.[70]

Enhancing Protection from Storms

As warmer oceans intensify storm systems, healthy coastal ecosystems, such as coastal wetlands, play a critical role in absorbing wave energy, stabilizing shorelines, and accumulating sediment more effectively than dredged or fragmented wetlands. Marine reserves help protect these ecosystems from degradation and loss, ensuring they remain intact and functional as natural coastal defenses. At the same time, they support human livelihoods by maintaining ecosystem services that benefit fisheries, tourism, and local communities.[71]

Enhancing Livelihoods

As fisheries productivity declines, marine reserves help rebuild overexploited fish stocks and restore degraded habitats, increasing reproduction and replenishing fishing grounds. By preserving a diverse age structure within fish populations, marine reserves enhance population stability and resilience, reducing fluctuations in stock replenishment. Additionally, they benefit adjacent areas through spillover, as juvenile and adult fish migrate beyond reserve boundaries, supporting surrounding fisheries and ecosystems.[72]

Enhancing Resilience of Marine Organisms

Although marine reserves cannot entirely shield ecosystems from the effects of climate change, they create low-stress environments that enhance the ability of marine organisms to adapt. By providing protected areas, marine reserves allow species to relocate and escape environmental pressures, increasing their chances of survival. Additionally, they serve as valuable reference sites for monitoring the effects of climate change on marine ecosystems.[73]

Case 8.10: Community-Based Fisheries Management in Micronesia

The Micronesian Community-based Fisheries Management as a Nature-Based Solution for Coastal Resilience (MiCOAST) project is a regional initiative supporting community-based fisheries management as an NBS for coastal resilience across Micronesia. Implemented in the Federated States of Micronesia, the Marshall Islands, Nauru, and Palau, the project strengthens local fisheries management by integrating climate adaptation, biodiversity conservation, and sustainable resource use. MiCOAST collaborates with local communities, government agencies, and NGOs to enhance fisheries governance through participatory management approaches. By engaging fishers, traditional authorities, and state officials, the project promotes adaptive strategies that incorporate scientific knowledge and traditional ecological practices. A key component is the co-design of fisheries management plans that integrate climate-resilient measures, ensuring sustainability across diverse coastal ecosystems. To address declining fish stocks and habitat degradation, MiCOAST implements MPAs, species-specific conservation measures, and habitat restoration initiatives. The project also develops alternative livelihood strategies, such as small-scale aquaculture and sustainable resource-based enterprises, reducing pressure on marine resources while strengthening economic resilience. Recognizing gender equity and social inclusion as essential to sustainable fisheries, MiCOAST actively engages women, youth, and marginalized groups in decision-making processes. The project conducts gender and inclusion analysis, builds local capacity, and facilitates networking opportunities to ensure equitable participation in fisheries governance. Through multi-stakeholder collaboration, policy alignment, and capacity building, MiCOAST strengthens community resilience to climate change, supports food security, and enhances the long-term sustainability of coastal fisheries in Micronesia.[74]

Kiwa Initiative. "Micoast: Micronesian Community-Based Fisheries Management as a Nature-Based Solution for Coastal Resilience." (n.d.). https://kiwainitiative.org/en/projects/micoast-micronesian-community-based-fisheries-management-as-a-nature-based-solution-for-coastal-resilience.

Sustainable Fisheries Management

Fishing impacts marine ecosystems beyond just fish stocks, as it can lead to the bycatch of non-target species, habitat destruction, disruptions to food chains, and biodiversity loss. To address these challenges, the Food and Agriculture Organization developed the Ecosystem Approach to Fisheries Management (EAFM) to ensure healthy ecosystems, sustainable fish stocks, and the well-being of fishing communities and stakeholders. EAFM considers five key components: target species, non-target species, the ecosystem, community outcomes, and fishery administration.

The EAFM process involves four main stages:

1. Defining the scope of what needs to be managed or assessed.
2. Identifying all relevant issues across the five components.
3. Conducting a risk analysis to determine which issues require direct management.
4. Setting acceptable management targets, applying appropriate measures, and establishing a review process to assess effectiveness.[75,76]

Table 8.4 Ecosystem-Based Adaptation Approaches

Habitat	*Project Name*	*Activities*	*Co-Benefits*
Agriculture	Kenya Climate Smart Agriculture Project	Implemented across 24 counties, the project promotes climate-smart agriculture through soil and water conservation, agroforestry, crop diversification, and small-scale irrigation. Enhances livestock and rangeland management, seed systems, advisory services, and climate-responsive emergency support.	Improves productivity and climate resilience, enhances food security, supports drought response, and empowers smallholder farmers.
Forests	Protecting and Restoring Coastal Temperate Rainforests in Chile	Restores and protects 50,000 hectares of Valdivian rainforest, adjacent to Alerce Costero National Park. Involves native reforestation, invasive species removal, carbon credit generation, and community water provisioning.	Enhances biodiversity, protects ancient tree species, generates carbon revenue, and supports sustainable tourism and clean water access.
Rivers	Restoration of the Emscher River, Germany	Transforms a century-old wastewater canal into a resilient, natural river system using underground sewers, restructured river paths, native vegetation, and green infrastructure.	Improves water quality, enhances biodiversity, creates recreational spaces, boosts employment, and revitalizes industrial areas.
Wetlands/ floodplains	Victorian Murray Floodplain Restoration Project, Australia	Restores floodplain ecosystems across nine sites using environmental infrastructure like flow regulators and channels to mimic natural flooding cycles.	Improves biodiversity, supports wetland regeneration, enhances habitat resilience, and safeguards ecosystems from climate impacts.
Water resources	Central Arizona Project Recharge Program	Stores surface water underground through artificial recharge at six major sites to replenish aquifers, mitigate drought impacts, and stabilize groundwater levels.	Secures long-term water supply, prevents land subsidence, supports groundwater sustainability, and enhances drought resilience.

(*Continued*)

Table 8.4 (Continued)

Habitat	*Project Name*	*Activities*	*Co-Benefits*
Coastal wetlands	Coastal Wetland Restoration – Hunter River, Australia	Restores tidal exchange and hydrology in degraded wetlands by modifying floodgates and barriers. Projects include Hexham Swamp and Tomago Wetland Rehabilitation.	Restores saltmarsh and fish habitats, improves water quality, and enhances resilience to sea-level rise and biodiversity conservation.
Coastal protection	The Sand Motor, Netherlands	Constructed an artificial sand peninsula to reinforce the coastline through natural redistribution by wind and waves, minimizing seabed disruption.	Enhances coastal defense, supports biodiversity, reduces maintenance, and offers recreational and ecological benefits.
Mangroves	Community-Led Mangrove Restoration – Umm Al Quwain, UAE	Restores degraded mangroves through local nursery planting, scientific methods, and community engagement. Focuses on optimal spacing, hydrology, and monitoring.	Strengthens coastal resilience, improves biodiversity, engages communities, and promotes sustainable ecosystem stewardship.
Coral reefs	Coral Nurture Program – Great Barrier Reef	Uses Coralclip® technology for efficient coral out-planting. Implements thermally tolerant coral selection, with tourism operators supporting logistics and education.	Enhances reef resilience, increases coral cover, integrates science and tourism, and raises awareness among visitors.
Fisheries	Community-Based Fisheries Management – Micronesia	Supports locally driven fisheries governance and habitat restoration, co-designed with communities using traditional and scientific knowledge.	Promotes climate resilience, food security, biodiversity, and gender-inclusive coastal resource management.

Effective EAFM strengthens the resilience of coastal communities to climate change and functions as an EbA measure. Community-based EAFM engages local stakeholders in developing fisheries management plans, ensuring long-term sustainability. Additionally, EAFM complements other EbA strategies that enhance coastal ecosystems and increase the availability of ecosystem services, thereby further supporting environmental and community resilience.[77]

Case Study Summary

Various EbA approaches have been implemented across terrestrial and marine habitats to mitigate the impacts of climate change (Table 8.4).

Notes

1 World Bank and WWF, "Operational Framework for Ecosystem-Based Adaptation. Implementing and Mainstreaming Ecosystem-Based Adaptation Responses in the Greater Mekong Sub-Region," (2013), http://d2ouvy59p0dg6k.cloudfront.net/downloads/wwf_wb_eba_project_2014_gms_ecosystem_based_adaptation_general_framework.pdf.
2 Holly P. Jones, David G. Hole, and Erika S. Zavaleta, "Harnessing Nature to Help People Adapt to Climate Change," *Nature Climate Change* 2 (2012).
3 Convention on Biological Diversity, "Connecting Biodiversity and Climate Change Mitigation and Adaptation," (2009), https://cbd.int/doc/publications/ahteg-brochure-en.pdf.
4 Christine Wamsler et al., "Operationalizing Ecosystem-Based Adaptation: Harnessing Ecosystem Services to Buffer Communities against Climate Change," *Ecology and Society* 21, no. 1 (2016).
5 Jones, Hole, and Zavaleta, "Harnessing Nature to Help People Adapt to Climate Change."
6 Jones, Hole, and Zavaleta, "Harnessing Nature to Help People Adapt to Climate Change."
7 World Bank and WWF, "Operational Framework for Ecosystem-Based Adaptation. Implementing and Mainstreaming Ecosystem-Based Adaptation Responses in the Greater Mekong Sub-Region."
8 World Bank and WWF, "Operational Framework for Ecosystem-Based Adaptation. Implementing and Mainstreaming Ecosystem-Based Adaptation Responses in the Greater Mekong Sub-Region."
9 Richard Munang et al., "Climate Change and Ecosystem-Based Adaptation: A New Pragmatic Approach to Buffering Climate Change Impacts," *Current Opinion in Environmental Sustainability* 5, no. 1 (2013).
10 UNEP, "Making the Case for Ecosystem-Based Adaptation. Building Resilience to Climate Change," (2010), https://wedocs.unep.org/handle/20.500.11822/13737.
11 UNEP, "Options for Ecosystem-Based Adaptation in Coastal Environments: A Guide for Environmental Managers and Planners," (2016), https://unep-wcmc.org/system/dataset_file_fields/files/000/000/380/original/Options_for_Ecosystem_based_Adaptation_in_Coastal_Environments_low-res.pdf?1462462607.
12 UNEP, "Options for Ecosystem-Based Adaptation in Coastal Environments: A Guide for Environmental Managers and Planners."
13 World Bank and WWF, "Operational Framework for Ecosystem-Based Adaptation. Implementing and Mainstreaming Ecosystem-Based Adaptation Responses in the Greater Mekong Sub-Region."
14 World Bank and WWF, "Operational Framework for Ecosystem-Based Adaptation. Implementing and Mainstreaming Ecosystem-Based Adaptation Responses in the Greater Mekong Sub-Region."

15 World Bank and WWF, "Operational Framework for Ecosystem-Based Adaptation. Implementing and Mainstreaming Ecosystem-Based Adaptation Responses in the Greater Mekong Sub-Region."
16 UNEP, "Technologies for Climate Change Adaptation – Agriculture Sector," (2011), http://tech-action.org/Publications/TNA-Guidebooks.
17 World Bank and WWF, "Operational Framework for Ecosystem-Based Adaptation. Implementing and Mainstreaming Ecosystem-Based Adaptation Responses in the Greater Mekong Sub-Region."
18 UNEP, "Technologies for Climate Change Adaptation – Agriculture Sector."
19 UNEP, "Technologies for Climate Change Adaptation – Agriculture Sector."
20 World Bank and WWF, "Operational Framework for Ecosystem-Based Adaptation. Implementing and Mainstreaming Ecosystem-Based Adaptation Responses in the Greater Mekong Sub-Region."
21 World Bank and WWF, "Operational Framework for Ecosystem-Based Adaptation. Implementing and Mainstreaming Ecosystem-Based Adaptation Responses in the Greater Mekong Sub-Region."
22 Kenya Climate Smart Agriculture Project, "Kenya Climate Smart Agriculture Project (Kcsap)," (n.d.), https://kcsap.go.ke/home.
23 Rodney J. Keenan, "Climate Change Impacts and Adaptation in Forest Management: A Review," *Annals of Forest Science* 72, no. 2 (2015).
24 FAO, "Managing Forests for Climate Change," (2010), http://fao.org/3/i1960e/i1960e00.pdf.
25 FAO, "Managing Forests for Climate Change," (2010), http://fao.org/3/i1960e/i1960e00.pdf.
26 K. Swiderska, C. King-Okumu, and M. Monirul Islam, "Ecosystem-Based Adaptation: A Handbook for Eba in Mountain, Dryland and Coastal Ecosystems," (2018), https://pubs.iied.org/17460IIED/.
27 ODI, "Pastoralism and Climate Change: Enabling Adaptive Capacity," (2009), https://odi.org/sites/odi.org.uk/files/odi-assets/publications-opinion-files/4317.pdf.
28 FAO, "Managing Forests for Climate Change."
29 Conservation International, "Valdivian Coastal Reserve," (2025), https://conservation.org/docs/default-source/publication-pdfs/valdivia-coastal-reserve_factsheet.pdf?Status=Master&sfvrsn=40b2c330_3#:~:text=The%20Reserve%20is%20an%20important,back%20to%20native%20temperate%20rainforest.
30 Emilia Pramova et al., "Forests and Trees for Social Adaptation to Climate Variability and Change," *Wiley Interdisciplinary Reviews: Climate Change* 3, no. 6 (2012).
31 Climate-ADAPT, "Water Sensitive Forest Management," (2015), https://climate-adapt.eea.europa.eu/metadata/adaptation-options/water-sensitive-forest-management.
32 Ecofys, "Assessing Adaptation Knowledge in Europe: Ecosystem-Based Adaptation," (2016), https://ec.europa.eu/clima/sites/clima/files/adaptation/what/docs/ecosystem_based_adaptation_en.pdf.
33 USAID, "Ecosystem-Based Adaptation and Water Security," (2017), https://files.globalwaters.org/water-links-files/Ecosystem-based%20Adaptation%20and%20Water%20Security.pdf.
34 Ecofys, "Assessing Adaptation Knowledge in Europe: Ecosystem-Based Adaptation."

35 Climate-ADAPT, "Rehabilitation and Restoration of Rivers," (2015), https://climate-adapt.eea.europa.eu/metadata/adaptation-options/rehabilitation-and-restoration-of-rivers.
36 Ecofys, "Assessing Adaptation Knowledge in Europe: Ecosystem-Based Adaptation."
37 Urban Nature Atlas, "Restoration of the Emscher River," (n.d.), https://una.city/nbs/essen/restoration-emscher-river.
38 Urban Nature Atlas, "Restoration of the Emscher River."
39 State Government of Victoria, "Victorian Murray Floodplain Restoration Project," (n.d.), https://water.vic.gov.au/our-programs/murray-darling-basin/victorian-murray-floodplain-restoration-project.
40 R.C. Brears, *Climate Resilient Water Resources Management* (Cham, Switzerland: Palgrave Macmillan, 2018).
41 Central Arizona Project, "Central Arizona Project (Cap) Recharge Program," (n.d.), https://storymaps.arcgis.com/stories/70f626809fd84e228c0f32aef4222dda.
42 Ecofys, "Assessing Adaptation Knowledge in Europe: Ecosystem-Based Adaptation."
43 Ecofys, "Assessing Adaptation Knowledge in Europe: Ecosystem-Based Adaptation."
44 CoastAdapt, "A Case Study of Good Coastal Adaptation on the Hunter River, Nsw," (2016), https://coastadapt.com.au/sites/default/files/case_studies/CSO4_Coastal_adaptation_Hunter_River.pdf.
45 CoastAdapt, "A Case Study of Good Coastal Adaptation on the Hunter River, Nsw."
46 World Bank and WWF, "Operational Framework for Ecosystem-Based Adaptation. Implementing and Mainstreaming Ecosystem-Based Adaptation Responses in the Greater Mekong Sub-Region."
47 World Bank and WWF, "Operational Framework for Ecosystem-Based Adaptation. Implementing and Mainstreaming Ecosystem-Based Adaptation Responses in the Greater Mekong Sub-Region."
48 UNEP, "Options for Ecosystem-Based Adaptation in Coastal Environments: A Guide for Environmental Managers and Planners."
49 World Bank and WWF, "Operational Framework for Ecosystem-Based Adaptation. Implementing and Mainstreaming Ecosystem-Based Adaptation Responses in the Greater Mekong Sub-Region."
50 De Zandmotor, "About the Sand Motor," (n.d.), https://dezandmotor.nl/en/about-the-sand-motor/#:~:text=nl%20en-,About%20the%20Sand%20Motor,coast%20and%20into%20the%20dunes.
51 De Zandmotor, "About the Sand Motor."
52 De Zandmotor, "About the Sand Motor."
53 De Zandmotor, "About the Sand Motor."
54 Swiderska, "Ecosystem-Based Adaptation: A Handbook for Eba in Mountain, Dryland and Coastal Ecosystems."
55 World Bank and WWF, "Operational Framework for Ecosystem-Based Adaptation. Implementing and Mainstreaming Ecosystem-Based Adaptation Responses in the Greater Mekong Sub-Region."
56 UNEP, "Making the Case for Ecosystem-Based Adaptation. Building Resilience to Climate Change."
57 United Arab Emirates WWF, "Restoring Mangrove Ecosystems in the UAE," (n.d.), https://emiratesnaturewwf.ae/en/page/restoring-mangrove-ecosystems-in-the-uae.

58 UNEP, "Options for Ecosystem-Based Adaptation in Coastal Environments: A Guide for Environmental Managers and Planners."
59 Auckland Council, "Coastal Wetlands, Saltmarshes & Estuaries," (2013), https://naturespace.org.nz/sites/default/files/u4/BC2224-Coastal%20wetlands%2C%20saltmarshes%20%26%20estuaries%20A4%20%28Web%29.pdf.
60 UNEP, "Options for Ecosystem-Based Adaptation in Coastal Environments: A Guide for Environmental Managers and Planners."
61 UNEP, "Options for Ecosystem-Based Adaptation in Coastal Environments: A Guide for Environmental Managers and Planners."
62 Coral Nurture Program, "Boosting Resilience at High-Value Reef Sites for the Great Barrier Reef," (n.d.), https://coralnurtureprogram.org.
63 NOAA Fisheries, "Oyster Reef Habitat," (n.d.), https://fisheries.noaa.gov/national/habitat-conservation/oyster-reef-habitat.
64 Antonio B. Rodriguez et al., "Oyster Reefs Can Outpace Sea-Level Rise," *Nature Climate Change* 4 (2014).
65 NOAA Fisheries, "Oyster Reef Habitat."
66 UNEP, "Options for Ecosystem-Based Adaptation in Coastal Environments: A Guide for Environmental Managers and Planners."
67 USAID, "Ecosystem-Based Adaptation and Coastal Populations," (2018), https://climatelinks.org/resources/ecosystem-based-adaptation-and-coastal-populations.
68 FAO, "Integrated Management of Coastal Zones," (1994), https://fao.org/4/t0708e/T0708E00.htm#TOC.
69 Callum M. Roberts et al., "Marine Reserves Can Mitigate and Promote Adaptation to Climate Change," *Proceedings of the National Academy of Sciences* 114, no. 24 (2017).
70 Roberts et al., "Marine Reserves Can Mitigate and Promote Adaptation to Climate Change."
71 Roberts et al., "Marine Reserves Can Mitigate and Promote Adaptation to Climate Change."
72 Roberts et al., "Marine Reserves Can Mitigate and Promote Adaptation to Climate Change."
73 IUCN, "Marine Protected Areas and Climate Change," (2017), https://iucn.org/sites/dev/files/mpas_and_climate_change_issues_brief.pdf.
74 Kiwa Initiative, "Micoast: Micronesian Community-Based Fisheries Management as a Nature-Based Solution for Coastal Resilience," (n.d.), https://kiwainitiative.org/en/projects/micoast-micronesian-community-based-fisheries-management-as-a-nature-based-solution-for-coastal-resilience.
75 WWF, "Ecosystem Approach to Fisheries Management (Eafm)," (2011), http://d2ouvy59p0dg6k.cloudfront.net/downloads/eafm_revised.pdf.
76 Adel Heenan et al., "A Climate-Informed, Ecosystem Approach to Fisheries Management," *Marine Policy* 57 (2015).
77 UNEP, "Options for Ecosystem-Based Adaptation in Coastal Environments: A Guide for Environmental Managers and Planners."

References

Auckland Council. "Coastal Wetlands, Saltmarshes & Estuaries." (2013). https://naturespace.org.nz/sites/default/files/u4/BC2224-Coastal%20wetlands%2C%20saltmarshes%20%26%20estuaries%20A4%20%28Web%29.pdf.

Brears, R.C. *Climate Resilient Water Resources Management.* Cham, Switzerland: Palgrave Macmillan, 2018.

Climate-ADAPT. "Rehabilitation and Restoration of Rivers." (2015). https://climate-adapt.eea.europa.eu/metadata/adaptation-options/rehabilitation-and-restoration-of-rivers.

Climate-ADAPT. "Water Sensitive Forest Management." (2015). https://climate-adapt.eea.europa.eu/metadata/adaptation-options/water-sensitive-forest-management.

Convention on Biological Diversity. "Connecting Biodiversity and Climate Change Mitigation and Adaptation." (2009). https://cbd.int/doc/publications/ahteg-brochure-en.pdf.

Ecofys. "Assessing Adaptation Knowledge in Europe: Ecosystem-Based Adaptation." (2016). https://ec.europa.eu/clima/sites/clima/files/adaptation/what/docs/ecosystem_based_adaptation_en.pdf.

FAO. "Integrated Management of Coastal Zones." (1994). https://fao.org/4/t0708e/T0708E00.htm#TOC.

FAO. "Managing Forests for Climate Change." (2010). http://fao.org/3/i1960e/i1960e00.pdf.

Heenan, Adel, Robert Pomeroy, Johann Bell, Philip L. Munday, William Cheung, Cheryl Logan, Russell Brainard, et al. "A Climate-Informed, Ecosystem Approach to Fisheries Management." *Marine Policy* 57 (2015): 182–92.

IUCN. "Marine Protected Areas and Climate Change." (2017). https://iucn.org/sites/dev/files/mpas_and_climate_change_issues_brief.pdf.

Jones, Holly P., David G. Hole, and Erika S. Zavaleta. "Harnessing Nature to Help People Adapt to Climate Change." *Nature Climate Change* 2 (2012/06/26): 504.

Keenan, Rodney J. "Climate Change Impacts and Adaptation in Forest Management: A Review." *Annals of Forest Science* 72, no. 2 (2015/03/01): 145–67.

Munang, Richard, Ibrahim Thiaw, Keith Alverson, Musonda Mumba, Jian Liu, and Mike Rivington. "Climate Change and Ecosystem-Based Adaptation: A New Pragmatic Approach to Buffering Climate Change Impacts." *Current Opinion in Environmental Sustainability* 5, no. 1 (2013/03/01): 67–71.

NOAA Fisheries. "Oyster Reef Habitat." (n.d.). https://fisheries.noaa.gov/national/habitat-conservation/oyster-reef-habitat.

ODI. "Pastoralism and Climate Change: Enabling Adaptive Capacity." (2009). https://odi.org/sites/odi.org.uk/files/odi-assets/publications-opinion-files/4317.pdf.

Pramova, Emilia, Bruno Locatelli, Houria Djoudi, and Olufunso A. Somorin. "Forests and Trees for Social Adaptation to Climate Variability and Change." *Wiley Interdisciplinary Reviews: Climate Change* 3, no. 6 (2012/11/01): 581–96.

Roberts, Callum M., Bethan C. O'Leary, Douglas J. McCauley, Philippe M. Cury, Carlos M. Duarte, Jane Lubchenco, Daniel Pauly, et al. "Marine Reserves Can Mitigate and Promote Adaptation to Climate Change." *Proceedings of the National Academy of Sciences* 114, no. 24 (2017): 6167–75.

Rodriguez, Antonio B., F.J. Fodrie, Justin T. Ridge, Niels L. Lindquist, Ethan J. Theuerkauf, Sara E. Coleman, Jonathan H. Grabowski, et al. "Oyster Reefs Can Outpace Sea-Level Rise." *Nature Climate Change* 4 (2014/04/28): 493.

Swiderska, K., C. King-Okumu, and M. Monirul Islam. "Ecosystem-Based Adaptation: A Handbook for Eba in Mountain, Dryland and Coastal Ecosystems." (2018). https://pubs.iied.org/17460IIED/.

UNEP. "Making the Case for Ecosystem-Based Adaptation. Building Resilience to Climate Change." (2010). https://wedocs.unep.org/handle/20.500.11822/13737.

UNEP. "Options for Ecosystem-Based Adaptation in Coastal Environments: A Guide for Environmental Managers and Planners." (2016). https://unep-wcmc.org/system/dataset_file_fields/files/000/000/380/original/Options_for_Ecosystem_based_Adaptation_in_Coastal_Environments_low-res.pdf?1462462607.

UNEP. "Technologies for Climate Change Adaptation – Agriculture Sector." (2011). http://tech-action.org/Publications/TNA-Guidebooks.

USAID. "Ecosystem-Based Adaptation and Coastal Populations." (2018). https://climatelinks.org/resources/ecosystem-based-adaptation-and-coastal-populations.

USAID. "Ecosystem-Based Adaptation and Water Security." (2017). https://files.globalwaters.org/water-links-files/Ecosystem-based%20Adaptation%20and%20Water%20Security.pdf.

Wamsler, Christine, Lisa Niven, Thomas H. Beery, Torleif Bramryd, Nils Ekelund, K.I. Jönsson, Adelina Osmani, Thomas Palo, and Sanna Stålhammar. "Operationalizing Ecosystem-Based Adaptation: Harnessing Ecosystem Services to Buffer Communities against Climate Change." *Ecology and Society* 21, no. 1 (2016).

World Bank and WWF. "Operational Framework for Ecosystem-Based Adaptation. Implementing and Mainstreaming Ecosystem-Based Adaptation Responses in the Greater Mekong Sub-Region." (2013). http://d2ouvy59p0dg6k.cloudfront.net/downloads/wwf_wb_eba_project_2014_gms_ecosystem_based_adaptation_general_framework.pdf.

WWF. "Ecosystem Approach to Fisheries Management (Eafm)." (2011). http://d2ouvy59p0dg6k.cloudfront.net/downloads/eafm_revised.pdf.

9 Restoring Degraded Ecosystems through Nature-Based Solutions

Introduction

Ecosystem restoration involves helping degraded, damaged, or destroyed ecosystems recover their ecological functions while restoring the goods and services that benefit people. It takes a landscape-based approach, integrating ecological, social, and developmental priorities to balance conservation and sustainable land use. Restoration efforts focus on enhancing ecosystem health, promoting biodiversity, improving landscape connectivity, and supporting long-term ecosystem management.

Restoring Degraded Ecosystems

Restoring degraded ecosystems through nature-based solutions (NBS) strengthens ecosystem resilience, allowing them to continue providing essential services while addressing broader societal challenges.[1] Ecosystem restoration can improve food and water security, support climate change mitigation and adaptation, provide timber, firewood, and other forest products, enhance pollinator resources, and create habitat for wildlife while also helping to reduce conflict and migration risks.[2,3] Restoration differs from rehabilitation in its approach and methodology. Restoration aims to fully reinstate an ecosystem's pre-disturbance state, including all its biological complexity, whereas rehabilitation focuses on restoring specific, targeted functions or services. In cases where degradation has exceeded recovery thresholds, such as when species have gone extinct, or seed banks and soil biota have been lost, rehabilitation may be the only viable option.[4]

Attributes of Restored Ecosystems

An ecosystem is restored when it has sufficient biotic and abiotic resources to continue developing without ongoing human intervention. A restored ecosystem is self-sustaining, maintaining its structural and functional integrity while showing resilience to environmental stress and disturbances. It also interacts naturally with neighboring ecosystems, facilitating the flow of

DOI: 10.4324/9781003517948-9

Table 9.1 Attributes of Restored Ecosystems

Attribute	*Description*
Similar diversity and community structure compared to reference sites	The restored ecosystem contains a characteristic mix of species typical of the reference ecosystem, providing appropriate community structure and ecological function.
Presence of indigenous species	The ecosystem is composed primarily of native species. In cultural landscapes, some non-invasive domesticated or historically associated species may be acceptable.
Presence of functional groups necessary for long-term stability	All ecological functional groups essential for long-term development and stability are present, or likely to naturally recolonise over time.
Capacity of the physical environment to sustain reproducing populations	The physical conditions support self-sustaining, reproducing populations of key species needed for ecosystem continuity and resilience.
Normal functioning	The ecosystem functions appropriately for its current developmental stage, with no indicators of ecological dysfunction.
Integration with the landscape	The restored site is ecologically connected to the surrounding landscape, supporting interactions through energy, nutrient, and species flows.
Elimination of potential threats	External threats from surrounding land uses have been minimized or removed to protect the health and integrity of the ecosystem.
Resilience to natural disturbances	The ecosystem can withstand and recover from normal stress events such as fire, storms, or droughts without losing integrity.
Self-sustainability	The ecosystem can maintain its structure and function over time under current environmental conditions, adapting naturally to periodic stress and gradual change.

species, nutrients, and cultural connections. The completion of restoration is assessed based on nine key attributes, summarized in Table 9.1.[5]

Landscape Connectivity

When selecting areas for restoration, priority should be given to improving the extent, quality, and connectivity of high-biodiversity areas, particularly those that support threatened or endangered species and provide essential ecosystem services. Ecological restoration efforts that promote biodiversity conservation and landscape connectivity include protecting remaining forest fragments, restoring degraded native vegetation, and converting degraded or low-productivity agricultural lands into corridors and stepping-stones to enhance habitat connectivity. These actions help expand and strengthen ecological networks, supporting wildlife movement and ecosystem resilience.[6]

Co-Benefits of Restoring Degraded Ecosystems

Restoring degraded ecosystems using NBS provides multiple co-benefits, including habitat and species recovery, accelerated ecosystem regeneration, and a reduced need for interventions such as fertilization. It also enhances climate resilience, supports carbon sequestration, and improves air and water quality. Additional benefits include increased water retention, enhanced pollination services, soil erosion control, and expanded habitats for wildlife. Restored ecosystems also contribute to recreational opportunities, improved public health, and job creation, making them valuable for environmental sustainability and community well-being.[7,8]

Soils

Soil quality refers to the ability of soil to function within natural or managed ecosystems, supporting plant and animal productivity, maintaining water and air quality, and sustaining human health and habitation. Changes in soil function occur in response to management practices and climate conditions, which in turn affect key soil properties. The impact of soil quality on ecosystems includes:

- Regulating water availability for plant growth from rainfall and snowmelt.
- Influencing runoff, water infiltration, and erosion potential.
- Determining nutrient availability is crucial for plant growth and development.
- Providing suitable conditions for germination, seedling establishment, and root growth.
- Acting as a natural filter to protect water and air quality.[9]

To safeguard soil health, farmers can implement various NBS, including:

Stocking Rate

The stocking rate refers to the number of animals per unit of land over a specific period. Maintaining an appropriate stocking rate is essential to ensure forage availability, as overgrazing can cause long-term land degradation and reduce pasture quality and yield potential. Proper stocking management helps maintain pasture productivity and supports animal health. To achieve this, pasture size and stock density should be determined based on available grass cover, the number of grazing animals, and the duration of grazing in each area.[10]

Rotational Grazing

During the growing season, plants can begin regrowing immediately after grazing. Still, they direct much of their energy into leaf growth, which can slow or even stop root development if too much foliage is removed. When plants start to regrow, livestock often preferentially graze on the fresh shoots, weakening root systems and leading to plant death over time.[11] Rotational grazing helps prevent overgrazing by moving livestock through a series of paddocks, allowing pastures to recover between grazing periods. This practice enhances pasture management, restricts livestock access to streams, and minimizes environmental impacts. By resting paddocks, vegetation can re-establish itself, manure is more evenly distributed, and sediment runoff is minimized by reducing livestock traffic in fragile areas.[12]

Meadows and Pastures

Meadows are fields dominated by grasses and non-woody plants, primarily used for mowing and hay production, while pastures are grassed or wooded areas typically used for grazing livestock.[13] Restoring meadows and pastures to their original state may not always be feasible due to changes in climate, wildlife activity, and plant species composition. However, understanding the site's history, ecological processes, and past variability provides a foundation for developing a practical and effective restoration plan.[14] Meadows and pastures, with their deep-rooted vegetation and permanent ground cover, play a crucial role in temporarily storing water during floods, reducing flood impacts. They also enhance water quality by trapping sediments, absorbing nutrients, and preventing soil erosion, making them valuable for ecosystem health and sustainable land management.[15]

Crop Rotation

Crop rotation is growing different plant species in a defined sequence on the same land over a fixed period.[16] This method enhances soil structure and fertility by alternating between deep-rooted and shallow-rooted plants, thereby improving nutrient cycling, reducing erosion, increasing water infiltration, and lowering downstream flood risk.[17] Beyond soil health, crop rotation helps preserve the environment, control weeds, diseases, and pests, and diversify farm income by reducing reliance on a single crop. A diverse farm can also attract educational programs, agritourism, farm stays, and recreational activities such as hunting, adding economic value while maintaining the natural resources and rural landscape.[18,19,20]

Intercropping

Intercropping involves growing two or more crops together to maximize land use efficiency and enhance resource utilization. This method increases yields by using space, nutrients, and water that a single crop might not fully utilize. Intercropping improves soil structure, as shallow-rooted plants help aerate the soil, while leguminous crops enrich soil fertility by fixing atmospheric nitrogen, making essential nutrients more available to plants.[21] Additionally, using cover crops in otherwise bare soil – such as between rows or beneath other crops – reduces runoff and erosion while increasing water infiltration. Beyond soil benefits, intercropping enhances biodiversity by creating habitats for beneficial insects and soil organisms, which support natural pest control and reduce the spread of pathogens and insect pests in agricultural systems.[22,23]

No-Till Agriculture

No-till agriculture is a method that allows crops or pastures to grow without disturbing the soil through tillage. Traditional tillage mechanically modifies the soil, which can break down soil structure, increase erosion, reduce water retention, and deplete organic matter by compacting and altering soil pores.[24] By keeping crop residues on the soil surface, no-till farming helps prevent soil loss from wind and water erosion, reduces water evaporation and runoff, and improves precipitation infiltration, leading to healthier, more resilient soils.[25]

Low-Till Agriculture

Low-till agriculture, also known as conservation tillage or reduced tillage, is a farming practice applied to arable land that minimizes soil disturbance while maintaining soil health and fertility. It involves leaving at least 30% of crop residue on the soil surface when the soil is most vulnerable to erosion. Some light surface tillage may still be used, but the goal is to slow water movement, reduce soil erosion, and improve water infiltration, enhancing overall soil stability and moisture retention.[26,27]

Green Cover

Green cover, including cover crops, consists of crops planted in late summer or autumn to protect soil from erosion during the winter when fields would otherwise remain bare and vulnerable to wind and water damage. Cover crops help improve soil structure, enhance biodiversity within cropping systems, and reduce nutrient loss by preventing the leaching of soluble nutrients.[28]

Case 9.1: Minnesota Nutrient Management Initiative for Sustainable Farming

The Minnesota Nutrient Management Initiative (NMI) is a statewide program designed to help farmers and crop advisers evaluate and improve nutrient management practices through on-farm field trials. By testing various nitrogen application methods, timings, stabilizer use, and alternative nutrient sources, the initiative aims to reduce nitrate losses, enhance crop productivity, and promote sustainable agricultural practices. Farmers and crop advisers participate in Field Trials and Advanced Rate Trials, which compare existing nutrient management practices with alternative approaches. Field Trials focus on testing a new method against current practices, while Advanced Rate Trials assess the impact of different nitrogen rates on crop yield and environmental outcomes. To incentivize participation, farmers receive $1,300 for Field Trials and $1,500 for Advanced Rate Trials, while crop advisers are compensated $600 and $700, respectively. The trials follow strict guidelines to ensure accurate data collection and meaningful results. Farmers record field operations, submit soil test data, and monitor crop yields, while crop advisers verify and submit trial results. Advanced Rate Trials require Variable Rate Technology and GPS mapping, allowing for precise nutrient application strategies. Farmers can work with a crop adviser or a representative from the Minnesota Department of Agriculture to ensure trials are properly conducted and supported. NMI contributes to reducing agricultural runoff and improving water quality by providing real-world data to refine Minnesota's nitrogen fertilizer recommendations. Insights from these trials help identify best management practices that enhance efficiency while minimizing environmental impacts. By integrating scientific research with practical, on-farm experimentation, the program supports sustainable agriculture, climate resilience, and long-term environmental protection.[29]

Minnesota Department of Agriculture. "Nutrient Management Initiative (Nmi)." (n.d.). https://mda.state.mn.us/protecting/cleanwaterfund/onfarmprojects/nmi#:~:text=The%20Minnesota%20Nutrient%20Management%20Initiative,efficiency%20and%20improved%20water%20quality.

Forests

Forest restoration aims to return a degraded forest to its original structure, productivity, and species diversity, fully restoring its ecological functions. On the other hand, forest rehabilitation focuses on restoring the capacity of degraded forest land to provide products and ecosystem services. However, it may not fully recover all plant and animal species that were previously present. Both restoration and rehabilitation can be implemented in unproductive or abandoned agricultural land, deforested grasslands, brushlands, scrublands, barren areas, and degraded forests. A summary of forest restoration and rehabilitation measures is provided in Table 9.2.[30]

Forest Landscape Restoration

Forest Landscape Restoration is a holistic approach to forest restoration that engages all stakeholders across affected land-use sectors through participatory decision-making. It focuses on managing the complex interactions

Table 9.2 Forest Restoration and Rehabilitation Measures

Measure	*Description*
Protective measures	Implementing protections against threats such as fire, overgrazing, and soil erosion to create favourable conditions for natural regeneration.
Measures to accelerate recovery	Using techniques like direct seeding or planting seedlings in degraded primary or secondary forests to speed up the restoration process.
Measures to assist natural regeneration	Supporting natural recovery processes through actions such as weed control and site preparation on degraded lands and marginal agricultural areas.
Planting of native or introduced trees	Establishing trees in monoculture or mixed-species plantations, agroforestry systems, or as trees outside forests to restore ecosystem function and productivity.

between people, natural resources, and land use, ensuring that restoration efforts are sustainable, inclusive, and adaptable to the needs of the environment and local communities.[31]

Case 9.2: Conservation and Restoration of the Dry Forest Ecosystem in the Santa Rosa Watershed, Colombia

The Santa Rosa Watershed in the Dagua Municipality, Cauca Valley Department, Colombia, is a highly degraded tropical dry forest (TDF) ecosystem that has been impacted by land conversion, agriculture, and biodiversity loss. In response, Corporación Ambiental y Forestal del Pacífico implemented a conservation and restoration project to enhance ecosystem services and promote sustainable production practices. The project integrated landscape management strategies that combined ecological restoration with local economic activities. Key interventions included the protection of remaining forest patches by fencing off 1.1 kilometer of TDF remnants and expanding conservation zones by restoring 1 hectare of protective forest. Additionally, the project introduced sustainable production practices by converting 1 hectare of pasture into a silvopastoral system, which integrates trees, forage, and livestock to improve land productivity while maintaining ecological balance. To support local engagement, community workshops were held on ecological restoration, sustainable agriculture, and organic fertilizers, fostering knowledge sharing and long-term stewardship of the land. To address water scarcity, a mist-catchment system was installed to provide a reliable water source for silvopastoral areas, reducing dependence on conventional water sources. Conservation agreements with local landowners ensured continued protection of restored areas, reinforcing the long-term sustainability of the initiative. By integrating conservation with rural livelihoods, the project aimed to halt biodiversity loss, restore ecosystem services, and improve soil and water retention.[32]

Satoyama Initiative. "Conservation and Restoration Actions in a Dry Forest Ecosystem in the Santa Rosa Watershed, Colombia." (n.d.). https://satoyamainitiative.org/case_studies/conservation-and-restoration-actions-in-a-dry-forest-ecosystem-in-the-santa-rosa-watershed-colombia/.

Silviculture and Management of Planted Forests

Planted forests are established by planting or deliberately seeding native or introduced species. This can occur on previously unforested land (afforestation) or on land that was once forested (reforestation). Planted forests play a key role in restoring and rehabilitating fragile ecosystems, combating desertification, reducing erosion, and absorbing wastewater. They also contribute to soil and water conservation, helping to maintain environmental stability. Additionally, fast-growing forests efficiently produce wood fuel as a by-product of other forest plantations or through dedicated bioenergy plantations. Beyond these benefits, planted forests act as significant carbon sinks, making them an effective tool for climate change mitigation by capturing and storing atmospheric carbon dioxide.[33]

Silviculture in Natural Forests

Natural forest silviculture involves managing the establishment, growth, composition, health, and quality of natural forests to meet various ecological, economic, and social objectives. This practice involves targeted interventions to preserve or enhance forest functions, including biodiversity conservation, wood and forest product production, recreation, and environmental services.

Silvicultural decisions in natural forests are made at different levels, outlined in Table 9.3.[34]

Land-Use Planning

Land-use planning is the process of allocating land for different uses in a way that balances economic, social, and environmental priorities. Its goal is to determine the optimal combination of land uses within a landscape to meet stakeholder needs while ensuring sustainable resource management for the future. Effective land-use planning guides land-use activities, promotes user synergies, and helps manage competing demands. In forestry, it involves assessing the potential of forestland, evaluating the suitability of various land uses, and considering economic, social, and environmental factors to identify the most sustainable land-use options in forested or partly forested areas.[35]

Table 9.3 Decision-Levels of Silviculture in Natural Forests

Decision-Level	*Description*
1. Silviculture system	The overarching approach for managing forest crops, including how stands are established, maintained, and replaced over time – resulting in forest stands with specific structural characteristics.
2. Silviculture treatment	A long-term plan comprising a sequence of silvicultural actions applied over the entire or part of a forest rotation to achieve desired stand conditions.
3. Silviculture operations	Site-specific activities such as canopy thinning, tree harvesting, planting, or selective pruning, implemented to meet defined management goals such as regeneration, growth enhancement, or timber improvement.

Table 9.4 Nature-Based Solutions to Protect Dryland Soils

Measure	*Description*
Dune fixation	Dune fixation reduces sand and dust erosion using natural vegetation or mechanical barriers. It includes: 1. Primary fixation: mechanically stabilizing sand by slowing movement, e.g. with barriers or vegetative debris. 2. Biological fixation: establishing permanent vegetation cover through seeding grasses or planting trees.
Conserving soil fertility	Soil fertility can be maintained by integrating leguminous species, applying compost, and avoiding practices like burning that deplete soil nutrients.
Controlling erosion caused by rainfall	Erosion control includes reducing runoff speed and volume via increased vegetation cover, vegetated contour bunds (biological method), or mechanical cross-slope barriers.
Using vegetative strips in farmland	Strips of unploughed land planted with trees, shrubs, or grasses serve as cross-slope barriers, improving water retention, preventing erosion, enriching soil nutrients, and enhancing on-farm biodiversity.
Sustainable grazing management	Managing livestock pressure includes rotating grazing zones, restricting water access in overgrazed areas, and empowering community institutions to oversee grazing rules and enforcement.

Drylands

Drylands are crucial in global food production, offering pharmaceutical resources, raw materials, cultural and aesthetic value, and economic opportunities through ecotourism.[36] However, soil erosion – driven by wind and water – is a significant threat to drylands worldwide, making soil protection essential for maintaining their productivity and ecosystem health. A range of NBS can help prevent further erosion and enhance soil resilience in dryland environments. Table 9.4 provides a summary of these solutions.[37]

Case 9.3: Enhancing Dryland Resilience through the Drylands Development Program

The Drylands Development Programme (DryDev) was a six-year initiative implemented from August 2013 to July 2019 across Burkina Faso, Ethiopia, Kenya, Mali, and Niger. With a budget of USD 50 million, the program was funded by the Ministry of Foreign Affairs of the Netherlands, with additional contributions from World Vision Australia. World Agroforestry (ICRAF) served as the overall implementing agency, working with five National Lead Organizations (NLOs) and 13 Implementation Partners (IPs) to support over 227,000 farmers in transitioning from subsistence farming to sustainable rural development. DryDev operated through a dual approach, providing direct support to farmers while also working to influence policy and institutional frameworks. Eight Work Packages structured the implementation, addressing sub-catchment-level natural resource management, on-farm water and soil management, agricultural commodity production, market access, financial services, local governance, monitoring and evaluation, and policy analysis. Through these components, the program promoted improved soil and water conservation, diversified agricultural production, and strengthened local governance structures. Farmers received training in agroforestry, water harvesting, and conservation agriculture to enhance food security and resilience to climate variability. Market access and financial inclusion were key areas of intervention, enabling farmers to integrate into value chains and access essential financial services, including credit and insurance. Local governance mechanisms were strengthened to ensure sustainable management of natural resources. DryDev worked with policymakers to integrate lessons from implementation into broader policy frameworks. World Agroforestry (ICRAF), which merged with the Center for International Forestry Research (CIFOR) in 2019, played a central role in managing partnerships and ensuring evidence-based approaches guided decision-making.[38]

World Agroforestry. "Drylands Development Programe (Drydev)." https://worldagroforestry.org/project/drylands-development-programe-drydev.

Grasslands

Grassland restoration has become increasingly important in areas where native grasslands have been lost due to land-use changes, fragmentation, and biodiversity loss. Restoration efforts often focus on formerly cultivated lands, where native grassland plant and animal species are reintroduced to restore ecological balance. Additional restoration measures include reducing woody plant encroachment, removing invasive species and replacing them with native grasses, and implementing sustainable pasture management that considers seasonal cycles to prevent overgrazing. Restoring grasslands enhances key ecosystem services, such as:

- Maintaining water quantity and quality for rivers, streams, and aquifers
- Providing high-quality forage for livestock
- Enhancing nutrient cycling and carbon sequestration
- Supporting pollination services essential for biodiversity
- Sustaining traditional livelihoods such as thatching and weaving
- Preserving medicinal and food plant resources
- Protecting cultural, heritage, and recreational sites, many of which have tourism value
- Improving deep, nutrient-rich soils, which contribute to long-term ecosystem health[39,40,41,42,43]

Case 9.4: Conservation of Natural Temperate Grassland in Crace, Australian Capital Territory, Australia

In late 2024, the Australian Capital Territory (ACT) Government established a new conservation zone in Crace to protect the region's remaining Natural Temperate Grassland. This initiative followed conservation mapping that identified the area between the Gungaderra Grasslands Nature Reserve and the bicycle path on Drooka Crescent as critical for biodiversity. The conservation zone plays a vital role in preserving habitat for several threatened species, including the Striped Legless Lizard, Superb Parrot, Gang-gang Cockatoo, and Box Gum Woodland. To enhance ecological resilience, the management plan includes a restricted mowing schedule and intensified control of invasive plants. These measures aim to maintain the habitat's ecological integrity while ensuring the survival of its native species. Despite its conservation status, the new zone remains accessible for public recreation. Pathways and connections to the existing nature reserve remain unchanged, allowing the community to enjoy the area while ensuring minimal disturbance to the ecosystem. Natural Temperate Grasslands in the ACT are critically endangered, with less than 5% of their original extent remaining. These ecosystems face ongoing threats from urban development, weed invasion, and poor land management practices. The Crace conservation zone aligns with broader conservation efforts under the Environment Protection and Biodiversity Conservation Act 1999 and the Nature Conservation Act 2014, reinforcing the government's commitment to protecting these ecosystems. This initiative is part of a broader strategy to enhance the condition and functionality of Natural

Temperate Grasslands through ecological restoration, improved land management, and increased community awareness of their ecological importance.[44]

ACT Government. "Natural Temperate Grassland." (n.d.). https://act.gov.au/environment/animals-and-plants/act-threatened-species/natural-temperate-grassland.

Freshwater

NBS for water resources management utilizes ecosystem services to enhance water availability, improve water quality, and strengthen resilience to climate change. These approaches provide multiple benefits, including provisioning services such as drinking water and fisheries, regulating services like flood and drought management, and cultural services that support recreation and ecotourism. Different NBS strategies can be applied to manage water resources effectively, integrating natural processes into water conservation, filtration, and flood mitigation efforts while promoting sustainable land use and biodiversity conservation.[45,46,47]

Riparian Buffers

Riparian buffers, such as wetlands and forests, act as natural barriers that protect waterways from pollutants originating from agricultural runoff, industrial activities, and urban development. While these buffers often occur naturally, they are increasingly degraded due to urbanization and changes in land-use patterns. Restoring riparian buffers involves re-establishing vegetation through carefully planned projects that select appropriate plant species and recreate habitat conditions for wildlife. Restoration efforts should also consider the existing uses of waterways, such as fishing, water collection, and recreation, and ensure that these activities are preserved through community consultation.[48] When implemented across a watershed, riparian buffers provide multiple environmental benefits, including maintaining stream base flows and creating an interconnected network of habitats. Riparian buffers can enhance wildlife diversity by preserving these habitat corridors, particularly in urban areas, while offering recreational opportunities such as cycling and walking trails.[49,50]

Floodplains

A floodplain is a naturally occurring area adjacent to a river that serves as a storage space for floodwater and rainwater, helping to regulate water flow and reduce flood risks. Floodplain soils are typically highly fertile, making them attractive for agriculture, which often leads to their drainage and conversion for farming purposes. In their natural state, floodplains are biodiversity hotspots, supporting rare species adapted to periodic flooding. However, many floodplains have been separated from rivers by dikes, berms, and other flood-control structures or buried under legacy sediment, leading to widespread loss of floodplains due to land drainage, urban expansion, and river channelization. Floodplain restoration seeks to restore retention capacity and ecosystem functions by reconnecting floodplains to rivers. Restoration efforts may involve modifying river channels, removing legacy sediment, creating lakes or ponds, adapting agricultural practices, afforestation, planting native grasses, shrubs, and trees, establishing wetlands, removing invasive species, and installing riparian buffers. Beyond flood retention, floodplains provide numerous ecosystem services, including improving river water quality by filtering out nutrients. They also contribute to climate change mitigation, as their soils and vegetation store carbon, reducing atmospheric carbon dioxide levels.[51,52]

Case 9.5: Restoring the Sacramento River's Riparian Habitat

The Sacramento River, stretching 384 miles, supplies 35% of California's developed water while supporting critical ecosystems. However, decades of agricultural expansion, water diversions, and flood control infrastructure have significantly degraded its riparian habitats. With 90% of forests cleared and levees restricting the river's natural course, fish and bird populations have declined, and the ecological functions of the river have been compromised. The Nature Conservancy (TNC) has led extensive efforts to restore the river's floodplains, using scientific research to demonstrate both ecological and economic benefits. Studies have revealed that riparian restoration supports special-status species, such as the western yellow-billed cuckoo, while reducing regulatory constraints for farmers. Research also dispelled concerns that restored habitats would increase agricultural pests and highlighted how reconnected floodplains provide natural flood mitigation, improving water infiltration and soil health. A key strategy has been the construction of setback levees, allowing the river to reclaim sections of its historical floodplain. This approach reduces flood risks for nearby communities, enhances groundwater recharge, and expands habitat connectivity. More than 6,000 acres have been restored, creating habitats for migratory birds, bats, and aquatic species such as salmon. Restoration has also boosted recreation opportunities, benefiting local economies. By applying the "Ecological Flows Tool" to assess the impacts of river flow and tracking long-term progress, TNC continues to refine its conservation strategies.[53,54]

The Nature Conservancy. "Restoring California's Largest River." (n.d.). https://scienceforconservation.org/science-in-action/restoring-californias-largest-river.

U.S. Fish and Wildlife Service. "Revitalizing California's Floodplains Benefits People and Wildlife." (n.d.). https://fws.gov/story/2021-04/benefits-floodplain-restoration.

Lakes

A lake is a natural or artificial water retention system that helps store water for flood control and supports various uses, including water supply, irrigation, fisheries, and tourism. Lakes also function as carbon sinks and provide critical habitats for many plants and animals. However, many lakes have been drained for agricultural purposes or have become silted due to a lack of maintenance, thereby reducing their ecological and functional capacity. Lake restoration involves rebuilding their structure and functionality, particularly in areas where they were previously drained, to restore water storage, improve ecosystem health, and enhance biodiversity.[55]

Wetlands

Wetland restoration is restoring a wetland's health and function, enabling it to provide essential ecosystem services once again. This involves re-establishing its natural extent and ecological processes by removing non-wetland features and recreating lost wetland areas. Restoration efforts may include excavation, removal of hydraulic structures, clearing invasive species, redirecting polluted runoff, and controlled re-flooding.[56] Restoring wetlands helps improve water quality, providing clean water for ecosystems, drinking supplies, and the habitats of wildlife. Wetlands also play a crucial role in mitigating droughts and controlling floods, as they store large volumes of water and release it gradually, thereby reducing flood peaks and sustaining water availability during dry periods. Additionally, wetlands support microorganisms that filter pollutants, such as nitrogen and phosphorus, by trapping them in soils or absorbing them through wetland plants and organisms.[57] Given their multifunctional nature, wetland restoration also brings social and economic benefits, including opportunities for recreation, fishing, and harvesting reeds, as well as fodder and fruits.[58]

Case 9.6: Comprehensive Everglades Restoration Plan

The Everglades, a vital wetland ecosystem spanning from central Florida to Florida Bay, has experienced significant degradation due to urban expansion, agricultural development, and extensive flood control measures. Over the past century, the region has lost more than half of its original size, with over 70% of its water flow redirected or diminished. This has led to the decline of native wildlife populations and disrupted the natural water cycle that sustains the ecosystem. To address these challenges, the Comprehensive Everglades Restoration Plan (CERP) was introduced in 2000 as a multi-billion-dollar initiative aimed at restoring the ecosystem's hydrology. Managed through a 50/50 partnership between the South Florida Water Management District and the U.S. Army Corps of Engineers, CERP encompasses 68 projects designed to improve water storage, distribution, and quality. Among these, the C-43 West Basin Storage Reservoir and the Everglades Agricultural Area (EAA) Reservoir play pivotal roles. The C-43 Reservoir, set for completion in 2025, spans an area equivalent to the City of Naples and will store 170,000 acre-feet of water. This reservoir is designed to regulate water availability for the Caloosahatchee River, mitigating extreme fluctuations between wet and dry seasons. Meanwhile, the EAA Reservoir, a 10,500-acre site south of Lake Okeechobee, is projected to store 240,000 acre-feet of water and treat nutrient-laden runoff before it flows into the Everglades. Its stormwater treatment area, spanning 6,500 acres, utilizes aquatic vegetation to remove pollutants, thereby improving water quality before it reaches the wetlands. The Picayune Strand Restoration Project, a key CERP initiative, is focused on reversing the hydrological damage caused by failed development in eastern Collier County. This project has made significant progress, with 90% of its restoration work completed. Remaining efforts include the construction of the Southwest Protection Feature, a levee and canal system designed to prevent flooding in adjacent agricultural areas. Until this final component is in place, only partial ecological benefits can be realized. Efforts under CERP continue to advance, with ongoing infrastructure projects designed to restore water flow, reduce pollution, and enhance ecosystem resilience.[59]

Conservancy of Southwest Florida. "Everglades Restoration." (n.d.). https://conservancy.org/everglades-restoration/#:~:text=In%202000%2C%20Congress%20approved%20the,ecosystems%20of%20the%20Greater%20Everglades.

Managed Aquifer Recharge

Managed aquifer recharge is designed to enhance natural groundwater storage and increase water availability during periods of low flow and high seasonal variability. It involves intentionally recharging aquifers during wet periods to replenish water supplies for future use. The socioeconomic benefits of managed aquifer recharge include enhancing water supply resilience, reducing evaporation-related water losses, facilitating access to alternative water sources, and increasing freshwater availability for households and economic activities. The environmental benefits include maintaining healthy water flows, preventing over-extraction and degradation of water sources, sustaining groundwater-dependent ecosystems such as wetlands, reducing saltwater intrusion and land subsidence, and improving water quality through natural filtration.[60]

Terrestrial Protected Areas

Protected areas include national parks, nature reserves, wilderness areas, and indigenous and community-conserved areas, all of which have clearly defined geographical boundaries and are legally or effectively managed to ensure the long-term conservation of nature, ecosystem services, and cultural values. However, even well-managed protected areas face threats like previous environmental degradation, climate change, invasive species, and changes in surrounding landscapes and seascapes. Illegal encroachment, poaching, and weak management can also lead to severe ecosystem degradation. Restoration efforts within protected areas may be undertaken for various reasons, including recovering individual species, enhancing ecosystem connectivity at a landscape or seascape scale, improving visitor experiences, and restoring critical ecosystem services. Successful restoration requires collaboration between protected area managers, stakeholders, and partners within and beyond protected area boundaries to ensure effective and sustainable conservation outcomes.[61]

Coastal and Marine Systems

Coastal and marine ecosystems have suffered significant degradation, which has negatively impacted biodiversity and the livelihoods of local communities. However, restoration is feasible for key ecosystems such as sand dunes, beaches, mangrove forests, and coral reefs. These ecosystems serve as natural coastal defenses, reducing the impacts of tidal and storm surges by dissipating wave energy, breaking offshore waves, and minimizing inland flooding. They also provide diverse habitats for wildlife, support both wild and farmed seafood fisheries, and sustain recreational and commercial fishing industries. Beyond their ecological functions, coastal and marine ecosystems hold cultural significance, promote human well-being, support tourism, and contribute to climate regulation by absorbing greenhouse gases.[62,63]

Sand Dunes and Beaches

Human activities often deplete sand dunes, reducing their ability to buffer coastal storms and flooding. Beach erosion is driven by river damming, which limits sediment flow to estuaries and shorelines, as well as by urbanization and agricultural practices. Dune restoration can involve trapping sand with fences or stabilizing dunes by planting vegetation. Native plants can be transplanted from intact dune systems or sourced from nurseries. In some cases, artificial dunes are constructed to enhance coastal defenses.[64] Beach nourishment is another strategy involving pumping or transporting sand onto eroded beaches to prevent shoreline loss, protect beachfront structures, and enhance recreational areas. While dune building is often integrated into beach nourishment projects, it can also be implemented separately. Along coastlines, sand dunes and beaches typically sit at higher elevations than inland areas, acting as natural barriers against flooding and wave energy. Dunes also play a key role in coastal sediment replenishment, as wind and tides redistribute sand to surrounding beaches, helping to reduce erosion and maintain coastal stability.[65]

Mangrove Restoration

Mangrove restoration is a cost-effective strategy for strengthening coastal defenses against natural hazards while enhancing human well-being and livelihoods. However, restoration success is often measured by the number of seedlings planted or short-term survival rates, which can be misleading. Many projects show initial high survival rates but experience long-term failures once monitoring ends. Several recurring factors contribute to unsuccessful mangrove restoration, including:

- Lack of community involvement or support for conservation due to socio-economic challenges, such as reliance on aquaculture, leading to mangrove areas being converted into fish or shrimp ponds.
- Planting monocultures results in non-functional, low-resilience mangroves with limited ecological benefits.
- Incorrect species selection can lead to high mortality or slow growth, especially when plants are planted in areas exposed to waves and erosion.
- Blocking natural sediment and water flow can disrupt the recovery of larger-scale ecosystems.
- Planting in unsuitable locations, such as open intertidal mudflats, where mangroves did not naturally exist.[66]

Successful Restoration

Successful mangrove restoration should establish a large, diverse, functional, and self-sustaining mangrove forest that provides multiple benefits. These include reducing coastal storm and cyclone risks, enhancing water security by preventing saltwater intrusion, supporting food security by restoring fish nurseries and improving energy security by increasing fuelwood availability.

To ensure effective mangrove restoration, two key principles should be followed:

1 *Ensure biophysical conditions are suitable*: Mangrove loss or degradation may be linked to land-use changes, disruptions to freshwater supply, sediment loss, or coastal infrastructure development. Sometimes, the original site conditions may no longer support the regrowth of mangroves. Restoration can only be successful if the necessary biophysical conditions – such as water flow, sediment availability, and salinity levels – are restored to support healthy mangrove regeneration.
2 *Ensure socio-economic conditions support long-term recovery*: If mangroves were removed in the past due to human activities, they are at risk of being cleared again unless local communities see value in their conservation. Sustainable economic activities that benefit from restored mangroves should be developed. Additionally, land ownership and use rights must be established, and local communities and governments must be actively involved to ensure long-term protection and sustainable management.[67]

Coastal Saltmarsh

Coastal saltmarshes have historically been undervalued, often regarded as boggy swamps or wastelands. As a result, many have been drained, reclaimed, or degraded due to human activities, leading to significant habitat loss. However, saltmarshes play critical ecological roles, including providing habitat and shelter for fish, birds, and invertebrates, serving as a food source for various species, acting as a natural buffer against shoreline erosion, improving water quality, and functioning as carbon sinks that store large amounts of carbon in both plant biomass and sediments. Saltmarsh restoration can be achieved through:

- Restoring tidal flow by removing barriers such as dykes, levees, and poorly designed water-control structures that restrict natural tidal hydrodynamics.
- Removing invasive species can increase marsh elevation through the accumulation of excessive sediment and biomass, thereby reducing tidal flooding and altering the marsh's function.
- Excavating dredged material from marshes impacted by fill allows for the re-establishment of historical intertidal elevations needed for a healthy salt marsh ecosystem.[68,69]

Case 9.7: Collaborative Protection of Coastal Saltmarsh at Corner Inlet, Victoria, Australia

Saltmarsh and intertidal habitats along the Corner Inlet and Nooramunga coastlines support a variety of threatened bird species. However, since 1750, these ecosystems have undergone significant decline, with a 45%

loss of saltmarsh, mangroves, and estuarine wetlands. Currently, 8,200 hectares of saltmarsh and associated vegetation remain, 1,100 hectares of which are located on private land. Both climate-induced pressures, such as sea level rise, and property-scale activities, including livestock grazing, pose threats to these areas. Over the past 15 years, targeted protection efforts have been led under the Corner Inlet Connections initiative, supported by the West Gippsland Catchment Management Authority (WGCMA). These efforts have included community engagement, on-the-ground conservation actions, and collaboration with private landholders. Trust for Nature has played a crucial role in supporting landholders in establishing Conservation Covenants on private property, thereby ensuring long-term protection of habitats. Conservation Covenants are legally binding agreements registered on property titles, preserving ecological values while maintaining private ownership of the land. In 2019, the Trust for Nature conducted a desktop study identifying 139 private properties with saltmarshes. Of these, 66 were prioritized based on their conservation potential. Currently, four properties with saltmarsh habitats are already protected under covenant, aiming to increase coverage from 70 to 250 hectares over the next three years. On-ground habitat enhancement and a monitoring program are underway, alongside projects by WGCMA, private landholders, the Gunaikurnai Land and Waters Aboriginal Corporation, and philanthropic partners. This initiative is supported by funding from the Australian Government's National Landcare Program.[70]

Commissioner for Environmental Sustainability Victoria. "Collaborating to Protect Coastal Saltmarsh at Corner Inlet." (n.d.). https://ces.vic.gov.au/state-of-reports/state-marine-and-coastal-environment-2021-report/stories/collaborating-protect.

Seagrass

Seagrass restoration aims to transform unvegetated seafloors into diverse, vegetated ecosystems, increasing structural complexity and ecological function. Seagrasses evolved approximately 100 million years ago and are flowering marine plants characterized by their roots, stems, leaves, and seed production. There are approximately 72 species classified into four major groups, which can form dense underwater meadows; some of these meadows are visible from space. As one of the world's most productive ecosystems, seagrasses provide food and shelter for many species, including invertebrates, large fish, crabs, turtles, marine mammals, and birds. Beyond their ecological importance, they offer valuable services to humans. Seagrass meadows reduce bacteria levels harmful to humans and marine life by up to 50%, improving water quality. Additionally, seagrasses contribute to sustainable aquaculture, supporting the growing demand for food as global populations increase.[71,72,73]

Seagrass restoration can be achieved through natural recolonization or proactive restoration methods. Encouraging natural recolonization relies on improving surface water quality, assuming that once conditions are suitable, seagrasses will regenerate naturally. This requires a long-term effort involving municipal sewage system upgrades and programs to identify and reduce pollution sources in coastal areas, including those from industrial, residential, and agricultural runoff.[74]

Proactive restoration involves transplanting seagrass from healthy donor beds or using seedlings cultivated in laboratory conditions. This method involves relocating viable seedlings or mature plants to restoration sites where conditions have been optimized for their survival and growth. However, transplantation is labor-intensive, as divers must plant individual seagrass units by hand. Another restoration technique involves seagrass propagation through direct seed application in estuarine waters, providing a less invasive method for re-establishing seagrass meadows.[75]

Case 9.8: Eelgrass Revival in Burtons Bay, Virginia

The Submerged Aquatic Vegetation Program at the Virginia Institute of Marine Science has completed a significant eelgrass restoration initiative in Burtons Bay on Virginia's Eastern Shore. Funded by a $2,259,633 grant from the Bipartisan Infrastructure Law's Coastal Zone Management Habitat Protection and Restoration Awards, the project initially aimed to restore 60 acres of eelgrass and introduce six million scallops to the bay over three years. In its first year, however, the project team exceeded expectations by planting 800,000 eelgrass seeds across 80 acres, setting a new state record and surpassing the initial target of 15 acres. As the first project completed under this particular funding stream, the restoration in Burtons Bay builds upon earlier work in South Bay, where more than 10,000 acres of eelgrass have been restored. These established meadows have played a crucial role in enhancing water quality, mitigating shoreline erosion, and providing a natural buffer against storm surges. They also provide habitat for marine species, support commercial and recreational fisheries, and contribute to local ecotourism economies. Partners in the project include the National Oceanic and Atmospheric Administration, The Nature Conservancy, and the Coastal and Estuarine Ecology Lab at the Virginia Institute of Marine Science. Seed distribution was led by team members aboard boats, manually broadcasting eelgrass seeds over targeted restoration areas. This hands-on approach reflects the collaborative nature of the effort and the technical expertise required to select optimal planting sites. Restoration activities are expected to continue, with long-term monitoring planned to assess ecological outcomes.[76]

NOAA. "Record-Breaking Eelgrass Restoration in Virginia's Eastern Shore." (n.d.). https://coast.noaa.gov/states/stories/eelgrass-restoration.html.

Coral Reefs

Coral reef restoration involves growing coral fragments in nurseries until they are suitable for outplanting onto reefs. These fragments can be attached directly to the reef using epoxy, cement, or ties or secured to platforms that are then fixed to the reef. An alternative approach, larval seeding, involves releasing large quantities of coral larvae onto reefs or settling them onto structures before placing them within the reef. However, success rates for

Table 9.5 Examples of Ecosystem Services Coral Reefs Provide

Category	*Ecosystem Service*	*Examples*
Supporting	Biodiversity benefit	• Coral reefs are among the most biodiverse ecosystems globally. • Their species richness supports genetic diversity critical for resilience and adaptation.
	Habitat	• Coral structures create habitat complexity, offering refuge and shelter. • Reefs support species across various life stages. • Corals interact with adjacent ecosystems like seagrasses and mangroves.
Regulating	Coastal protection	• Reefs absorb up to 97% of wave energy, protecting coastal communities. • They reduce storm damage and erosion risks along shorelines.
	Water quality and biogeochemical cycling	• Coral reefs disperse sediment loads affected by land reclamation. • Coral mucus transfers energy to nearby ecosystems. • Sponges play a key role in nutrient cycling across trophic levels.
Provisioning	Fishery	• Reefs support diverse fisheries vital to coastal nutrition and livelihoods. • Provide subsistence and commercial income. • Fishing contributes to cultural identity, social prestige, and a valued lifestyle.
Cultural	Cultural	• Coral reefs hold value for tourism (diving, snorkeling, reef tours). • Support coastal attractions such as calm lagoons, beaches, and seafood cuisine. • Source of marine compounds with pharmaceutical and biotechnological potential.

larval seeding remain low, making fragment outplanting the most widely used method in coral restoration efforts. The primary goals of coral reef restoration include re-establishing foundational coral species, increasing live coral biomass, enhancing biodiversity, improving reef structural complexity, providing habitat for marine life, strengthening coastal protection, and improving connectivity between coral colonies. Coral reefs provide significant ecological and economic benefits, as illustrated in Table 9.5.[77,78,79]

Case 9.9: Coral Restoration in Seychelles

In response to the severe coral bleaching events of 1998 and 2016, which drastically reduced coral cover in Seychelles, Nature Seychelles launched the Reef Rescuers (RR) project in 2010. The initiative aimed to accelerate reef recovery through active restoration methods, particularly "coral gardening," a technique that involves cultivating coral fragments in underwater nurseries before transplanting them to degraded sites. Supported by USAID, the Adaptation Fund, and various partners, the project has pioneered large-scale reef restoration in the region. Between 2011 and 2015, RR established 12 underwater nurseries within the Cousin Island Special Reserve, cultivating 40,000 coral fragments across 34 species. Over 24,000 corals were subsequently transplanted, covering 0.52 hectares of degraded reef. Following the 2016 bleaching event, which reduced coral cover in restored areas from 40% to less than 10%, the project expanded to include additional restoration techniques. By 2022, RR had deployed four mid-water floating rope nurseries, stocked with over 8,000 corals from the Acropora and Pocillopora genera, and outplanted more than 3,000 coral colonies using cementing and metal structures. The project integrates scientific research, training, and community engagement, collaborating with universities to enhance restoration efficiency. Recent efforts have included assessing artificial reef structures and testing new stocking methods to enhance survival rates. Plans involve establishing a large-scale land-based nursery capable of producing 10,000 coral colonies annually. Ongoing monitoring and refinement of restoration techniques aim to ensure the long-term resilience of the Seychelles' reefs, supporting biodiversity and safeguarding the economic benefits tied to marine ecosystems.[80]

The Nature Conservancy. "Seychelles – Coral Restoration." (n.d.). https://reefresilience.org/case-studies/seychelles-coral-restoration/.

Marine Protected Areas

Marine Protected Areas (MPAs) preserve ecological functions and marine landscapes. They can be designed to protect entire ecosystems or specific habitats, such as seamounts, and may also safeguard particular species, including marine mammals and seabirds. MPAs vary in their levels of protection. Fully protected MPAs prohibit all potentially harmful activities, such as fishing, mining, aquaculture, and dredging. Partially protected MPAs, on the other hand, may regulate activities, for example, by banning dredging and trawling to prevent damage to sensitive seabed habitats. The benefits of MPAs include biodiversity conservation, particularly for critical habitats and threatened species, protection of natural areas that support sustainable tourism, increased fisheries productivity, and enhanced food security and fisheries management. Several factors and criteria must be considered when designating an MPA or defining its boundaries. Table 9.6 provides a summary of these considerations.[81,82,83,84,85]

Table 9.6 Criteria for Selecting Marine Protected Areas

Criteria	*Description*
Biogeographic	• Area represents rare biogeographic types or features • Presence of unique geological formations or characteristics
Ecological	• Supports essential ecological processes or life-support systems • High ecosystem integrity, potentially part of a larger, connected system • Diverse habitats, including areas for rare/endangered species • Critical zones such as nurseries, feeding, breeding, or resting grounds • Unique or rare habitats and genetic diversity among species
Naturalness	• Area remains relatively undisturbed or minimally affected by human activities, maintaining its natural condition
Economic importance	• Offers existing or potential economic value through activities like ecotourism, traditional uses, fisheries, or as breeding grounds for commercially important species
Social importance	• Holds cultural, traditional, aesthetic, educational, recreational, or historical value to local, national, or international communities
Scientific importance	• High potential for use in scientific research and long-term environmental monitoring
International or national significance	• Area is nationally or internationally designated, or holds strong potential for inclusion in official protected area systems
Practicality or feasibility	• Logistically viable: accessible, manageable, compatible with current land or sea use • Supported socially and politically, including by local communities • Insulated from external threats where possible
Duality or replication	• Replication of key ecosystems is encouraged, especially for smaller MPAs, to reduce vulnerability to disturbance and ensure representativeness across regions

Case 9.10: Isla Ají Marine Protected Area Safeguards Coastal Ecosystems

The Colombian government has designated Isla Ají as a new MPA, spanning 24,600 hectares (246 square kilometers). Located in the Delta of the Naya River on the central Pacific coast, the MPA includes tropical wet forests, mangrove forests, beaches, mudflats, and tidal wetlands. These ecosystems serve as critical blue carbon sinks while providing essential habitat for threatened species, including the critically endangered largetooth sawfish, humpback whales, migratory shorebirds, and sea turtles such as the green sea turtle and the olive ridley sea turtle. The MPA also plays a crucial role in sustaining local fisheries, which are vital to the livelihoods of Afro-Colombian coastal communities. Isla Ají serves as a nursery ground for commercially important fish species, including Pacific sierra and armed snook, as well as Pacific blue crabs and ark clams. The designation process was led by the Afro-Colombian community council of the Naya River, with support from the Regional Environmental Authority of Valle del Cauca and the Wildlife Conservation Society (WCS). The initiative included participatory zoning, governance development, and strategic planning to address threats such as mangrove deforestation, solid waste mismanagement, and unsustainable coastal development. The establishment of Isla Ají contributes to Colombia's conservation goals, aligning with the Convention on Biological Diversity targets and the global 30 × 30 initiative. The project received financial backing from the WCS MPA Fund, Arcadia, and the Santo Domingo Foundation, with additional support from the Bezos Earth Fund to advance conservation efforts in the region.[86]

Wildlife Conservation Society. "New Marine Protected Area Isla Ají Safeguards Important Coastal Ecosystems (English and Spanish)." (n.d.). https://newsroom.wcs.org/News-Releases/articleType/ArticleView/articleId/17153/New-Marine-Protected-Area-Isla-Aji-Safeguards-Important-Coastal-Ecosystems-English-and-Spanish.aspx#:~:text=The%20Isla%20Ají%20MPA%20provides,and%20many%20other%20threatened%20species.

Case Summary

Various NBS approaches have been implemented across terrestrial and marine habitats to restore degraded ecosystems (Table 9.7).

Table 9.7 Restoring Degraded Ecosystems

Habitat	*Project Name*	*Activities*	*Co-Benefits*
Agriculture	Minnesota Nutrient Management Initiative	Uses on-farm field trials to test nitrogen application strategies, improve nutrient efficiency, and reduce nitrate runoff. Includes Field Trials and Advanced Rate Trials with GPS-based variable rate technology.	Reduces agricultural runoff, enhances productivity, informs nitrogen guidelines, and promotes sustainable farming practices.
Forests	Dry Forest Restoration–Santa Rosa Watershed, Colombia	Combines forest protection with silvopastoral systems and mist-catchment installations. Implements community workshops and conservation agreements for long-term ecosystem restoration.	Improves soil and water retention, halts biodiversity loss, and promotes sustainable rural livelihoods.
Drylands	Drylands Development Programme (DryDev)	Supports over 227,000 farmers in five African countries with agroforestry, soil and water conservation, market access, and financial inclusion. Operates through eight work packages and policy engagement.	Enhances food security, builds resilience to climate variability, and improves local governance and value chain participation.
Grasslands	Natural Temperate Grassland Conservation – Crace, Australia	Establishes a conservation zone in Crace to protect endangered grasslands and habitat for threatened species. Includes invasive species control, restricted mowing, and public access pathways.	Protects biodiversity, preserves habitat connectivity, and raises awareness of ecological value.
Riparian zones	Sacramento River Riparian Habitat Restoration	Reconstructs floodplains through setback levees and vegetation restoration. Applies ecological tools to evaluate river flow and benefits to agriculture and biodiversity.	Improves flood mitigation, water infiltration, species habitat, and recreational access.
Wetlands	Comprehensive Everglades Restoration Plan	Restores hydrology across the Everglades through reservoir construction, water treatment areas, and canal systems. Includes over 68 projects under a U.S. federal-state partnership.	Improves water storage and quality, restores ecosystem flow, and enhances resilience of native species and communities.

(*Continued*)

Table 9.7 (Continued)

Habitat	*Project Name*	*Activities*	*Co-Benefits*
Saltmarsh	Corner Inlet Saltmarsh Protection – Victoria,Australia	Protects saltmarsh on public and private land using Conservation Covenants. Engages local landholders and First Nations through land agreements and monitoring.	Secures habitat for threatened species, improves ecological integrity, and builds private-public partnerships for conservation.
Seagrass	Eelgrass Revival–Burtons Bay, Virginia	Restores eelgrass across 80 acres using manual seeding and long-term monitoring. Builds on prior South Bay restoration and sets state restoration records.	Improves water quality, reduces erosion, enhances habitat for fisheries, and supports coastal economies.
Coral reefs	Coral Restoration – Seychelles	Implements coral gardening using underwater nurseries, cementing, and rope-based methods. Expands research and develops land-based coral propagation nursery.	Restores reef biodiversity, builds restoration capacity, and enhances reef resilience to climate change.
Marine protected areas	Isla Ají MPA–Colombia	Designates a 24,600-hectare MPA with participatory zoning and governance for mangroves, wetlands, and nursery grounds. Led by Afro-Colombian communities with NGO and government support.	Safeguards coastal ecosystems, supports fisheries, empowers communities, and advances biodiversity targets.

Notes

1 European Commission, "Towards an Eu Research and Innovation Policy Agenda for Nature-Based Solutions & Re-Naturing Cities," (2015), https://publications.europa.eu/en/publication-detail/-/publication/fb117980-d5aa-46df-8edc-af367cddc202.

2 Government of El Salvador, "Un Decade of Ecosystem Restoration 2020–2030. Communiqué Issued by El Salvador with the Support of Countries of the Central American Integration System (Sica)," (2018), http://wedocs.unep.org/handle/20.500.11822/26027?show=full.

3 Forest Ecosystem Restoration Initiative, "Partnering with Nature: The Case for Natural Regeneration in Forest and Landscape Restoration," (2017), https://cdn.ymaws.com/www.ser.org/resource/resmgr/custompages/publications/ser_publications/feri_single_page.pdf.

4 Sasha Alexander et al., "The Relationship between Ecological Restoration and the Ecosystem Services Concept," *Ecology and Society* 21, no. 1 (2016).

5 Society for Ecological Restoration International, "The Ser International Primer on Ecological Restoration," (2004), https://cdn.ymaws.com/www.ser.org/resource/resmgr/custompages/publications/SER_Primer/ser_primer.pdf.
6 L. Janishevski et al., "Ecosystem Restoration, Protected Areas and Biodiversity Conservation," *Unasylva* 245, no. 3 (2015).
7 Asa L. Aradottir and Dagmar Hagen, "Chapter Three – Ecological Restoration: Approaches and Impacts on Vegetation, Soils and Society," in *Advances in Agronomy*, ed. Donald L. Sparks (Cambridge, Massachusetts: Academic Press, 2013).
8 T. Elmqvist et al., "Benefits of Restoring Ecosystem Services in Urban Areas," *Current Opinion in Environmental Sustainability* 14 (2015).
9 USDA, "Grazing Management and Soil Health," (2016), https://nrcs.usda.gov/Internet/FSE_DOCUMENTS/nrcseprd621806.pdf.
10 USDA, "Grazing Management and Soil Health."
11 USDA, "Grazing Management and Soil Health."
12 Iowa State University Extension and Outreach, "Managing Pasture to Reduce Soil Erosion," (n.d.), https://crops.extension.iastate.edu/managing-pasture-reduce-soil-erosion.
13 Natural Water Retention Measures, "Meadows and Pastures," (2015), http://nwrm.eu/sites/default/files/nwrm_ressources/a1_-_meadows_and_pastures_0.pdf.
14 Stillwater Sciences, "A Guide for Restoring Functionality to Mountain Meadows of the Sierra Nevada," (2012), http://stillwatersci.com/resources/2012meadowrestguide.pdf.
15 Natural Water Retention Measures, "Meadows and Pastures."
16 Roberta Farina et al., "Chapter 15 – Do Crop Rotations Improve the Adaptation of Agricultural Systems to Climate Change? A Modeling Approach to Predict the Effect of Durum Wheat-Based Rotations on Soil Organic Carbon and Nitrogen," in *Soil Management and Climate Change*, ed. María Á. Muñoz and Raúl Zornoza (Cambridge, Massachusetts: Academic Press, 2018).
17 Natural Water Retention Measures, "Crop Rotation," (2015), http://nwrm.eu/sites/default/files/nwrm_ressources/a3_-_crop_rotation_0.pdf.
18 Abdullah Alhameid et al., "Chapter 9 – Intensified Agroecosystems and Changes in Soil Carbon Dynamics," in *Soil Health and Intensification of Agroecosytems*, ed. Mahdi M. Al-Kaisi and Birl Lowery (Cambridge, Massachusetts: Academic Press, 2017).
19 C.A. Francis, "Crop Rotations," in *Encyclopedia of Soils in the Environment*, ed. Daniel Hillel (Oxford: Elsevier, 2005).
20 Natural Water Retention Measures, "Crop Rotation."
21 Natural Water Retention Measures, "Intercropping," (2015), http://nwrm.eu/sites/default/files/nwrm_ressources/a5_-_intercropping.pdf.
22 Beata Gabryś and Bożena Kordan, "Chapter 18 – Cultural Control and Other Non-Chemical Methods," in *Insect Pests of Potato*, ed. Andrei Alyokhin, Charles Vincent, and Philippe Giordanengo (San Diego, CA: Academic Press, 2013).
23 Natural Water Retention Measures, "Intercropping."
24 Natural Water Retention Measures, "No Till Agriculture," (2015), http://nwrm.eu/sites/default/files/nwrm_ressources/a6_-_no_till_agriculture.pdf.
25 David L. Regehr and Charles A. Norwood, "Chapter 14 – Benefits of Triazine Herbicides in Ecofallow," in *The Triazine Herbicides*, ed. Homer M. LeBaron, Janis E. McFarland, and Orvin C. Burnside (San Diego, CA: Elsevier, 2008).

26 J.T. Sims and P.A. Vadas, "Phosphorus in Soils | Overview," in *Encyclopedia of Soils in the Environment*, ed. Daniel Hillel (Oxford: Elsevier, 2005).
27 Natural Water Retention Measures, "Low Till Agriculture," (2015), http://nwrm.eu/sites/default/files/nwrm_ressources/a7_-_low_till_agriculture_0.pdf.
28 Natural Water Retention Measures, "Green Cover," (2015), http://nwrm.eu/sites/default/files/nwrm_ressources/a8_-_green_cover_0.pdf.
29 Minnesota Department of Agriculture, "Nutrient Management Initiative (Nmi)," (n.d.), https://mda.state.mn.us/protecting/cleanwaterfund/onfarmprojects/nmi#:~:text=The%20Minnesota%20Nutrient%20Management%20Initiative,efficiency%20and%20improved%20water%20quality.
30 FAO, "Forest Restoration and Rehabilitation," (2017), http://fao.org/sustainable-forest-management/toolbox/modules/forest-restoration-and-rehabilitation/basic-knowledge/en/?type=111.
31 FAO, "Forest Restoration and Rehabilitation."
32 Satoyama Initiative, "Conservation and Restoration Actions in a Dry Forest Ecosystem in the Santa Rosa Watershed, Colombia," (n.d.), https://satoyamainitiative.org/case_studies/conservation-and-restoration-actions-in-a-dry-forest-ecosystem-in-the-santa-rosa-watershed-colombia/.
33 FAO, "Silviculture and Management of Planted Forests," (2017), http://fao.org//sustainable-forest-management/toolbox/modules/silviculture-and-management-of-planted-forests/basic-knowledge/en/?type=111.
34 FAO, "Silviculture in Natural Forests," (2017), http://fao.org//sustainable-forest-management/toolbox/modules/silviculture-in-natural-forests/basic-knowledge/en/?type=111http://fao.org//sustainable-forest-management/toolbox/modules/silviculture-in-natural-forests/basic-knowledge/en/?type=111.
35 FAO, "Land-Use Planning," (2017), http://fao.org//sustainable-forest-management/toolbox/modules/land-use-planning/basic-knowledge/en/?type=111.
36 UNEP, "Global Drylands: A Un System-Wide Response," (2011), https://unep-wcmc.org/resources-and-data/global-drylands--a-un-system-wide-response.
37 FAO, "Global Guidelines for the Restoration of Degraded Forests and Landscapes in Drylands. Building Resilience and Benefiting Livelihoods," (2015), http://fao.org/3/a-i5036e.pdf.
38 World Agroforestry, "Drylands Development Programe (Drydev)," https://worldagroforestry.org/project/drylands-development-programe-drydev.
39 John Blair, Jesse Nippert, and John Briggs, "Grassland Ecology," in *Ecology and the Environment*, ed. Russell K. Monson (New York: Springer New York, 2014).
40 Uta Schirpke et al., "Future Impacts of Changing Land-Use and Climate on Ecosystem Services of Mountain Grassland and Their Resilience," *Ecosystem Services* 26 (2017).
41 SANBI, "Grasslands Ecosystem Guidelines: Landscape Interpretation for Planners and Managers," (2013), http://biodiversityadvisor.sanbi.org/wp-content/uploads/2015/06/2013_Grassland-Ecosystem-Guidelines.pdf.
42 SANBI, "Grasslands Ecosystem Guidelines: Landscape Interpretation for Planners and Managers."
43 IUCN, "Drylands and Land Degradation," (2017), https://iucn.org/sites/dev/files/drylands_and_land_degradation_issues_brief_0.pdf.
44 ACT Government, "Natural Temperate Grassland," (n.d.), https://act.gov.au/environment/animals-and-plants/act-threatened-species/natural-temperate-grassland.

45 Camino Liquete et al., "Integrated Valuation of a Nature-Based Solution for Water Pollution Control. Highlighting Hidden Benefits," *Ecosystem Services* 22 (2016).
46 Zahra Kalantari et al., "Nature-Based Solutions for Flood-Drought Risk Mitigation in Vulnerable Urbanizing Parts of East-Africa," *Current Opinion in Environmental Science & Health* 5 (2018).
47 R.C. Brears, *Blue and Green Cities: The Role of Blue-Green Infrastructure in Managing Urban Water Resources* (London, UK: Palgrave Macmillan UK, 2018).
48 WWF, "Natural and Nature-Based Flood Management: A Green Guide," (2017), https://worldwildlife.org/publications/natural-and-nature-based-flood-management-a-green-guide.
49 R.C. Brears, *Climate Resilient Water Resources Management* (Cham, Switzerland: Palgrave Macmillan, 2018).
50 Brears, *Blue and Green Cities: The Role of Blue-Green Infrastructure in Managing Urban Water Resources*.
51 Natural Water Retention Measures, "Floodplain Restoration and Management," (2015), http://nwrm.eu/sites/default/files/nwrm_ressources/n3_-_floodplain_restoration_and_management.pdf.
52 BMU, "Strategic Framework for Setting Priorities for Restoring Degraded Ecosystems in Germany," (2015), https://bmu.de/en/download/strategic-framework-for-setting-priorities-for-restoring-degraded-ecosystems-in-germany/.
53 U.S. Fish and Wildlife Service, "Revitalizing California's Floodplains Benefits People and Wildlife," (n.d.), https://fws.gov/story/2021-04/benefits-floodplain-restoration.
54 The Nature Conservancy, "Restoring California's Largest River," (n.d.), https://scienceforconservation.org/science-in-action/restoring-californias-largest-river.
55 Natural Water Retention Measures, "Lake Restoration," (2015), http://nwrm.eu/sites/default/files/nwrm_ressources/n12_-_lake_restoration_0.pdf.
56 WWF, "Natural and Nature-Based Flood Management: A Green Guide."
57 Brears, *Blue and Green Cities: The Role of Blue-Green Infrastructure in Managing Urban Water Resources*.
58 WWF, "Natural and Nature-Based Flood Management: A Green Guide."
59 Conservancy of Southwest Florida, "Everglades Restoration," (n.d.), https://conservancy.org/everglades-restoration/#:~:text=In%202000%2C%20Congress%20approved%20the,ecosystems%20of%20the%20Greater%20Everglades.
60 Brears, *Climate Resilient Water Resources Management*.
61 IUCN, "Ecological Restoration for Protected Areas: Principles, Guidelines and Best Practices," (2012), https://iucn.org/content/ecological-restoration-protected-areas-principles-guidelines-and-best-practices.
62 Cheruvathoor L. Libini et al., "Chapter 27 – Marine Biodiversity – Strategies for Conservation, Management and Ecological Restoration," in *Biodiversity and Climate Change Adaptation in Tropical Islands*, ed. Chandrakasan Sivaperuman et al. (Cambridge, Massachusetts: Academic Press, 2008).
63 Living with Environmental Change, "New Tools to Improve the Management of the Coastal Environment," (2014), https://nerc.ukri.org/research/partnerships/ride/lwec/ppn/ppn10/.
64 WWF, "Natural and Nature-Based Flood Management: A Green Guide."
65 WWF, "Natural and Nature-Based Flood Management: A Green Guide."
66 Wetlands International, "Mangrove Restoration: To Plant or Not to Plant?," (2018).

67 Society for Ecological Restoration International, "Investing in Our Ecological Infrastructure," (2011), https://cdn.ymaws.com/www.ser.org/resource/resmgr/custompages/publications/ser_publications/economic_rationale_for_resto.pdf.
68 NSW Department of Primary Industries, "Coastal Saltmarsh," (2013), http://dpi.nsw.gov.au/__data/assets/pdf_file/0007/459628/Coastal-Saltmarsh-Primefact.pdf.
69 Coastal Resources Management Council Narragansett Bay Estuary Program, "Restoration Methods," (n.d.), http://edc.uri.edu/restoration/html/tech_sci/restsalt.htm.
70 Commissioner for Environmental Sustainability Victoria, "Collaborating to Protect Coastal Saltmarsh at Corner Inlet," (n.d.), https://ces.vic.gov.au/state-of-reports/state-marine-and-coastal-environment-2021-report/stories/collaborating-protect.
71 Diogo Paulo et al., "Open Coast Seagrass Restoration. Can We Do It? Large Scale Seagrass Transplants," *Frontiers in Marine Science* 6, no. 52 (2019).
72 Smithsonian Ocean, "Seagrass and Seagrass Beds," (n.d.), https://ocean.si.edu/ocean-life/plants-algae/seagrass-and-seagrass-beds.
73 Joleah B. Lamb et al., "Seagrass Ecosystems Reduce Exposure to Bacterial Pathogens of Humans, Fishes, and Invertebrates," *Science* 355, no. 6326 (2017).
74 Coastal Resources Management Council Narragansett Bay Estuary Program, "Restoration Methods," (n.d.), http://edc.uri.edu/restoration/html/tech_sci/restsea.htm.
75 Coastal Resources Management Council Narragansett Bay Estuary Program, "Restoration Methods."
76 NOAA, "Record-Breaking Eelgrass Restoration in Virginia's Eastern Shore," (n.d.), https://coast.noaa.gov/states/stories/eelgrass-restoration.html.
77 Shawna A. Foo and Gregory P. Asner, "Scaling up Coral Reef Restoration Using Remote Sensing Technology," *Frontiers in Marine Science* 6, no. 79 (2019).
78 Mark C. Ladd et al., "Harnessing Ecological Processes to Facilitate Coral Restoration," *Frontiers in Ecology and the Environment* 16, no. 4 (2018).
79 Anna J. Woodhead et al., "Coral Reef Ecosystem Services in the Anthropocene," *Functional Ecology* 33, no. 6 (2019).
80 The Nature Conservancy, "Seychelles – Coral Restoration," (n.d.), https://reefresilience.org/case-studies/seychelles-coral-restoration/.
81 Fei Li et al., "Ecological Restoration Zoning for a Marine Protected Area: A Case Study of Haizhouwan National Marine Park, China," *Ocean & Coastal Management* 98 (2014).
82 Environment Foundation, "Marine Protected Areas," 2019. http://environmentguide.org.nz/issues/marine/marine-protected-areas/.
83 Dan Laffoley et al., "Chapter 29 – Marine Protected Areas," in *World Seas: An Environmental Evaluation (Second Edition)*, ed. Charles Sheppard (Cambridge, Massachusetts: Academic Press, 2019).
84 IUCN, "Guidelines for Marine Protected Areas," (1999), https://portals.iucn.org/library/sites/library/files/documents/PAG-003.pdf.
85 IUCN, "Guidelines for Marine Protected Areas."
86 Wildlife Conservation Society, "New Marine Protected Area Isla Ají Safeguards Important Coastal Ecosystems (English and Spanish)," (n.d.), https://newsroom.wcs.org/News-Releases/articleType/ArticleView/articleId/17153/New-Marine-Protected-Area-Isla-Aji-Safeguards-Important-Coastal-Ecosystems-English-and-Spanish.aspx#:~:text=The%20Isla%20Ají%20MPA%20provides,and%20many%20other%20threatened%20species.

References

Alexander, Sasha, James Aronson, Oliver Whaley, and David Lamb. "The Relationship between Ecological Restoration and the Ecosystem Services Concept." *Ecology and Society* 21, no. 1 (2016).

Alhameid, Abdullah, Colin Tobin, Amadou Maiga, Sandeep Kumar, Shannon Osborne, and Thomas Schumacher. "Chapter 9 – Intensified Agroecosystems and Changes in Soil Carbon Dynamics." In *Soil Health and Intensification of Agroecosytems*, edited by Mahdi M. Al-Kaisi, and Birl Lowery, 195–214. Cambridge, Massachusetts: Academic Press, 2017.

Aradottir, Asa L., and Dagmar Hagen. "Chapter Three – Ecological Restoration: Approaches and Impacts on Vegetation, Soils and Society." In *Advances in Agronomy*, edited by Donald L. Sparks, 173–222. Cambridge, Massachusetts: Academic Press, 2013.

Blair, John, Jesse Nippert, and John Briggs. "Grassland Ecology." In *Ecology and the Environment*, edited by Russell K. Monson, 389–423. New York: Springer New York, 2014.

BMU. "Strategic Framework for Setting Priorities for Restoring Degraded Ecosystems in Germany." (2015). https://bmu.de/en/download/strategic-framework-for-setting-priorities-for-restoring-degraded-ecosystems-in-germany/.

Brears, R.C. *Blue and Green Cities: The Role of Blue-Green Infrastructure in Managing Urban Water Resources*. London, UK: Palgrave Macmillan UK, 2018.

Brears, R.C. *Climate Resilient Water Resources Management*. Cham, Switzerland: Palgrave Macmillan, 2018.

Coastal Resources Management Council Narragansett Bay Estuary Program. "Restoration Methods." (n.d.). http://edc.uri.edu/restoration/html/tech_sci/restsalt.htm.

Elmqvist, T., H. Setälä, S.N. Handel, S. van der Ploeg, J. Aronson, J.N. Blignaut, E. Gómez-Baggethun, et al. "Benefits of Restoring Ecosystem Services in Urban Areas." *Current Opinion in Environmental Sustainability* 14 (2015/06/01): 101–08.

Environment Foundation. "Marine Protected Areas." (n.d.). http://environmentguide.org.nz/issues/marine/marine-protected-areas/.

European Commission. "Towards an Eu Research and Innovation Policy Agenda for Nature-Based Solutions & Re-Naturing Cities." (2015). https://publications.europa.eu/en/publication-detail/-/publication/fb117980-d5aa-46df-8edc-af367cddc202.

FAO. "Forest Restoration and Rehabilitation." (2017). http://fao.org/sustainable-forest-management/toolbox/modules/forest-restoration-and-rehabilitation/basic-knowledge/en/?type=111.

FAO. "Global Guidelines for the Restoration of Degraded Forests and Landscapes in Drylands. Building Resilience and Benefiting Livelihoods." (2015). http://fao.org/3/a-i5036e.pdf.

FAO. "Land-Use Planning." (2017). http://fao.org//sustainable-forest-management/toolbox/modules/land-use-planning/basic-knowledge/en/?type=111.

FAO. "Silviculture and Management of Planted Forests." (2017). http://fao.org//sustainable-forest-management/toolbox/modules/silviculture-and-management-of-planted-forests/basic-knowledge/en/?type=111.

FAO. "Silviculture in Natural Forests." (2017). http://fao.org//sustainable-forest-management/toolbox/modules/silviculture-in-natural-forests/basic-knowledge/en/?type=111http://fao.org//sustainable-forest-management/toolbox/modules/silviculture-in-natural-forests/basic-knowledge/en/?type=111.

Farina, Roberta, Claudia Di Bene, Chiara Piccini, Alessandro Marchetti, Antonio Troccoli, and Rosa Francaviglia. "Chapter 15 – Do Crop Rotations Improve the Adaptation of Agricultural Systems to Climate Change? A Modeling Approach to Predict the Effect of Durum Wheat-Based Rotations on Soil Organic Carbon and Nitrogen." In *Soil Management and Climate Change*, edited by María Á. Muñoz, and Raúl Zornoza, 221–36. Cambridge, Massachusetts: Academic Press, 2018.

Foo, Shawna A., and Gregory P. Asner. "Scaling up Coral Reef Restoration Using Remote Sensing Technology." [In English]. *Frontiers in Marine Science* 6, no. 79 (2019/03/13).

Forest Ecosystem Restoration Initiative. "Partnering with Nature: The Case for Natural Regeneration in Forest and Landscape Restoration." (2017). https://cdn.ymaws.com/www.ser.org/resource/resmgr/custompages/publications/ser_publications/feri_single_page.pdf.

Francis, C.A. "Crop Rotations." In *Encyclopedia of Soils in the Environment*, edited by Daniel Hillel, 318–22. Oxford: Elsevier, 2005.

Gabryś, Beata, and Bożena Kordan. "Chapter 18 – Cultural Control and Other Non-Chemical Methods." In *Insect Pests of Potato*, edited by Andrei Alyokhin, Charles Vincent, and Philippe Giordanengo, 517–41. San Diego, CA: Academic Press, 2013.

Government of El Salvador. "Un Decade of Ecosystem Restoration 2020 – 2030. Communiqué Issued by El Salvador with the Support of Countries of the Central American Integration System (Sica)." (2018). http://wedocs.unep.org/handle/20.500.11822/26027?show=full.

Iowa State University Extension and Outreach. "Managing Pasture to Reduce Soil Erosion." (n.d.). https://crops.extension.iastate.edu/managing-pasture-reduce-soil-erosion.

IUCN. "Drylands and Land Degradation." (2017). https://iucn.org/sites/dev/files/drylands_and_land_degradation_issues_brief_0.pdf.

IUCN. "Ecological Restoration for Protected Areas: Principles, Guidelines and Best Practices." (2012). https://iucn.org/content/ecological-restoration-protected-areas-principles-guidelines-and-best-practices.

IUCN. "Guidelines for Marine Protected Areas." (1999). https://portals.iucn.org/library/sites/library/files/documents/PAG-003.pdf.

Janishevski, L., C. Santamaria, Sarat Gidda, H. Cooper, and Pedro Brancalion. "Ecosystem Restoration, Protected Areas and Biodiversity Conservation." *Unasylva* 245, no. 3 (2015): 19–28.

Kalantari, Zahra, Carla S.S. Ferreira, Saskia Keesstra, and Georgia Destouni. "Nature-Based Solutions for Flood-Drought Risk Mitigation in Vulnerable Urbanizing Parts of East-Africa." *Current Opinion in Environmental Science & Health* 5 (2018/10/01): 73–78.

Ladd, Mark C., Margaret W. Miller, John H. Hunt, William C. Sharp, and Deron E. Burkepile. "Harnessing Ecological Processes to Facilitate Coral Restoration." *Frontiers in Ecology and the Environment* 16, no. 4 (2018/05/01): 239–47.

Laffoley, Dan, John M. Baxter, Jon C. Day, Lauren Wenzel, Paula Bueno, and Katherine Zischka. "Chapter 29 – Marine Protected Areas." In *World Seas: An Environmental Evaluation (Second Edition)*, edited by Charles Sheppard, 549–69. Cambridge, Massachusetts: Academic Press, 2019.

Lamb, Joleah B., Jeroen A.J.M. van de Water, David G. Bourne, Craig Altier, Margaux Y. Hein, Evan A. Fiorenza, Nur Abu, Jamaluddin Jompa, and C. Drew Harvell.

"Seagrass Ecosystems Reduce Exposure to Bacterial Pathogens of Humans, Fishes, and Invertebrates." *Science* 355, no. 6326 (2017): 731.

Li, Fei, Min Xu, Qing liu, Zaifeng Wang, and Wenjian Xu. "Ecological Restoration Zoning for a Marine Protected Area: A Case Study of Haizhouwan National Marine Park, China." *Ocean & Coastal Management* 98 (2014/09/01): 158–66.

Libini, Cheruvathoor L., K.A. Albert Idu, C.C. Manjumol, Vasant Kripa, and Kolliyil Sunil Mohamed. "Chapter 27 – Marine Biodiversity – Strategies for Conservation, Management and Ecological Restoration." In *Biodiversity and Climate Change Adaptation in Tropical Islands*, edited by Chandrakasan Sivaperuman, Ayyam Velmurugan, Awnindra Kumar Singh, and Iyyappan Jaisankar, 745–62. Cambridge, Massachusetts: Academic Press, 2008.

Liquete, Camino, Angel Udias, Giulio Conte, Bruna Grizzetti, and Fabio Masi. "Integrated Valuation of a Nature-Based Solution for Water Pollution Control. Highlighting Hidden Benefits." *Ecosystem Services* 22 (2016/12/01): 392–401.

Living with Environmental Change. "New Tools to Improve the Management of the Coastal Environment." (2014). https://nerc.ukri.org/research/partnerships/ride/lwec/ppn/ppn10/.

Natural Water Retention Measures. "Crop Rotation." (2015). http://nwrm.eu/sites/default/files/nwrm_ressources/a3_-_crop_rotation_0.pdf.

Natural Water Retention Measures. "Floodplain Restoration and Management." (2015). http://nwrm.eu/sites/default/files/nwrm_ressources/n3_-_floodplain_restoration_and_management.pdf.

Natural Water Retention Measures. "Green Cover." (2015). http://nwrm.eu/sites/default/files/nwrm_ressources/a8_-_green_cover_0.pdf.

Natural Water Retention Measures. "Intercropping." (2015). http://nwrm.eu/sites/default/files/nwrm_ressources/a5_-_intercropping.pdf.

Natural Water Retention Measures. "Lake Restoration." (2015). http://nwrm.eu/sites/default/files/nwrm_ressources/n12_-_lake_restoration_0.pdf.

Natural Water Retention Measures. "Low Till Agriculture." (2015). http://nwrm.eu/sites/default/files/nwrm_ressources/a7_-_low_till_agriculture_0.pdf.

Natural Water Retention Measures. "Meadows and Pastures." (2015). http://nwrm.eu/sites/default/files/nwrm_ressources/a1_-_meadows_and_pastures_0.pdf.

Natural Water Retention Measures. "No Till Agriculture." (2015). http://nwrm.eu/sites/default/files/nwrm_ressources/a6_-_no_till_agriculture.pdf.

NSW Department of Primary Industries. "Coastal Saltmarsh." (2013). http://dpi.nsw.gov.au/__data/assets/pdf_file/0007/459628/Coastal-Saltmarsh-Primefact.pdf.

Paulo, Diogo, Alexandra H. Cunha, Joana Boavida, Ester A. Serrão, Emanuel J. Gonçalves, and Mark Fonseca. "Open Coast Seagrass Restoration. Can We Do It? Large Scale Seagrass Transplants." [In English]. *Frontiers in Marine Science* 6, no. 52 (2019/03/07).

Regehr, David L., and Charles A. Norwood. "Chapter 14 – Benefits of Triazine Herbicides in Ecofallow." In *The Triazine Herbicides*, edited by Homer M. LeBaron, Janis E. McFarland, and Orvin C. Burnside, 175–83. San Diego, CA: Elsevier, 2008.

SANBI. "Grasslands Ecosystem Guidelines: Landscape Interpretation for Planners and Managers." (2013). http://biodiversityadvisor.sanbi.org/wp-content/uploads/2015/06/2013_Grassland-Ecosystem-Guidelines.pdf.

Schirpke, Uta, Marina Kohler, Georg Leitinger, Veronika Fontana, Erich Tasser, and Ulrike Tappeiner. "Future Impacts of Changing Land-Use and Climate on Ecosystem

Services of Mountain Grassland and Their Resilience." *Ecosystem Services* 26 (2017/08/01): 79–94.
Sims, J.T., and P.A. Vadas. "Phosphorus in Soils | Overview." In *Encyclopedia of Soils in the Environment*, edited by Daniel Hillel, 202–10. Oxford: Elsevier, 2005.
Smithsonian Ocean. "Seagrass and Seagrass Beds." (n.d.). https://ocean.si.edu/ocean-life/plants-algae/seagrass-and-seagrass-beds.
Society for Ecological Restoration International. "Investing in Our Ecological Infrastructure." (2011). https://cdn.ymaws.com/www.ser.org/resource/resmgr/custompages/publications/ser_publications/economic_rationale_for_resto.pdf.
Society for Ecological Restoration International. "The Ser International Primer on Ecological Restoration." (2004). https://cdn.ymaws.com/www.ser.org/resource/resmgr/custompages/publications/SER_Primer/ser_primer.pdf.
Stillwater Sciences. "A Guide for Restoring Functionality to Mountain Meadows of the Sierra Nevada." (2012). http://stillwatersci.com/resources/2012meadowrestguide.pdf.
UNEP. "Global Drylands: A Un System-Wide Response." (2011). https://unep-wcmc.org/resources-and-data/global-drylands--a-un-system-wide-response.
USDA. "Grazing Management and Soil Health." (2016). https://nrcs.usda.gov/Internet/FSE_DOCUMENTS/nrcseprd621806.pdf.
Wetlands International. "Mangrove Restoration: To Plant or Not to Plant?" (2018).
Woodhead, Anna J., Christina C. Hicks, Albert V. Norström, Gareth J. Williams, and Nicholas A.J. Graham. "Coral Reef Ecosystem Services in the Anthropocene." *Functional Ecology* 33, no. 6 (2019/06/01): 1023–34.
WWF. "Natural and Nature-Based Flood Management: A Green Guide." (2017). https://worldwildlife.org/publications/natural-and-nature-based-flood-management-a-green-guide.

10 Enhancing Urban Sustainability with Nature-Based Solutions

Introduction

In urban areas, nature-based solutions (NBS) are often implemented as green infrastructure (GI) – a strategically planned network of natural and semi-natural areas designed to provide environmental, economic, and social benefits.[1] GI is a key approach to addressing urban challenges, including reducing ecological footprints, enhancing public health and well-being, increasing climate resilience, and fostering social cohesion. GI helps create sustainable, livable, and climate-adaptive urban environments by integrating natural elements into city planning.[2]

Green Infrastructure

GI varies in scale and scope, encompassing buildings, public spaces, water bodies, drainage systems, and green corridors. It provides three key benefits. First, GI supports economic development, as urban economies depend on natural resources, such as clean water for sanitation and manufacturing. By enhancing resource availability, GI contributes to long-term economic sustainability. Second, GI has positive environmental impacts, helping mitigate the effects of climate change by storing carbon, reducing emissions, and increasing resilience to floods, droughts, and heatwaves. Third, GI improves social well-being, as access to green spaces has been shown to boost happiness, improve health, and enhance overall quality of life in urban areas.[3] By working with nature, cities can provide essential ecosystem services that make them more competitive, resource-efficient, climate-resilient, and healthier, while enhancing natural capital rather than depleting it. A summary of these ecosystem services is provided in Table 10.1, while Table 10.2 outlines the specific environmental, economic, and social benefits of GI at the urban scale.[4,5,6,7,8,9,10,11]

Principles Informing Green Infrastructure Solutions

Six key principles guide the planning and design of GI in urban areas:

DOI: 10.4324/9781003517948-10

Table 10.1 Ecosystem Services Provided by Urban Nature-Based Solutions

Ecosystem Service	*Examples*
Provisioning	• Supply of raw materials (e.g. timber from urban forests) • Access to fresh water sources • Urban agriculture and food production • Medicinal plants and natural remedies
Regulating	• Moderation of local temperatures and improvement in air quality • Carbon capture and storage by urban trees and vegetation • Reduction of flood risk and extreme weather impacts • Natural wastewater filtration and stormwater regulation
Cultural	• Recreational spaces such as parks and greenways • Benefits to physical and mental well-being • Cultural tourism and nature-based activities • Visual inspiration for art, architecture, and design • Spaces fostering community identity, pride, and spiritual reflection
Supporting	• Creation of habitats for urban wildlife and pollinators • Contribution to the conservation of species and genetic diversity • Ecological connectivity across fragmented urban landscapes

Table 10.2 Environmental, Economic, and Social Benefits of Nature-Based Solutions

Benefit	*Description*
Environmental	• Reduce stormwater runoff, flood risk, and erosion through natural water absorption • Improve air and water quality by filtering harmful pollutants • Regulate local temperatures and mitigate the urban heat island effect, supporting energy efficiency • Conserve and restore natural ecosystems, providing habitats for biodiversity • Contribute to climate change mitigation by sequestering carbon and reducing energy demand
Economic	• Create employment and business opportunities in green infrastructure, tourism, and recreation • Increase foot traffic and stimulate spending in local commercial areas • Enhance property values in areas with accessible green space • Attract new residents, visitors, and businesses • Reduce energy, healthcare, and infrastructure maintenance costs
Social	• Support active lifestyles by offering access to outdoor recreation • Improve public health by enhancing environmental conditions • Advance environmental justice by increasing equitable access to green space • Foster social cohesion by providing gathering spaces and cultural expression venues • Beautify urban spaces and enhance community pride • Promote connections with nature that support mental and emotional well-being, especially for children and elderly • Improve educational outcomes and reduce crime through increased community engagement

1. *Multifunctionality*: By leveraging natural processes, GI provides multiple benefits within the same space. For example, integrating vegetation into the urban environment can regulate the water cycle, improve air quality, and enhance sustainable development while delivering essential ecosystem services.
2. *Connectivity*: GI is more effective when part of a network across the landscape. A nature reserve linked by green corridors to other reserves offers greater ecosystem benefits than an isolated one, as it facilitates wildlife movement, habitat continuity, and ecological resilience.
3. *Habitability*: GI creates functional and visually appealing outdoor spaces that benefit both people and nature. It improves air and water quality, enhances human health, increases opportunities for outdoor recreation, and restores native habitats, making urban environments more livable.
4. *Resiliency*: GI strengthens communities' ability to withstand and adapt to climate change and environmental disturbances. For example, trees and green roofs reduce the urban heat island effect, lowering cooling demands at both the building and city scales while providing shade and managing stormwater.
5. *Identity*: GI helps define a community's character and sense of place. It can reflect local cultural and ecological significance by incorporating native vegetation, recreational spaces, and artistic elements that enhance a location's visual appeal and social value.
6. *Return on Investment*: GI delivers economic benefits by reducing costs and generating financial gains for governments, businesses, and citizens. It can increase property values, support economic development, lower energy consumption, and reduce the need for expensive grey infrastructure. Planners should utilize cost-benefit analyses, establish clear objectives, and monitor measurable outcomes to ensure long-term effectiveness and efficiency.[12,13]

Buildings

A variety of GI solutions can be implemented at the building scale to restore and enhance the natural environment in urban areas while delivering environmental, economic, and social benefits, including:

Green Roofs

A green roof is a vegetated space on top of a building, partially or entirely covered with plants growing in a specialized growing medium. While 'green' refers to vegetation, green roofs vary in color and texture, incorporating different plant species, soil types, and features such as stones and sandy areas to support wildlife habitats. Advancements in design and technology have led to various types of green roofs, outlined in Table 10.3.[14] These roofs provide multiple environmental benefits, including reducing heat impacts, managing

Table 10.3 Typologies of Green Roofs

Green Roof Type	*Description*
Green roofs	Roofs where vegetation is deliberately established. Two main types: • Intensive green roofs: feature deep substrates (>200 millimeter), formal landscaping, irrigation, and regular maintenance. Often used for rooftop gardens or urban agriculture. • Extensive green roofs: use shallow substrates (40–150 millimeter), low-maintenance, drought-tolerant vegetation. Typically not irrigated and require minimal upkeep.
Biodiverse green roofs	Designed to replicate native habitats or support target species. These roofs typically have deeper soils than extensive roofs, can include a wide range of plants such as shrubs, and may or may not require irrigation.
Bio solar roofs	Combine green roofs with solar photovoltaic (PV) systems. The green roof substrate supports PV frames and enhances their efficiency by reducing overheating through evaporative cooling.
Blue-green roofs	Integrate rainwater storage and vegetation in a single system. A void below the vegetation layer stores water, supporting irrigation and delaying runoff. Often equipped with smart sensors for controlled water release and capable of supporting wetland habitats for biodiversity and amenity purposes.

stormwater, improving air quality, lowering fine dust pollution, cutting energy costs for heating and cooling, and serving as carbon sinks. Additionally, green roofs contribute to urban biodiversity, creating habitat for birds, insects, and other species.[15,16,17,18]

Case 10.1: Hamburg's Green Roof Strategy

Hamburg was the first major German city to implement a comprehensive Green Roof Strategy, launched in 2014. The strategy set a target to green at least 70% of new buildings and suitable flat or gently pitched roofs undergoing renovation. By the end of 2024, the city aimed to establish 100 hectares of green roof area, supported by a €3 million funding program. The strategy operates across three levels: incentive, dialogue, and regulation. Financial incentives cover up to 60% of the installation costs per square meter, capped at €50,000, with additional subsidies available for city center projects. Rainwater fee reductions of up to 50% are offered to encourage uptake. Projects must meet minimum criteria, including a net vegetation area of 20 square meters and substrate thickness of at least 8 centimeters for extensive greening. Green roofs are to be voluntary initiatives, not resulting from binding legal requirements. In parallel, the city has incorporated green roofs into building regulations and urban planning codes. Scientific research led by HafenCity University supports the strategy through monitoring and evaluation of water retention and ecosystem performance. This evidence base informs the design of optimal roof systems and provides data for national scaling. Green roofs in Hamburg serve multifunctional purposes, including biodiversity conservation, stormwater retention, thermal regulation, and urban agriculture. Types range from low-maintenance extensive roofs to fully accessible intensive roof gardens with lawns, shrubs, and trees. Many projects, such as the green roof atop the Congress Center Hamburg, now serve both ecological and recreational purposes.[19]

City of Hamburg, Ministry of Environment and Energy. "More Green Roofs for Hamburg: Green Roof Strategy for Hamburg." (n.d.). https://hamburg.de/resource/blob/281688/a5544533af6cba7305c0d8c58946fa96/d-more-green-roofs-for-hamburg-data.pdf.

Urban Rooftop Farming

Urban rooftop farms are cultivated spaces on building rooftops that utilize raised beds, row farming, and hydroponic systems to grow a diverse range of vegetables, fruits, and nuts. Raised beds involve growing crops in shallow, soil-based beds placed on a waterproof membrane and may include root barriers, drainage layers, and irrigation systems. Row farming, where crops are grown in contiguous linear beds, requires strong structural support due to the weight of deep soil, high organic matter content, and water retention. Hydroponic rooftop systems replace soil with water-based nutrient solutions, enabling plants to grow in open-air settings or controlled environments, such as greenhouses, which increases yields and extends growing seasons. Urban rooftop farms provide a range of benefits, including enhanced social interaction, stronger community engagement, higher property values, improved food security, stormwater retention, better air quality, and carbon sequestration (Table 10.4).[20,21,22]

Rainwater Harvesting

Rainwater harvesting collects and stores precipitation from roof surfaces into cisterns or other storage systems. While harvested rainwater can be treated for potable use, it is primarily used for non-potable applications, such as toilet flushing, laundry, irrigation, fire suppression, and building cooling. By capturing and utilizing rainwater, these systems help reduce reliance on municipal water supplies, lower water bills, manage stormwater runoff, and promote sustainable water use practices. A summary of the specific benefits of rainwater harvesting systems is provided in Table 10.5.[23]

Table 10.4 Benefits of Rainwater Harvesting Systems

Benefit	*Description*
Reduced stormwater runoff	Captures rainfall at the source, reducing surface runoff and the risk of localized flooding.
Increased water availability	Reduces reliance on potable water by providing an alternative source for irrigation and non-potable uses.
Increased groundwater recharge	Supports aquifer replenishment by directing harvested rainwater into the ground during irrigation.
Reduced energy use	Minimizes the energy required for pumping, treating, and distributing municipal water supplies.
Improved air quality	By reducing demand on treatment plants, emissions from water and wastewater processing facilities are lowered, benefiting air quality.
Reduced atmospheric carbon emissions	Decreases the need for energy-intensive water treatment processes, thereby lowering greenhouse gas emissions from power plants.
Enhanced public education	Serves as a community engagement tool to raise awareness of green infrastructure and sustainable water resource practices.

Table 10.5 Opportunities to Enhance Park Features with Nature-Based Solutions

Park Infrastructure	*Nature-Based Solutions*
Parking lots	• Bioretention in landscaped areas, medians, and roundabouts • Permeable pavement in parking stalls and pedestrian zones • Tree planting in landscape islands to provide shade and absorb runoff
Visitor centres	• Demonstration gardens with bioretention features • Pollinator gardens designed to manage stormwater • Green roofs to reduce runoff and improve insulation • Rain barrels or cisterns for rainwater harvesting • Planter boxes for stormwater treatment in space-limited areas
Playing fields	• Use of fields as temporary stormwater detention areas during heavy rainfall • Permeable pavement in adjacent parking areas and access paths
Wetlands and drainage systems	• Restoration of natural buffers along stream channels • Creation of constructed wetlands within existing drainage corridors • Integration of wildlife habitat areas that also function as stormwater basins
Paved trails, walkways, and roads	• Use of permeable materials for trails, pathways, and emergency access lanes • Tree planting along routes to provide shade, enhance aesthetics, and manage runoff

Case 10.2: City of Tucson, Arizona's Rainwater Harvesting Rebate Program

The City of Tucson, Arizona, operates a rainwater harvesting rebate program that supports residential water conservation through on-site collection and storage of rainwater. Administered by Tucson Water, the program offers up to $2,000 per property for the installation of qualifying rainwater harvesting systems. Multiple system components across a property can be combined toward the rebate limit. To qualify for the maximum rebate rate of $1.00 per gallon, systems must be designed to capture at least one inch of rainfall from a designated drainage area, typically a rooftop. If the system captures less than this threshold or involves basins that do not receive direct roof runoff, the rebate is set at $0.50 per gallon. Properties with more than one drainage area can incorporate each into the total rebate calculation. Applicants must be active Tucson Water customers and attend an approved workshop before submitting a pre-approval form. From January 2024 onwards, only workshops completed after July 2023 are accepted. Pre-approval requires submitting an online form and a site plan that details the location and features of the proposed system. Approval must be obtained before any purchase or installation; otherwise, the system will not qualify for a rebate. Once installation is complete, the applicant must schedule an on-site inspection before submitting the rebate application. Following the inspection, the applicant submits the rebate application along with supporting documents such as itemized receipts and proof of payment. Additional documentation is required for higher-value rebates in line with local administrative procedures.[24]

City of Tucson. "Residential Customer Rebates." (n.d.). https://tucsonaz.gov/Departments/Water/Conservation/Residential-Customer-Rebates#:~:text=All%20water%20harvesting%20features%20on,full%20rebate%20amount%20%241.00%2Fgal.

Green Facades and Walls

Green facades and walls enhance urban spaces by integrating vegetation into building exteriors and interiors, providing ecological, environmental, and social benefits. Green facades use pot-bound plants arranged in rows along a supporting structure, which can be installed at any height or size on a building.[25] These facades meet residents' demands for nature, create habitats for insects and birds, and contribute to indoor and outdoor climate regulation. Green walls consist of plants grown on a vertical structure, which can be attached to a building or freestanding. These structures range from modular systems to broad-based designs with felt pockets for soil or other growing media. Both green facades and walls offer multiple benefits, including improving air quality by capturing fine particulate matter and reducing carbon dioxide levels, mitigating the urban heat island effect, lowering building energy consumption, and creating habitats to support biodiversity. Additionally, they contribute to psychological and physical well-being, making urban environments healthier and more sustainable.[26,27]

Case 10.3: City of Sydney Green Roofs and Walls Policy

The City of Sydney has adopted a formal policy to promote the installation of green roofs and walls throughout its local government area, aiming to increase urban climate resilience, enhance livability, and contribute to improved environmental quality. Approved by the Council in 2014, the policy outlines the role of green roofs and walls in stormwater management, urban heat mitigation, carbon absorption, biodiversity support, and improving both building efficiency and public amenity. The policy defines a green roof as vegetation covering at least 30% of the usable rooftop space, excluding areas occupied by mechanical equipment or stairwell access structures. A green roof typically consists of a growing medium, a waterproof membrane, and may also incorporate renewable energy systems, walkways, or furnishings. A green wall may be part of a building or freestanding structure and consists of vegetation supported by soil or inorganic material. Two main types are recognized: green façades and living walls. To support implementation, the city has developed a range of measures. These include education, technical guidance, stakeholder engagement, and incorporation of green infrastructure into planning instruments and sustainability rating tools. The city also leads by example, promoting green roofs and walls on its buildings and exploring demonstration projects. Monitoring of installations is part of the ongoing policy assessment process.[28]

City of Sydney. "Green Roofs and Walls." (n.d.). https://cityofsydney.nsw.gov.au/environmental-support-funding/green-roofs-and-walls.

Community Courtyards

Community courtyards between buildings offer ample space for integrating various GI elements, enhancing environmental quality, urban resilience, and community well-being. These elements include:

- Pervious pavements reduce surface runoff and help prevent flooding by allowing rainwater to infiltrate the ground.
- Vegetation in plant pots, planters, and green spaces improves air quality, mitigates the urban heat island effect, reduces noise pollution, and promotes ecological connectivity.
- Fountains and water features refresh the atmosphere, enhance aesthetic appeal, and help cool urban areas by reducing heat buildup.

- Vegetable gardens and fruit trees support local food production, increase biodiversity, and provide shade and recreational benefits for the community.[29]

Public Spaces

GI can be implemented in public spaces, including urban parks, plazas, and communal living areas, to achieve multiple benefits. These include restoring biodiversity, improving human health and well-being, and reducing the impacts of climate change.

Green Street Furniture

Incorporating biomaterials into benches, canopies, other street furniture, and pervious pavements can help moderate extreme summer temperatures and improve surface water drainage. Public transportation infrastructure, such as bus stops and shelters, can be "greened" by integrating vegetation and water retention systems to enhance cooling, water management, and urban biodiversity.[30,31]

Case 10.4: Utrecht's Bee-Friendly Bus Stops

The City of Utrecht, Netherlands, has implemented an innovative urban greening initiative by transforming 316 bus stops into bee-friendly green infrastructure. Each bus shelter is equipped with a sedum-covered roof, which contributes to improved air quality, enhanced rainwater retention, and cooling during the summer heat. These green roofs also serve as habitats for pollinators, such as bees, bumblebees, and butterflies, thereby enhancing urban biodiversity. The green roofs are primarily composed of sedum, a low-maintenance plant species well-suited to dry conditions. Utrecht is now experimenting with additional plant varieties at select shelters to evaluate their ability to attract more pollinators. These changes are being monitored to assess the broader ecological benefits of alternative plantings. Beyond the green roofs, the shelters incorporate a range of sustainable features. Ninety-six shelters have integrated solar panels to power efficient LED lighting. Benches are made from sustainable bamboo, and the shelter bases are constructed using recycled concrete. Maintenance of the shelters is conducted using electric vehicles to reduce carbon emissions. The project was implemented through a public-private partnership with RBL Outdoor and Clear Channel, funded entirely by advertising revenue at no cost to the municipality. This approach reflects the city's commitment to "healthy urban living," which was outlined in the original tender for bus stop installation and maintenance.[32]

ArchDaily. "Utrecht Creates 300 Bee-Friendly Bus Stops." (n.d.). https://archdaily.com/921850/utrecht-creates-300-bee-friendly-bus-stops.

Urban Parks

Parks typically feature open spaces that absorb stormwater, improving water quality, habitat conditions, and aesthetics. However, infrastructure such as car parks, roads, buildings, playing fields, and courts can be built or retrofitted with GI to manage stormwater more effectively by directing it to enhanced pervious surfaces (Table 10.5). Integrating GI into park infrastructure provides multiple benefits:

- *Educational opportunities*: Restored natural areas can serve as wildlife habitats while offering hands-on outdoor learning experiences.
- *Social and environmental equity*: Parks designed with GI improve public well-being, particularly in underserved communities, by fostering healthier environments, opportunities for physical activity, and spaces for community engagement.
- *Reduced maintenance costs*: Improved drainage and vegetation health reduce the upkeep required for open spaces, lowering long-term maintenance expenses.[33]

Case 10.5: La Marjal Floodable Park, Alicante, Spain

La Marjal Floodable Park in Alicante, Spain, was developed as a response to persistent flooding in the San Juan Beach area, a coastal zone historically characterized as a natural floodplain. Conceived in 2011 and opened in 2015, the park occupies a 36,700 square meters municipal plot designated as green space, operating as both a recreational amenity and a stormwater management system. It is designed to store up to 45,000 cubic meters of stormwater during intense rainfall events, mitigating urban flooding while mimicking the ecological functions of natural wetlands. The park's hydraulic design integrates two large collectors that channel overflow water from the stormwater network into the park once the drainage system's capacity is exceeded. Water retained in the park is later redirected to a treatment facility for reuse in urban irrigation and street cleaning, contributing to Alicante's circular water strategy. Ecologically, the park supports a diverse range of vegetation, including aquatic, marshland, and Mediterranean species. It has become a biodiverse habitat for fauna, with more than 18 bird species observed, 58% of which are migratory. Educational signage throughout the park facilitates environmental learning, while its trails offer accessible points of interest, including a viewpoint, waterfall, ponds, and diverse vegetation zones. To commemorate its 10th anniversary in 2025, the city announced plans for three additional floodable parks and up to twelve new stormwater retention basins. These will be in flood-prone urban zones, with a combined future capacity of 150,000 cubic meters, reinforcing La Marjal's model as a scalable nature-based solution for urban resilience.[34]

Cadena SER. "Décimo Aniversario De La Marjal Cuyo éxito El Ayuntamiento Quiere Replicar En "Zonas Sensibles" De Alicante." (n.d.). https://cadenaser.com/comunitat-valenciana/2025/02/28/decimo-aniversario-de-la-marjal-cuyo-exito-el-ayuntamiento-quiere-replicar-en-zonas-sensibles-de-alicante-radio-alicante/.

Urban Micro-Climates

Decommissioning and unsealing impervious surfaces, as well as incorporating water elements into squares and public spaces, can help moderate urban temperatures and create microclimates that reduce the urban heat island effect, making cities more resilient to climate change. Features such as allotment gardens, pocket parks, water playgrounds, fountains, and medium-scale retention or detention basins provide cooling benefits and encourage social interaction by creating welcoming public spaces that serve as community gathering points.[35,36]

Allotment Gardens

Allotment gardens provide urban food production, help mitigate the urban heat island effect, and control flooding areas during heavy rainfall. They are beneficial for regenerating degraded areas or repurposing wastelands, offering temporary and productive land use while enhancing urban sustainability. These gardens foster ecological connectivity, linking urban spaces with natural and peri-urban environments. They also promote community engagement, creating a sense of belonging, raising awareness of biodiversity conservation, and encouraging recycling and responsible consumption. Beyond environmental benefits, urban allotments contribute to public health by encouraging the consumption of fresh, organic produce, free from pesticides and synthetic chemicals, while promoting physical activity for those tending to the gardens.[37]

Urban Forests

Urban forests and vegetation provide a range of environmental benefits, including intercepting and filtering stormwater runoff, absorbing air pollutants, and acting as windbreaks to protect buildings from damage. They help regulate the urban heat island effect through shading and evaporation, offer habitats for wildlife, deliver valuable ecosystem services, and contribute to climate change mitigation by storing carbon dioxide. These benefits are summarized in Table 10.6.[38] Urban forest programs involve planting and

Table 10.6 Benefits of Urban Forests

Benefit	*Description*
Reduced stormwater runoff	Urban trees intercept rainfall, enhance infiltration, and reduce soil moisture through transpiration. They also help filter stormwater by removing nutrients and heavy metals.
Increased groundwater recharge	Tree roots and canopy structures promote percolation, helping to replenish underground aquifers.
Reduced energy use	Trees cool the environment by providing shade and reducing wind speeds, which decreases energy demand for air conditioning in summer and heating in winter.
Improved air quality	Trees absorb air pollutants, trap particulate matter, and reduce carbon emissions by decreasing energy use in nearby buildings.
Reduced atmospheric carbon emissions	Urban trees sequester carbon dioxide directly from the atmosphere, contributing to climate change mitigation.
Reduced urban heat island effect	Trees cool urban environments through shading and evapotranspiration, reducing localized heat buildup.
Improved community liveability	Urban forests enhance public spaces, promote well-being, reduce noise and pollution, and contribute to recreational and aesthetic value.
Improved habitat	Native tree species provide habitat for urban wildlife, increasing local biodiversity.
Enhanced public education	Tree planting programs offer educational opportunities and foster community engagement in environmental stewardship.

maintaining trees in public spaces such as parks, streets, alleys, and other government-managed areas, including stream corridors, public buildings, and vacant lots. They also extend to green belts that buffer waterways and regulate urban development. Additionally, urban forest programs may include acquiring and managing land to preserve urban watersheds, ensuring the protection of drinking water supply and quality.[39]

Case 10.6: Frankfurt's Green Belt, Germany

Frankfurt am Main, widely known for its financial district and modern skyline, is also home to one of the most expansive urban nature initiatives in Germany: The GrünGürtel (Green Belt). Established in 1991 by city council resolution, the Green Belt encompasses approximately one-third of Frankfurt's total area, amounting to 80 square kilometers. It serves as a multifunctional green infrastructure network supporting biodiversity, recreation, and climate resilience. The Green

Belt encompasses a diverse range of landscapes, including meadows, forests, orchards, agricultural fields, and historic landmarks. It is part of the larger Rhine-Main Regional Park, which connects Frankfurt to surrounding regions through a 550-kilometre trail network. The circular hiking route within the Green Belt spans approximately 68 kilometers, divided into nine segments, each marked with a unique stele for visitors to stamp as they complete each section. Among the Green Belt's notable features is the transformation of former industrial or degraded sites, such as the Weilbacher Gruben, which was once a gravel pit and dumping ground. Through local community efforts, the area has been rehabilitated into a nature reserve with gardens, educational signage, and a regional cuisine restaurant. The area also supports regional food systems. Orchards and organic farms, such as Dottenfelderhof and operations in Nieder-Erlenbach, provide produce, including apples used for cider production and ingredients for local specialities like Frankfurt's "Grüne Soße." Cultural and culinary traditions are integrated into the landscape, reinforcing the Green Belt's role in preserving ecological, social, and regional heritage. To secure its future, the Green Belt is legally protected as a landscape conservation area, with replacement requirements if land use changes occur.[40]

Stadt Frankfurt am Main. "Der Frankfurter Grüngürtel." (n.d.). https://frankfurt.de/themen/umwelt-und-gruen/orte/gruenguertel.

Renaturing Abandoned Areas

Underutilized public spaces, abandoned lots, and neglected areas present opportunities for urban regeneration and social revitalization. GI can transform these spaces into urban farms, community gardens, or green spaces on post-industrial sites, enhancing environmental sustainability and community engagement. Additionally, GI can help cities adapt to climate extremes, such as heatwaves, flooding, and poor air quality, by incorporating vegetation, permeable surfaces, and water management systems. These adaptive strategies improve urban resilience, create equitable economic opportunities, and enhance the overall quality of life for residents.[41]

Case 10.7: Transforming Underutilized Urban Space into Functional Green Infrastructure at First Creek Wetland, Adelaide

The First Creek Wetland, located within the Adelaide Botanic Garden, exemplifies the integration of green infrastructure into an urban public space. Developed on a previously underutilized area near Hackney Road, the site incorporates stormwater treatment, aquifer storage, and public engagement in a multifunctional landscape. The project is designed to harvest stormwater from First Creek, whose catchment extends 23 square kilometers into the Adelaide Hills, and treat it through a series of engineered and ecological processes to support the long-term irrigation of the gardens. The wetland system includes three ponds – sedimentation, macrophyte, and balance – connected via a staged flow path. After mechanical pre-treatment using a gross pollutant trap and a Hydrobrake device to manage flow, water is directed into the sedimentation pond, followed by filtration in the macrophyte pond. Treated water is then passed through a sand filter before being injected into a fractured rock aquifer. In summer, this water is extracted for irrigation via a balance pond. The system supports up to 100 million liters of stormwater reuse annually and features over 90 native aquatic and terrestrial plant species. Interactive design elements, including a sunken amphitheater, stepping stones, and interpretive signage, facilitate public and school engagement. Approximately 20,000 students visit annually for environmental education programs aligned with the Australian curriculum. Operational challenges, including sediment load, plant establishment, and algal growth, have informed adaptive management strategies. Maintenance is carried out by garden staff and contractors, with technical data provided for regulatory compliance. The project is supported by state and federal environmental funding.[42]

Government of South Australia, Department of Environment, Water and Natural Resources. "Green Infrastructure Case Study: Wetland and Stormwater Reuse – First Creek Wetland, Adelaide Botanic Garden." (2017). https://cdn.environment.sa.gov.au/landscape/docs/hf/green-infrastructure-first-creek-case-study.pdf.

Water Bodies and Drainage Systems

Water bodies and drainage systems can integrate GI to enhance water quality, regulate water flow, and restore natural hydrological processes in urban landscapes. By doing so, GI supports ecosystem services, improves biodiversity, and strengthens climate resilience, making cities more sustainable and environmentally balanced.

Sustainable Urban Drainage Systems

Sustainable Urban Drainage Systems (SUDS) are environmentally friendly drainage solutions that minimize long-term damage to water systems and the environment. They use a combination of management practices, control structures, and strategies to efficiently drain surface water while reducing pollution and protecting local waterways. By holding back rainwater, treating contaminants, and releasing water gradually, SUDS help prevent flooding and overburdening of sewer systems. Compared to traditional drainage methods, SUDS offer greater sustainability by:

- Managing runoff volumes and flow rates from hard surfaces to reduce urban flooding.
- Enabling rainwater reuse where it falls.
- Improving water quality by filtering pollutants from runoff.
- Preserving natural flow patterns in waterways.
- Integrating environmental and community needs into their design.
- Creating urban habitats that support wildlife.
- Enhancing evapotranspiration through vegetation and surface water.
- Encouraging groundwater recharge where appropriate.

SUDS come in various forms, as summarized in Table 10.7, and offer a holistic approach to urban water management that benefits both people and the environment.[43]

Stormwater Detention and Retention Systems

Stormwater detention and retention systems help manage urban runoff by capturing water from buildings and roads before it reaches stormwater sewers. These systems interrupt traditional stormwater pathways, reducing flood risks and improving water quality. Detention basins temporarily store

Table 10.7 Sustainable Urban Design Systems Features

SUDS Feature	*Description*
Green roofs	Vegetated roof systems that absorb rainwater, reduce runoff, and filter pollutants before they reach drainage systems.
Permeable pavement	Paving materials that allow water to pass through the surface, enabling natural filtration and reducing surface runoff.
Raingardens and bioretention	Planted depressions that temporarily store and filter rainwater. Bioretention systems include engineered soil and gravel layers to treat polluted runoff from impervious surfaces.
Filter strips	Gently sloping grass or vegetated areas next to hard surfaces that trap and filter pollutants from runoff.
Swales	Shallow, vegetated channels that slow water flow, promote infiltration, and transport runoff through the landscape while filtering pollutants.
Basins, ponds, and wetlands	Water-holding features that manage stormwater by capturing, storing, and gradually releasing water, while supporting biodiversity and improving water quality.
Filter drains	Trenches filled with gravel that filter runoff, promote infiltration, and convey water through the system to downstream SUDS features.

stormwater during heavy rainfall and gradually release it at a controlled rate, preventing the overwhelming of drainage systems. Retention basins, on the other hand, hold a permanent pool of water designed to control peak runoff while providing pre-treatment to remove pollutants before water is discharged. Additional features such as bioretention basins and swales enhance water infiltration and filtration, making these systems essential for sustainable urban water management.[44]

Rain Gardens

Rain gardens are vegetated depressions designed to capture, filter, and treat stormwater runoff from rooftops, footpaths, and streets. They help manage urban drainage, reduce flood risks, and improve water quality by naturally filtering pollutants. A rain garden consists of three main components: a drainage area, which collects rainwater from surrounding surfaces; a distribution system, which channels the collected water into the receiving area; and a receiving area, which retains and infiltrates rainwater, allowing for gradual absorption. As runoff enters the rain garden, it is first filtered through surface vegetation and then through an engineered soil layer, which removes contaminants. A perforated pipe at the base collects and transports the filtered water to a downstream detention system or designated discharge point. This process helps reduce stormwater pollution, enhances groundwater recharge, and supports urban biodiversity.[45]

Case 10.8: Rain Garden Implementation in New York City

The New York City Department of Environmental Protection (DEP) has implemented a widespread rain garden program as part of the city's broader green infrastructure strategy to manage stormwater and improve urban resilience. The initiative focuses on retrofitting dense urban streetscapes with rain gardens to reduce runoff entering the city's combined sewer system. Site selection begins with visual markers and subsequent infiltration testing to determine suitability. Only locations with adequate soil permeability progress to construction. Installation involves excavating the footpath or roadside to a depth of approximately 1.5 meters, followed by backfilling with graded layers of stone and engineered soil. Curbs are reinstalled with inlet cuts to direct stormwater into the garden. Each rain garden is planted with a mix of hardy vegetation suited to stormwater conditions. These vegetated depressions allow runoff to infiltrate through the soil profile, filtering pollutants and reducing peak flow volumes. DEP conducts inspections upon completion to confirm functionality and assumes long-term maintenance responsibilities, including sediment removal, replanting, and litter clearance. Designs are carefully planned to avoid pedestrian and vehicular disruptions. Gardens are engineered to fully drain within 48 hours, preventing mosquito breeding. Thousands of these systems have been installed across New York City, contributing to local cooling through evapotranspiration, supporting biodiversity, and enhancing streetscape aesthetics. Community members can participate in ongoing maintenance through the city's Rain Garden Stewardship Program.[46]

NYC DEP. "Rain Gardens." (n.d.). https://nyc.gov/site/dep/water/rain-gardens.page.

Riparian Buffers

Riparian buffers serve as natural filters between catchments and receiving environments, intercepting and absorbing excess nutrients before they reach water bodies. They slow and filter stormwater runoff, allowing plants to uptake and transform contaminants, improving water quality. The vegetation and humus layers in riparian buffers retain significant amounts of water, promoting infiltration into the soil and releasing it gradually over time, reducing the risk of flooding and erosion. In addition to improving water management, riparian buffers help preserve the natural form and function of waterways by restricting development along creeks, streams, and rivers. When implemented across a watershed, they maintain stream base flows, support interconnected habitats, and enhance urban biodiversity. By providing continuous natural corridors, riparian buffers benefit wildlife and offer recreational opportunities, such as walking and cycling trails, enriching urban green spaces.[47]

Floodplain Restoration and Management

Transforming floodplains to reduce flood risks can be achieved by increasing water storage capacity or improving water conveyance through the floodplain. Implementing GI in floodplains helps manage floodwaters by enlarging retention areas, which increases storage capacity and prevents water from flooding areas used for human activities. Various GI measures support floodplain restoration and flood risk reduction, including:

- Creating green rivers or 'flood bypasses' to divert excess water.
- Removing or setting back embankments to allow natural expansion of the floodplain.
- Connecting isolated water bodies to improve hydrological connectivity.
- Excavating floodplains (lowering or dredging) to increase floodwater capacity.
- Restoring vegetation to enhance water absorption and slow runoff.
- Changing land use to promote natural water retention and infiltration.

Table 10.8 Effects of Floodplain Restoration Projects

Effect Type	*Examples*
Direct effects	• Storage and attenuation of runoff • Regulation of river flow • Reduction of surface runoff • Temporary or permanent water retention
Indirect effects (ecosystem benefits)	• Enhancement of fish populations and recruitment • Support for natural biomass productivity • Conservation of biodiversity • Improved resilience to climate change impacts • Replenishment of groundwater reserves • Soil stabilization and erosion prevention • Natural filtration of pollutants • Opportunities for recreation and tourism • Preservation of aesthetic and cultural landscape values

The impacts of floodplain restoration projects are detailed in Table 10.8, demonstrating how GI solutions can enhance flood resilience while supporting biodiversity and ecosystem health.[48]

Case 10.9: Floodplain Buyout Program, Charlotte-Mecklenburg, USA

Mecklenburg County's Storm Water Services has implemented a voluntary floodplain buyout program aimed at reducing long-term flood risk and restoring natural floodplain functions. Since 1999, over 490 homes have been purchased and demolished, with more than 785 families and businesses relocated to safer areas. The program was not initiated in response to a catastrophic event, but rather emerged from a strategic overhaul of floodplain management following several moderate floods. Properties are selected for buyout based on risk and cost-effectiveness, using a scoring system established in the county's Risk Assessment, Risk Reduction Plan. The criteria include flood likelihood and estimated damage, with annual updates guiding prioritization. Buyouts are primarily funded through a mix of local fees and federal grants. Federal Emergency Management Agency mitigation grants contribute when properties meet national eligibility requirements. At the same time, local risk-based and "Quick Buy" initiatives offer flexibility to address high-risk structures more swiftly using local resources. After acquisition, properties are cleared and converted into green infrastructure features. These include reforested areas, stormwater retention zones, floodplain restoration sites, and greenways. Interim maintenance is provided until long-term community use is finalized through public engagement. In some instances, public safety agencies use the acquired buildings for training before demolition. The program integrates with broader urban planning by regulating upstream development and updating future floodplain maps to prevent new buildings from becoming future buyout candidates. With a total investment exceeding $90 million, Charlotte-Mecklenburg's approach combines risk mitigation with environmental enhancement, supporting resilient urban water management through permanent land-use change.[49]

Mecklenburg County Storm Water Services. "Floodplain Buyout Program." (n.d.). https://stormwaterservices.mecknc.gov/floodplain-buyout-program.

Transport Linear Infrastructure

Minimizing the fragmentation of natural ecosystems caused by transportation infrastructure is a GI strategy in the transportation sector. To reduce habitat

Table 10.9 Greening Streets

Purpose of Nature-Based Solution	*Nature-Based Solution*	*Description*
Capturing stormwater runoff	Buffer vegetation and lawns	Vegetated strips or lawns that absorb stormwater from adjacent roads and pavement, promoting evaporation and infiltration while reducing local flooding.
	Stormwater planters	Rectangular systems installed at street level to collect runoff. These planters store, filter, and slowly release water through infiltration and evapotranspiration.
	Stormwater bump-outs	Kerb extensions planted with vegetation that intercept street runoff. Positioned mid-block or at intersections to reduce stormwater flows and enhance streetscape greening.
	Stormwater tree trenches	Underground systems connecting street trees with a shared infiltration zone. Runoff enters the trench and slowly infiltrates while watering the trees.
Allowing water to percolate into the ground	Permeable pavement	Pavement system that allows rainwater to filter through its surface into a subsurface reservoir for temporary storage and soil infiltration.
	Depaved areas	Removal of unnecessary hard surfaces to allow vegetation and soil to naturally absorb stormwater, improving infiltration and supporting biodiversity.
	Gravel trenches	Stone-filled underground trenches that collect runoff and allow gradual infiltration into surrounding soil, reducing surface water accumulation.

disruption and maintain ecological connectivity, various GI solutions can be implemented, including the following:

Greening Streets

Greening the streets is a district-level intervention integrating vegetation and permeable surfaces to manage stormwater while enhancing urban environments. The primary goals of GI in streets are to capture stormwater runoff in vegetated areas and allow water to percolate into the ground, reducing flood risks and improving water management. Stormwater capture can be achieved through buffer vegetation, lawns, stormwater planters, bump-outs, and tree trenches, while permeable pavements, depaved areas, and gravel trenches facilitate water infiltration (see Table 10.9).[50] Beyond stormwater management, greening the streets offers numerous co-benefits, including air pollution filtration, noise reduction, shade provision, and mitigation of the urban heat island effect. Additionally, integrating greenery into urban streetscapes encourages active transportation, such as walking and cycling, contributing to healthier and more sustainable cities.[51]

Case 10.10: Toronto's Green Streets Initiative

The City of Toronto's Green Streets program integrates green infrastructure into the design of public roadways to address urban stormwater challenges while enhancing the public realm. Replacing traditional street designs that rapidly channel rainwater into storm sewers, Green Streets introduce planted areas and permeable surfaces that slow, capture, and filter stormwater before it reaches local waterways. This shift supports the city's resilience strategy by reducing pollutant loads, mitigating urban flooding, and promoting groundwater recharge. By redirecting runoff through vegetation and engineered soil systems, Green Streets contribute to improved water quality and enhanced erosion control in Toronto's rivers and lakes. The approach also enhances the growing conditions for urban trees, contributing to the longevity and health of the city's urban forest. Cooler surface temperatures are promoted through increased shade and evapotranspiration, mitigating the urban heat island effect. Additionally, the introduction of native vegetation provides habitat opportunities and supports local biodiversity goals outlined in the Toronto Biodiversity Strategy. Projects span across various neighborhoods and street types, guided by the City's Green Streets Technical Guidelines. These guidelines provide standardized design practices and selection tools for integrating green infrastructure in both new and retrofit street projects. The selection of street-specific interventions, such as bioswales, tree trenches, or permeable pavements, is based on conditions like road width, traffic patterns, and adjacent land use. The city also maintains an online Green Streets map, which displays both completed and ongoing projects. Ongoing maintenance is managed by City staff, with public involvement encouraged through stewardship opportunities.[52]

City of Toronto. "Green Streets." (n.d.). https://toronto.ca/services-payments/streets-parking-transportation/enhancing-our-streets-and-public-realm/green-streets/.

Permeable Pavements

Permeable pavement is a porous urban surface made of open-pore pavers, concrete, or asphalt with an underlying stone reservoir that allows water to infiltrate the ground instead of running off into storm drains. Typical applications include carparks, low-traffic roads, footpaths, and driveways. Permeable pavement offers several environmental and urban benefits:

- Reduces runoff volume by trapping and slowly releasing precipitation into the ground, preventing excess water from entering storm drains.
- Filters pollutants through physical, chemical, and biological processes, such as trapping contaminants in pavement or soil, microbial breakdown of pollutants, and plant uptake in vegetated pavers.
- Lowers urban runoff temperature, reducing thermal stress on aquatic ecosystems.
- Mitigates the urban heat island effect by lowering surrounding air temperatures.
- Enhances community livability by reducing noise pollution and improving aesthetic and environmental quality in urban areas.[53,54,55]

Naturing of High-Capacity Transport Linear Infrastructure

Integrating GI into high-capacity transport corridors can support urban revitalization and enhance climate resilience in growing, densely populated cities. Planting vegetation along roads, railways, and bridges creates new habitats for flora and fauna while establishing ecological corridors that improve connectivity between urban green spaces. This approach enhances biodiversity and contributes to improved air quality, reduced noise, and increased urban cooling, making cities more livable and environmentally sustainable.[56]

Green Corridors

Green corridors are natural pathways that facilitate the movement of people, animals, seeds, and water while enhancing urban biodiversity and ecological connectivity. These corridors serve multiple functions, including providing habitat, supporting the daily and seasonal movement of wildlife,

Table 10.10 Benefits of Green Corridors in Urban Settings

Benefit	*Description*
Ecological	• Enhances habitat connectivity and supports biodiversity, especially in natural corridors like rivers, wetlands, and aquatic systems • Reduces soil erosion and improves stormwater drainage by absorbing rainfall • Mitigates urban heat island effects by cooling the built environment
Social	• Offers spaces for recreation, relaxation, and social interaction • Strengthens local identity and cultural expression • Encourages community engagement and fosters a sense of belonging
Environmental	• Integrates nature into urban design, improving urban livability and well-being • Enhances accessibility through networks of trails and greenways • Influences city form and land use planning by defining spatial boundaries • Supports sustainable transport by offering green routes for walking and cycling

and facilitating species dispersal. By linking trees, parks, green spaces, and drainage areas with vegetated pedestrian walkways, cycle paths, and transport routes, green corridors create thriving ecosystems within cities. They also play a crucial role in building climate resilience, helping to mitigate floods, droughts, heatwaves, and other impacts of climate change.[57,58] The specific benefits of green corridors in urban environments are detailed in Table 10.10, demonstrating their role in creating healthier, more sustainable, and livable cities.[59]

Case 10.11: Kingfisher Creek Corridor Concept Plan, Brisbane

Brisbane City Council has developed the Kingfisher Creek corridor concept plan as the final of three large-scale integrated projects under the Norman Creek 2012–2031 Master Plan. Focused on the inner-city suburbs of Woolloongabba and East Brisbane, the concept plan outlines a long-term vision to revitalize a largely underground creek system by creating a connected sequence of green open spaces. The project prioritizes resilience, sustainability, and accessibility in areas experiencing significant urban growth. The corridor connects locations such as Watt Park, Woolloongabba Rotary Park, and Moorhen Flats. These spaces will be enhanced with features including improved active travel infrastructure, expanded vegetation, and interpretative signage. Public feedback collected over multiple engagement stages informed the plan's priorities, which include protecting and celebrating the unique bushland setting at Moorhen Flats, improving park amenities and safety at Woolloongabba Rotary Park, and enhancing connectivity through Watt Park. The key objectives of the project are to strengthen biodiversity and ecological resilience, foster community connections, and enhance accessibility to nature. The plan also seeks to address challenges such as arterial road barriers, limited signage, and heat island effects by introducing shade trees, wayfinding systems, and green infrastructure. The corridor will support broader transformation in the precinct ahead of the Brisbane 2032 Olympic and Paralympic Games, as Woolloongabba and East Brisbane continue to evolve as cultural, transport, and employment hubs. Community consultation also informed updates to the local road network, aiming to enhance safety, reduce through-traffic, and maintain community cohesion.[60]

Brisbane City Council. "Waterway Projects." (n.d.). https://brisbane.qld.gov.au/environment-and-water/creeks-and-waterways/waterway-projects.

Case Summary

Various GI approaches have been implemented to enhance sustainable urbanization (Table 10.11).

Notes

1 R.C. Brears, *Blue and Green Cities: The Role of Blue-Green Infrastructure in Managing Urban Water Resources* (London, UK: Palgrave Macmillan UK, 2018).

2 Stephan Pauleit et al., "Urban Landscapes and Green Infrastructure," in *Oxford Research Encyclopedia of Environmental Science* (Oxford, UK: Oxford University Press, 2017).

3 European Commission, "Towards an Eu Research and Innovation Policy Agenda for Nature-Based Solutions & Re-Naturing Cities," (2015), https://publications.europa.eu/en/publication-detail/-/publication/fb117980-d5aa-46df-8edc-af367cddc202.

4 Green Surge, "What Benefits Does Green Infrastructure in Cities Offer?," (2017), https://greensurge.eu/pdfer/05_Benefits._A_functional_perspective.pdf.

5 S. Lehmann, "Reconnecting with Nature: Developing Urban Spaces in the Age of Climate Change," *Emerald Open Research* 1, no. 2 (2019).

6 Laura A. Wendling et al., "Benchmarking Nature-Based Solution and Smart City Assessment Schemes against the Sustainable Development Goal Indicator Framework," *Frontiers in Environmental Science* 6, no. 69 (2018).

7 APA, "Green Infrastructure: A Landscape Approach," (2013), http://caeau.com.ar/wp-content/uploads/2018/11/46.GREEN-INFRAESTRUCTURE.pdf.

8 M. van den Bosch and Å. Ode Sang, "Urban Natural Environments as Nature-Based Solutions for Improved Public Health – A Systematic Review of Reviews," *Environmental Research* 158 (2017).

Table 10.11 Green Infrastructure Enhancing Sustainable Urbanization

Habitat	*Project Name*	*Activities*	*Co-Benefits*
Green roofs	Hamburg's Green Roof Strategy	Implements financial incentives, building code integration, and scientific monitoring to promote 100 hectares of green roof area by 2024, targeting 70% coverage for new and renovated roofs.	Enhances stormwater retention, biodiversity, thermal regulation, and public amenity through multifunctional rooftop spaces.
Water harvesting	Tucson Rainwater Harvesting Rebate Program	Offers up to $2,000 in rebates for installing residential rainwater systems that meet specific runoff and design criteria, requiring workshops and pre-approval.	Promotes household-level water conservation, reduces urban water demand, and supports climate adaptation.
Green roofs and walls	Sydney's Green Roofs and Walls Policy	Establishes definitions, design standards, and implementation tools to promote green infrastructure on rooftops and façades across the city.	Supports stormwater management, biodiversity, thermal regulation, and green building performance.
Urban biodiversity	Utrecht's Bee-Friendly Bus Stops	Transforms 316 bus stops with sedum roofs, solar lighting, bamboo seating, and recycled materials, implemented through advertising revenue.	Improves pollinator habitat, air quality, rainwater retention, and public space sustainability.
Flood management	La Marjal Floodable Park – Alicante, Spain	Designs a 36,700 square meters park with 45,000 cubic meters flood storage capacity integrated into a public green space, mimicking natural wetlands.	Provides flood protection, biodiversity habitat, public recreation, and urban cooling.
Urban green space	Frankfurt's Green Belt	Covers 80 square kilometers with meadows, forests, orchards, and restored industrial areas, forming part of the Rhine-Main Regional Park.	Enhances biodiversity, food systems, recreation, and regional identity while legally protecting green space.
Stormwater wetlands	First Creek Wetland – Adelaide	Converts underutilized urban land into stormwater harvesting and treatment wetlands with aquifer storage and educational features.	Supports irrigation reuse, environmental education, urban biodiversity, and public engagement.

Table 10.11 (Continued)

Habitat	*Project Name*	*Activities*	*Co-Benefits*
Rain gardens	New York City Rain Garden Program	Retrofits sidewalks with rain gardens to manage runoff and improve infiltration, using layered soil media and vegetation.	Reduces combined sewer overflow, supports biodiversity, cools urban streetscapes, and engages local residents.
Floodplain restoration	Charlotte-Mecklenburg Floodplain Buyout Program	Voluntarily acquires flood-prone properties and restores them to green infrastructure for flood mitigation and habitat creation.	Reduces flood risk, restores ecological functions, and integrates urban planning with climate resilience.
Green streets	Toronto's Green Streets Initiative	Integrates bioswales, permeable pavements, and vegetated tree trenches in street design to manage stormwater and enhance public space.	Improves water quality, reduces flooding, supports biodiversity, and mitigates heat island effects.
Urban green corridors	Kingfisher Creek Corridor Concept Plan – Brisbane	Revitalizes an underground creek system through a network of open spaces with biodiversity corridors, park upgrades, shade trees, signage, and improved connectivity, shaped by community input.	Strengthens urban biodiversity, mitigates heat island effects, improves access to nature, and fosters resilient, connected communities.

9 Nadja Kabisch, Matilda van den Bosch, and Raffaele Lafortezza, "The Health Benefits of Nature-Based Solutions to Urbanization Challenges for Children and the Elderly – A Systematic Review," *Environmental Research* 159 (2017).

10 Joana Vieira et al., "Green Spaces Are Not All the Same for the Provision of Air Purification and Climate Regulation Services: The Case of Urban Parks," *Environmental Research* 160 (2018).

11 T. Elmqvist et al., "Benefits of Restoring Ecosystem Services in Urban Areas," *Current Opinion in Environmental Sustainability* 14 (2015).

12 APA, "Green Infrastructure: A Landscape Approach."

13 Brears, *Blue and Green Cities: The Role of Blue-Green Infrastructure in Managing Urban Water Resources.*

14 European Federation of Green Roof and Green Wall Associations and Livingroofs.org, "Living Roofs and Walls: From Policy to Practice. 10 Years of Urban Greening in London and Beyond," (2019), https://livingroofs.org/wp-content/uploads/2019/04/LONDON-LIVING-ROOFS-WALLS-REPORT-2019.pdf.

15 Ahmet B. Besir and Erdem Cuce, "Green Roofs and Facades: A Comprehensive Review," *Renewable and Sustainable Energy Reviews* 82 (2018).

16 Reeman M. Rehan, "Cool City as a Sustainable Example of Heat Island Management Case Study of the Coolest City in the World," *HBRC Journal* 12, no. 2 (2016).
17 Daniel Roehr and Yuewei Kong, "Runoff Reduction Effects of Green Roofs in Vancouver, Bc, Kelowna, Bc, and Shanghai, P.R. China," *Canadian Water Resources Journal / Revue canadienne des ressources hydriques* 35, no. 1 (2010).
18 Lehmann, "Reconnecting with Nature: Developing Urban Spaces in the Age of Climate Change."
19 Ministry of Environment and Energy City of Hamburg, "More Green Roofs for Hamburg: Green Roof Strategy for Hamburg," (n.d.), https://hamburg.de/resource/blob/281688/a5544533af6cba7305c0d8c58946fa96/d-more-green-roofs-for-hamburg-data.pdf.
20 Nadia Sabeh, "Chapter 6 – Rooftop Plant Production Systems in Urban Areas," in *Plant Factory*, ed. Toyoki Kozai, Genhua Niu, and Michiko Takagaki (San Diego, CA: Academic Press, 2016).
21 RUAF Foundation, "Rooftop Agriculture – A Climate Change Perspective," (2012), https://ruaf.org/sites/default/files/RUAF%20Manual%20on%20rooftopgardening%20-a%20climate%20change%20perspective%20%28Final%29.pdf.
22 Mastura Safayet, Md F. Arefin, and Md M.U. Hasan, "Present Practice and Future Prospect of Rooftop Farming in Dhaka City: A Step Towards Urban Sustainability," *Journal of Urban Management* 6, no. 2 (2017).
23 K. Exall and T.D. Vassos, "14 – Integrated Urban Water Management: Water Use and Reuse," in *Metropolitan Sustainability*, ed. Frank Zeman (Woodhead Publishing, 2012).
24 City of Tucson, "Residential Customer Rebates," (n.d.), https://tucsonaz.gov/Departments/Water/Conservation/Residential-Customer-Rebates#:~:text=All%20water%20harvesting%20features%20on,full%20rebate%20amount%20%241.00%2Fgal.
25 E. Morello, I. Mahmoud, and N. Colaninno, "Catalogue of Nature-Based Solutions for Urban Regeneration," (2019), http://labsimurb.polimi.it/nbs-catalogue/.
26 Morello, Mahmoud, and Colaninno, "Catalogue of Nature-Based Solutions for Urban Regeneration."
27 I. Susorova, "5 – Green Facades and Living Walls: Vertical Vegetation as a Construction Material to Reduce Building Cooling Loads," in *Eco-Efficient Materials for Mitigating Building Cooling Needs*, ed. F. Pacheco-Torgal et al. (Oxford: Woodhead Publishing, 2015).
28 City of Sydney, "Green Roofs and Walls," (n.d.), https://cityofsydney.nsw.gov.au/environmental-support-funding/green-roofs-and-walls.
29 Ihobe Environmental Management Agency, "Nature-Based Solutions for Local Climate Adaptation in the Basque Country," (2017), http://growgreenproject.eu/wp-content/uploads/2018/05/NBS-Climate-Adaptation-Basque-Country.pdf.
30 Ihobe Environmental Management Agency, "Nature-Based Solutions for Local Climate Adaptation in the Basque Country."
31 Morello, "Catalogue of Nature-Based Solutions for Urban Regeneration."
32 ArchDaily, "Utrecht Creates 300 Bee-Friendly Bus Stops," (n.d.), https://archdaily.com/921850/utrecht-creates-300-bee-friendly-bus-stops.
33 US EPA, "Green Infrastructure in Parks: A Guide to Collaboration, Funding, and Community Engagement," (2017), https://epa.gov/sites/production/files/2017-05/documents/gi_parksplaybook_2017-05-01_508.pdf.

34 Cadena SER, "Décimo Aniversario De La Marjal Cuyo éxito El Ayuntamiento Quiere Replicar En "Zonas Sensibles" De Alicante," (n.d.), https://cadenaser.com/comunitat-valenciana/2025/02/28/decimo-aniversario-de-la-marjal-cuyo-exito-el-ayuntamiento-quiere-replicar-en-zonas-sensibles-de-alicante-radio-alicante/.

35 Martina Winker et al., "Chapter 21 – Greening and Cooling the City Using Novel Urban Water Systems: A European Perspective," in *Approaches to Water Sensitive Urban Design*, ed. Ashok K. Sharma, Ted Gardner, and Don Begbie (Cambridge, United Kingdom: Woodhead Publishing, 2019).

36 Ihobe Environmental Management Agency, "Nature-Based Solutions for Local Climate Adaptation in the Basque Country."

37 Ihobe Environmental Management Agency, "Nature-Based Solutions for Local Climate Adaptation in the Basque Country."

38 Brears, *Blue and Green Cities: The Role of Blue-Green Infrastructure in Managing Urban Water Resources*.

39 Josh Foster, Ashley Lowe, and Steve Winkelman, "The Value of Green Infrastructure for Urban Climate Adaptation," *Center for Clean Air Policy* (2011), https://nacto.org/wp-content/uploads/2017/08/The-value-of-green-infrastructure-for-urban-climate-adaptation.pdf.

40 Stadt Frankfurt am Main, "Der Frankfurter Grüngürtel," (n.d.), https://frankfurt.de/themen/umwelt-und-gruen/orte/gruenguertel.

41 Angelo Panno et al., "Nature-Based Solutions to Promote Human Resilience and Wellbeing in Cities during Increasingly Hot Summers," *Environmental Research* 159 (2017).

42 Department of Environment Government of South Australia, Water and Natural Resources, "Green Infrastructure Case Study: Wetland and Stormwater Reuse – First Creek Wetland, Adelaide Botanic Garden," (2017), https://cdn.environment.sa.gov.au/landscape/docs/hf/green-infrastructure-first-creek-case-study.pdf.

43 URBAN GreenUP, "D1.1 – Nbs Catalogue," (2018), https://urbangreenup.eu/insights/.

44 Brears, *Blue and Green Cities: The Role of Blue-Green Infrastructure in Managing Urban Water Resources*.

45 Brears, *Blue and Green Cities: The Role of Blue-Green Infrastructure in Managing Urban Water Resources*.

46 NYC DEP, "Rain Gardens," (n.d.), https://nyc.gov/site/dep/water/rain-gardens.page.

47 NYC DEP, "Rain Gardens."

48 EEA, "Green Infrastructure and Flood Management. Promoting Cost-Efficient Flood Risk Reduction Via Green Infrastructure Solutions," (2017), https://eea.europa.eu/publications/green-infrastructure-and-flood-management.

49 Mecklenburg County Storm Water Services, "Floodplain Buyout Program," (n.d.), https://stormwaterservices.mecknc.gov/floodplain-buyout-program.

50 Brears, *Blue and Green Cities: The Role of Blue-Green Infrastructure in Managing Urban Water Resources*.

51 Ihobe Environmental Management Agency, "Nature-Based Solutions for Local Climate Adaptation in the Basque Country."

52 City of Toronto, "Green Streets," (n.d.), https://toronto.ca/services-payments/streets-parking-transportation/enhancing-our-streets-and-public-realm/green-streets/.

53 USGS, "Evaluating the Potential Benefits of Permeable Pavement on the Quantity and Quality of Stormwater Runoff," https://usgs.gov/science/evaluating-potential-benefits-permeable-pavement-quantity-and-quality-stormwater-runoff?qt-science_center_objects=0#qt-science_center_objects.

54 Sam Kubba, "Chapter 6 – Choosing Materials and Products," in *Green Construction Project Management and Cost Oversight*, ed. Sam Kubba (Boston, MA: Architectural Press, 2010).
55 Brears, *Blue and Green Cities: The Role of Blue-Green Infrastructure in Managing Urban Water Resources*.
56 Ihobe Environmental Management Agency, "Nature-Based Solutions for Local Climate Adaptation in the Basque Country."
57 H.A. Aziz and M.H. Rasidi, "The Role of Green Corridors for Wildlife Conservation in Urban Landscape: A Literature Review," *IOP Conference Series: Earth and Environmental Science* 18 (2014).
58 ADB, "Nature-Based Solutions for Building Resilience in Towns and Cities: Case Studies from the Greater Mekong Subregion," (2016).
59 S.S.A. Aly and M.S.E. Amer, "Green Corridors as a Response for Nature: Greening Alexandria City by Creating a Green Infrastructure Network," *WIT Transactions on Ecology and the Environment* 130 (2010).
60 Brisbane City Council, "Waterway Projects," (n.d.), https://brisbane.qld.gov.au/environment-and-water/creeks-and-waterways/waterway-projects.

References

ADB. "Nature-Based Solutions for Building Resilience in Towns and Cities: Case Studies from the Greater Mekong Subregion." (2016).

Aly, S.S.A., and M.S.E. Amer. "Green Corridors as a Response for Nature: Greening Alexandria City by Creating a Green Infrastructure Network." *WIT Transactions on Ecology and the Environment* 130 (2010): 101–17.

APA. "Green Infrastructure: A Landscape Approach." (2013). http://caeau.com.ar/wp-content/uploads/2018/11/46.GREEN-INFRAESTRUCTURE.pdf.

Aziz, H.A., and M.H. Rasidi. "The Role of Green Corridors for Wildlife Conservation in Urban Landscape: A Literature Review." *IOP Conference Series: Earth and Environmental Science* 18 (2014/02/25): 012093.

Besir, Ahmet B., and Erdem Cuce. "Green Roofs and Facades: A Comprehensive Review." *Renewable and Sustainable Energy Reviews* 82 (2018/02/01): 915–39.

Brears, R.C. *Blue and Green Cities: The Role of Blue-Green Infrastructure in Managing Urban Water Resources*. London, UK: Palgrave Macmillan UK, 2018.

EEA. "Green Infrastructure and Flood Management. Promoting Cost-Efficient Flood Risk Reduction via Green Infrastructure Solutions." (2017). https://eea.europa.eu/publications/green-infrastructure-and-flood-management.

Elmqvist, T.H. Setälä, S.N. Handel, S. van der Ploeg, J. Aronson, J.N. Blignaut, E. Gómez-Baggethun, et al. "Benefits of Restoring Ecosystem Services in Urban Areas." *Current Opinion in Environmental Sustainability* 14 (2015/06/01): 101–08.

European Commission. "Towards an Eu Research and Innovation Policy Agenda for Nature-Based Solutions & Re-Naturing Cities." (2015). https://publications.europa.eu/en/publication-detail/-/publication/fb117980-d5aa-46df-8edc-af367cddc202.

European Federation of Green Roof and Green Wall Associations and Livingroofs.org. "Living Roofs and Walls: From Policy to Practice. 10 Years of Urban Greening in London and Beyond." (2019). https://livingroofs.org/wp-content/uploads/2019/04/LONDON-LIVING-ROOFS-WALLS-REPORT-2019.pdf.

Exall, K., and T.D. Vassos. "14 – Integrated Urban Water Management: Water Use and Reuse." In *Metropolitan Sustainability*, edited by Frank Zeman, 319–49. Cambridge, United Kingdom: Woodhead Publishing, 2012.

Foster, Josh, Ashley Lowe, and Steve Winkelman. "The Value of Green Infrastructure for Urban Climate Adaptation." *Center for Clean Air Policy* (2011). https://nacto.org/wp-content/uploads/2017/08/The-value-of-green-infrastructure-for-urban-climate-adaptation.pdf.

Green Surge. "What Benefits Does Green Infrastructure in Cities Offer?" (2017). https://greensurge.eu/pdfer/05_Benefits._A_functional_perspective.pdf.

Ihobe Environmental Management Agency. "Nature-Based Solutions for Local Climate Adaptation in the Basque Country." (2017). http://growgreenproject.eu/wp-content/uploads/2018/05/NBS-Climate-Adaptation-Basque-Country.pdf.

Kabisch, Nadja, Matilda van den Bosch, and Raffaele Lafortezza. "The Health Benefits of Nature-Based Solutions to Urbanization Challenges for Children and the Elderly – A Systematic Review." *Environmental Research* 159 (2017/11/01): 362–73.

Kubba, Sam. "Chapter 6 – Choosing Materials and Products." In *Green Construction Project Management and Cost Oversight*, edited by Sam Kubba, 221–66. Boston, MA: Architectural Press, 2010.

Lehmann, S. "Reconnecting with Nature: Developing Urban Spaces in the Age of Climate Change." *Emerald Open Research* 1, no. 2 (2019).

Morello, E., I. Mahmoud, and N. Colaninno. "Catalogue of Nature-Based Solutions for Urban Regeneration." (2019). http://labsimurb.polimi.it/nbs-catalogue/.

Panno, Angelo, Giuseppe Carrus, Raffaele Lafortezza, Luigi Mariani, and Giovanni Sanesi. "Nature-Based Solutions to Promote Human Resilience and Wellbeing in Cities during Increasingly Hot Summers." *Environmental Research* 159 (2017/11/01): 249–56.

Pauleit, Stephan, Rieke Hansen, Emily Rall, Teresa Zölch, Erik Andersson, Ana Luz, Luca Száraz, Iván Tosics, and Kati Vierikko. "Urban Landscapes and Green Infrastructure." In *Oxford Research Encyclopedia of Environmental Science*. Oxford, UK: Oxford University Press, 2017.

Rehan, Reeman M. "Cool City as a Sustainable Example of Heat Island Management Case Study of the Coolest City in the World." *HBRC Journal* 12, no. 2 (2016/08/01): 191–204.

Roehr, Daniel, and Yuewei Kong. "Runoff Reduction Effects of Green Roofs in Vancouver, Bc, Kelowna, Bc, and Shanghai, P.R. China." *Canadian Water Resources Journal / Revue canadienne des ressources hydriques* 35, no. 1 (2010/01/01): 53–68.

RUAF Foundation. "Rooftop Agriculture – A Climate Change Perspective." (2012). https://ruaf.org/sites/default/files/RUAF%20Manual%20on%20rooftopgardening%20-a%20climate%20change%20perspective%20%28Final%29.pdf.

Sabeh, Nadia. "Chapter 6 – Rooftop Plant Production Systems in Urban Areas." In *Plant Factory*, edited by Toyoki Kozai, Genhua Niu, and Michiko Takagaki, 105–11. San Diego, CA: Academic Press, 2016.

Safayet, Mastura, Md F. Arefin, and Md M.U. Hasan. "Present Practice and Future Prospect of Rooftop Farming in Dhaka City: A Step Towards Urban Sustainability." *Journal of Urban Management* 6, no. 2 (2017/12/01): 56–65.

Susorova, I. "5 – Green Facades and Living Walls: Vertical Vegetation as a Construction Material to Reduce Building Cooling Loads." In *Eco-Efficient Materials for Mitigating Building Cooling Needs*, edited by F. Pacheco-Torgal, J.A. Labrincha, L.F. Cabeza, and C.G. Granqvist, 127–53. Oxford: Woodhead Publishing, 2015.

URBAN GreenUP. "D1.1 – Nbs Catalogue." (2018). https://urbangreenup.eu/insights/.

US EPA. "Green Infrastructure in Parks: A Guide to Collaboration, Funding, and Community Engagement." (2017). https://epa.gov/sites/production/files/2017-05/documents/gi_parksplaybook_2017-05-01_508.pdf.

USGS. "Evaluating the Potential Benefits of Permeable Pavement on the Quantity and Quality of Stormwater Runoff." (n.d.). https://usgs.gov/science/evaluating-potential-benefits-permeable-pavement-quantity-and-quality-stormwater-runoff?qt-science_center_objects=0#qt-science_center_objects.

van den Bosch, M., and Å Ode Sang. "Urban Natural Environments as Nature-Based Solutions for Improved Public Health – A Systematic Review of Reviews." *Environmental Research* 158 (2017/10/01): 373–84.

Vieira, Joana, Paula Matos, Teresa Mexia, Patrícia Silva, Nuno Lopes, Catarina Freitas, Otília Correia, et al. "Green Spaces Are Not All the Same for the Provision of Air Purification and Climate Regulation Services: The Case of Urban Parks." *Environmental Research* 160 (2018/01/01): 306–13.

Wendling, Laura A., Aapo Huovila, Malin zu Castell-Rüdenhausen, Mari Hukkalainen, and Miimu Airaksinen. "Benchmarking Nature-Based Solution and Smart City Assessment Schemes against the Sustainable Development Goal Indicator Framework." [In English]. *Frontiers in Environmental Science* 6, no. 69 (2018/7/04).

Winker, Martina, Simon Gehrmann, Engelbert Schramm, Martin Zimmermann, and Annette Rudolph-Cleff. "Chapter 21 – Greening and Cooling the City Using Novel Urban Water Systems: A European Perspective." In *Approaches to Water Sensitive Urban Design*, edited by Ashok K. Sharma, Ted Gardner, and Don Begbie, 431–54. Cambridge, United Kingdom: Woodhead Publishing, 2019.

11 Nature-Based Solutions for Disaster Risk Management and Resilience

Introduction

Globally, disasters caused by hazard events, such as floods, droughts, heatwaves, tropical cyclones, volcanic eruptions, and earthquakes, disrupt livelihoods, hinder economic growth, and threaten overall human well-being. Over the past 20 years, 90% of disasters have been linked to weather-related events, including floods, storms, and heatwaves.[1] In the last decade, ecosystem-based approaches have gained recognition within disaster risk reduction (DRR) and environmental conservation communities. These approaches are increasingly valuable for mitigating hazards, securing livelihoods, and enhancing resilience to future disasters by leveraging natural systems to reduce risks and support sustainable development.[2]

Disaster Risk Reduction

Natural hazards are physical events such as cyclones, earthquakes, and tsunamis that have the potential to cause harm to people and property. However, these hazards only become disasters when they result in widespread disruption, human and material losses, and impacts that exceed a community's ability to cope using its resources, as defined by the United Nations International Strategy for Disaster Risk Reduction (UNISDR). Disasters are not just the result of natural events, but are also social constructs influenced by how societies manage their environment, their levels of vulnerability, their capacity to respond, and the available resources for recovery. Effective DRR, therefore, depends on both reducing exposure to hazards and strengthening societal resilience.[3]

While natural hazards cannot always be prevented, their potential to cause disasters can be avoided or mitigated through effective DRR strategies. One key approach is to leverage ecosystems to mitigate the damage caused by hazards. Ecosystems such as forests, wetlands, coastal and marine systems, and drylands play a crucial role in DRR, yet their

DOI: 10.4324/9781003517948-11

degradation increases vulnerability and amplifies disaster risks. Protecting and restoring these ecosystems can serve as a natural buffer against hazards, strengthening resilience and reducing the likelihood of severe impacts on communities.[4,5,6,7]

Ecosystem-Based Disaster Risk Reduction

Nature-based solutions (NBS) that focus on conserving and restoring ecosystems can significantly reduce disaster risk and enhance community resilience. In DRR, NBS are applied as ecosystem-based disaster risk reduction (Eco-DRR) strategies, which involve the sustainable management, conservation, and restoration of ecosystems to mitigate hazards and strengthen livelihood resilience. Well-managed ecosystems serve as natural infrastructure, reducing exposure to hazards while supporting socio-economic resilience by providing essential resources such as food and water. However, the ability of ecosystems to reduce disaster risk depends on their health and structure as well as the intensity of the hazard event.[8]

The health of ecosystems is crucial for disaster risk reduction as:

- Human well-being depends on ecosystems that help people withstand, cope with, and recover from disasters. There is a two-way relationship between poverty and disasters, with poor communities experiencing more disasters, especially in degraded environments.
- Ecosystems such as wetlands, forests, and coastal systems serve as cost-effective natural buffers, helping to mitigate the impacts of hazards and adapt to climate change.
- Healthy and diverse ecosystems are more resilient to extreme weather events. Intact ecosystems are less vulnerable to damage and more resilient in their ability to recover from disasters. However, disasters can still lead to habitat loss and species mortality.
- Ecosystem degradation reduces the ability of natural systems to sequester carbon, contributing to climate change and increasing the frequency and severity of climate-related disasters.[9]

Multiple Benefits of Ecosystem-Based Disaster Risk Reduction

Eco-DRR strategies provide environmental, economic, and social benefits that extend beyond disaster risk reduction. They are often considered low-regret or no-regret options because they provide value regardless of uncertainties in future climate conditions. For example, restoring mangroves protects coastlines from storm surges, supports fish habitats, enhances local livelihoods, and contributes to carbon storage. These strategies strengthen ecosystem resilience while simultaneously improving community well-being and economic stability. A range of co-benefits associated with Eco-DRR is outlined in Table 11.1.[10,11]

Table 11.1 Multiple Benefits of Ecosystem-Based Disaster Risk Reduction

Benefit	*Examples*
Economic	• Extends the lifespan of built flood protection infrastructure • Provides cost-effective natural barriers (e.g. wetlands, mangroves, forests) that reduce disaster risk and recovery costs
Climate	• Supports climate change mitigation through forest and coastal ecosystem conservation • Prevents emissions from deforestation and land degradation • Restores carbon-sequestering landscapes such as peatlands and mangroves
Social	• Enhances physical and mental health by improving environmental conditions • Builds community cohesion, trust, and shared responsibility through participation • Helps reduce conflict over natural resources by ensuring equitable access and sustainable management

Principles of Ecosystem-Based Disaster Risk Reduction

Eco-DRR should align with key principles of DRR to effectively enhance resilience in social-ecological systems. This includes fostering cross-sectoral collaboration among governments, local, and regional authorities, and ensuring inclusive participation, particularly for communities most vulnerable to disaster risks. Eco-DRR strategies should focus on building resilience, strengthening adaptive capacity, promoting inclusivity and equity, considering multiple scales, and ensuring effectiveness and efficiency. A detailed overview of these principles is provided in Table 11.2.[12]

Floods

Globally, communities are increasingly relying on NBS to manage flood risks. Eco-DRR strategies are being implemented to mitigate flood impacts, such as widening floodplains, protecting and expanding wetlands, and restoring coral reefs. Beyond flood risk reduction, Eco-DRR offers additional benefits including biodiversity conservation, job creation in agriculture and fisheries, enhanced recreation and tourism opportunities, and improved public health.[13,14]

Natural Drainage Path Restoration

Urbanization and agriculture often alter or eliminate natural drainage paths, such as small streams, disrupting water flow and increasing flood risks. Restoring these drainage paths involves identifying, monitoring, maintaining, and rehabilitating them to improve natural water management. In heavily modified landscapes, identifying these paths can be challenging, as they may

Table 11.2 Principles of Ecosystem-Based Disaster Risk Reduction

Principle	*Description*
Building resilience and adaptive capacity through Eco-DRR	Eco-DRR enhances the resilience and reduces vulnerability of both people and ecosystems. It addresses current and future disaster risks using biodiversity, ecosystem services, and landscape approaches without compromising ecological function, while providing social benefits and promoting sustainable development through equitable, transparent, and participatory processes.
Consider a full range of ecosystem-based approaches	Incorporate diverse ecosystem-based measures to strengthen the resilience of socio-ecological systems within overall disaster risk reduction strategies.
Use disaster response to build back better	Seize opportunities during recovery phases to integrate ecosystem considerations and strengthen resilience throughout all stages of disaster management.
Apply the precautionary principle	Exercise caution in Eco-DRR planning and implementation where scientific certainty is lacking to prevent harm to people, ecosystems, and future resilience.
Ensuring inclusivity and equity in planning and implementation	Ensure full participation of indigenous peoples and local communities with transparency and accountability at every stage of Eco-DRR. Prioritize interventions that protect vulnerable populations and ecosystems from disproportionate disaster impacts.
Achieving Eco-DRR on multiple scales	Design Eco-DRR actions to operate at appropriate spatial and temporal scales, recognizing that certain benefits only emerge over time or across wider geographic areas. Foster coordination and cross-sector collaboration.
Ensuring effectiveness and efficiency of Eco-DRR	Base Eco-DRR on sound science, diverse knowledge systems – including traditional and local knowledge – and practical experience. Integrate adaptive management, monitoring, and evaluation. Minimize trade-offs while maximizing co-benefits such as biodiversity conservation, sustainable development, gender equality, and climate adaptation.

remain dry for extended periods of time. However, expert assessments and local community knowledge can help determine historical flow patterns and discharge requirements. Remote sensing and Geographic Information Systems are valuable tools for identifying and mapping natural drainage paths, aiding in their effective restoration.[15]

Riparian Vegetation Restoration

Riparian plant species are well-adapted to withstand high water velocities and fluctuating water levels, making riparian habitats natural buffers against sudden surges in water flow. In upper watershed streams, these habitats help dissipate the energy of fast-moving water, reducing downstream risks of flash floods and erosion. Fallen trees and organic litter along riverbanks further slow water velocity and provide soil protection. Restoring riparian habitats plays a crucial role in reducing erosion and stabilizing riverbanks. The most critical aspects of riparian restoration are identifying priority areas for rehabilitation and selecting appropriate plant species for revegetation. These plants should stabilize the banks and support broader ecological functions, such as providing habitat for nesting, breeding, and feeding wildlife.[16]

Floodplains and Bypasses

Floodplain restoration re-establishes the natural connection between a river and its floodplain, allowing for more frequent inundation and improved floodwater storage. By restoring this hydrological connectivity, floodplain restoration helps reduce the intensity of flood peaks and lowers downstream flood depths, particularly for frequent, low-intensity floods. This process also enhances groundwater recharge, improves water quality by filtering pollutants, and increases biodiversity by creating habitats for aquatic and terrestrial species. Additionally, restored floodplains can support sustainable land uses such as agriculture, recreation, and ecotourism while providing climate resilience benefits by buffering extreme weather events.[17]

Integrated Flood Risk Management

Integrated flood management involves developing land and water resources within river basins to optimize floodplain use while minimizing loss of life and property damage. This approach works in harmony with nature by enhancing ecosystems and their services, such as restoring a river's capacity to store and slow down floodwaters. Key strategies include rehabilitating natural river features, such as floodplains and wetlands, which act as natural buffers by absorbing and retaining excess water, thereby reducing the speed and impact of flooding.[18] Effective integrated flood risk management requires coordination across multiple activities, including land-use planning, sustainable land management, early warning systems, community engagement, and the construction of protective infrastructure. Given that actions in one part of a river basin can affect other areas, flood risk management is most effective when implemented holistically and coordinated across the entire watershed.[19]

Case 11.1: Wairau Blue-Green Network Project, Auckland, New Zealand

In 2025, Auckland Council's Transport, Resilience and Infrastructure Committee approved the business case for Stages 1 and 2 of the Wairau Blue-Green Network under the Making Space for Water Programme.

The decision enables design, consenting, and early enabling works to proceed. The broader initiative aims to reduce flood risks across the Wairau catchment in Auckland's North Shore by restoring natural water flows and integrating flood management with public space development. Stage 1 focuses on AF Thomas Park, where the existing recreational area, including Takapuna Golf Course, is proposed for redevelopment into a flood storage wetland and detention basin. This would increase the park's flood storage capacity from 60 million to 550 million liters. The redevelopment is intended to protect downstream residential areas and key infrastructure, such as access roads to North Shore Hospital and nearby schools – stage 2 targets land acquired under the Category 3 buy-out program in Milford and Tōtara Vale. Initiatives include stream daylighting, channel widening, and enhancements to overland flow paths, alongside minor upgrades to existing detention facilities. Collectively, both stages aim to reduce flood risk for 261 dwellings, three retirement villages, and 3,900 square meters of commercial space, with 35 properties being removed from high-risk zones. Community engagement will continue throughout the planning and implementation process, with council staff working alongside mana whenua and stakeholders to balance recreational needs and flood resilience. The business case also incorporates a technical assessment of Takapuna Golf's proposal to ensure feasibility and cost-effectiveness ahead of final delivery approvals in 2026.[20]

Auckland Council. "Next Steps to Improve Flood Resilience for the Wairau Community." (n.d.). https://ourauckland.aucklandcouncil.govt.nz/news/2025/03/next-steps-to-improve-flood-resilience-for-the-wairau-community/.

Wetland Restoration

Healthy, well-managed wetlands serve as natural buffers against multiple hazards, reducing disaster risks while supporting local resilience. They sustain livelihoods by providing essential resources, including wild fruits, vegetables, fish, and rice. Wetlands also play a crucial regulatory role in mitigating the impacts of weather and climate-related hazards. For example, floodplains and valley-bottom wetlands absorb and disperse floodwaters, slowing their movement and reducing their destructive energy. During dry periods and droughts, wetlands provide critical water and grazing fodder, helping to maintain local ecosystems and livelihoods.

Wetland restoration involves restoring degraded wetlands to regain their ecological functions and enhance their role in flood management. This process includes re-establishing wetland extent and functionality by removing non-wetland features, recreating lost wetland areas, and implementing restoration measures such as excavation, clearing invasive species, removing

Table 11.3 Types of Natural Wetlands That Can Be Restored

Wetland Type	*Description*
Marshes	Typically found in floodplain areas near the lower reaches of rivers. Characterized by grassy vegetation and peaty soils capable of storing large volumes of water.
Estuarine or tidal wetlands	Located at river mouths or coastal lagoons, including reed beds, salt marshes, mangroves, and mudflats. These wetlands help buffer against tidal flooding and storm surges.
Riverine and forested wetlands	Wooded or shrub-dominated areas adjacent to rivers that can absorb minor increases in water flow and help prevent local flooding.
Shallow lakes and ponds	Naturally occurring in low-lying landscape depressions. They act as temporary reservoirs during storms, slowly releasing water to aquifers or downstream systems.

hydraulic structures, diverting polluted runoff, and reintroducing natural flooding cycles. Beyond flood mitigation, restored wetlands offer additional benefits, including recreational opportunities, fishing, and harvesting reeds, fodder, and fruits. Table 11.3 outlines various types of natural wetlands that can be restored.[21,22]

Case 11.2: Wetland Restoration in the New Forest, United Kingdom

Since 2010, a series of wetland restoration projects have been undertaken in the New Forest, southern England, as part of the Higher Level Stewardship (HLS) scheme. Led by Forestry England and supported by partner organizations, the initiative has improved the condition of over 5,000 hectares of designated wetland habitats, restoring approximately 20 miles of modified watercourses. The New Forest is home to some of the most ecologically significant freshwater habitats in Britain, including mires, bogs, streams, and wet woodlands. These ecosystems form part of the New Forest Site of Special Scientific Interest and are protected under international conservation frameworks. They support species such as the sundew, southern damselfly, tadpole shrimp, and bog pimpernel, many of which are rare or locally extinct elsewhere in England. Many of the Forest's watercourses had previously been straightened or deepened to expedite drainage. These modifications contributed to habitat degradation, increased erosion, and disrupted natural floodplain processes. The HLS scheme has reversed these impacts through interventions such as filling drainage ditches, re-meandering streams, and raising channel beds. These actions help restore natural hydrology, reduce downstream flood risk, and improve resilience to drought. The restorations also support broader environmental functions, including carbon storage in peat and improved water quality. The restored wetland areas now retain more water during heavy rainfall, allowing for seasonal flooding that benefits surrounding habitats. In recognition of its outcomes, the New Forest wetland restoration program received the UK River Prize in 2019.[23]

HLS New Forest Higher Level Stewardship Scheme. "Wetland Restoration." (n.d.). https://hlsnewforest.org.uk/about-us/.

Coastal Flooding

Eco-DRR measures are increasingly recognized in coastal management for their role in mitigating tropical cyclones, storm surges, waves, coastal flooding, and other hazards. Coastal ecosystems serve as natural buffers, stabilizing sediments and reducing wave energy, which helps prevent coastal erosion. In addition to protecting shorelines, Eco-DRR strategies support local economies by sustaining fisheries, tourism, and recreation. These nature-based solutions also contribute to freshwater storage, improve water quality, and enhance habitat and biodiversity, offering multiple environmental and socio-economic benefits.[24]

Beaches and Sand Dunes

Beaches and sand dunes, often elevated above the land behind them, act as natural barriers against coastal flooding and help dissipate wave energy. They also serve as sediment reservoirs, replenishing beaches through wind and tidal processes, thereby reducing coastal erosion. Dunes and beaches can be restored by trapping sand with fences or stabilizing them with vegetation. Native plants can be transplanted from intact dune systems or sourced from nurseries, while artificial dunes can be constructed to reinforce coastlines. Coastal restoration projects should consider the connection between the shoreline and adjacent back bays or lagoons to avoid unintended consequences, such as disrupting natural sediment replenishment or increasing flood risks in other areas.[25]

Beach Nourishment

Beaches serve as natural coastal defenses, but their effectiveness depends on maintaining sufficient width and elevation to withstand the forces of erosion. When beaches become depleted, their ability to mitigate flood risks is diminished. Beach nourishment restores or enhances beaches by adding sediment, reinforcing their role in coastal protection. The key considerations in beach nourishment include the type and quantity of sediment used. The particle size and composition should closely match the existing intertidal sediment to support natural sorting processes and maintain a stable beach profile. Using naturally occurring sediment ensures long-term stability and minimizes the risk of erosion or offshore loss.[26,27]

Case 11.3: City of Gold Coast, Queensland, Australia Beach Nourishment Program

The City of Gold Coast has implemented a comprehensive beach nourishment program to mitigate coastal erosion and maintain the condition of its beaches. Replenishment efforts respond to frequent erosion caused by storms, swells, and the disruption of natural sand movement from coastal infrastructure such as groynes and training walls. Key methods include offshore dredging, where over 3 million cubic meters of sand were added to the coastline in 2017 using a specialist vessel. This sand was placed along nearshore bar systems to protect vulnerable beaches. Additionally, annual creek dredging is conducted at Tallebudgera and Currumbin Creeks, contributing around 70,000 cubic meters of sand annually to Burleigh and Palm Beaches. This work is scheduled to ensure both beaches are restored in time for the summer season, with the city allocating over $1 million annually to support these efforts. The city also employs engineered solutions to maintain the natural flow of sand. The Gold Coast Seaway Sand Bypass System, operational since 1986, mechanically transfers sand around constructed barriers. A second bypassing system at the Tweed River was introduced in 2000 to restore sand supply to southern beaches. A long-term project, the Surfers Paradise Sand Backpass Pipeline, is under construction to redirect up to 120,000 cubic meters of sand annually from The Spit to northern beaches. The 7.8-kilometre pipeline enhances response capabilities after weather events. Excavated marine sand from coastal developments is reused under strict regulations, with over 2 million cubic meters returned to the foreshore since the 1970s, reinforcing beach resilience.[28]

City of Gold Coast. "Beach Nourishment." (n.d.). https://goldcoast.qld.gov.au/Environment-sustainability/Protecting-our-environment/Coastal-management/Beach-nourishment.

Living Shorelines

Coastal communities face ongoing challenges from shoreline erosion, threatening valuable coastal resources. Shorelines require protection from damage caused by storms, wave erosion, and sea-level rise. Traditional shoreline stabilization methods, such as seawalls and bulkheads, create a hard barrier between land and water; however, newer approaches, like living shorelines, offer a more sustainable alternative. Living shorelines help reduce erosion and storm damage while providing essential ecosystem services,

Table 11.4 Living Shoreline Composition

Type	*Composition*
Vegetation-only	Uses plants to buffer upland areas and reduce small wave impacts. Most effective in low wave energy environments.
Edging	Structural additions that stabilize the base of existing or planted slopes. Suitable for most settings except high-energy coastal areas.
Sills	Structures placed parallel to vegetated shorelines to reduce wave energy and prevent erosion. Best used in moderate to low wave energy areas.
Breakwater	Offshore structures that absorb and reduce wave force, promoting sediment buildup and shoreline stabilization. Appropriate for a range of coastal conditions.

including food production, nutrient and sediment removal, and improved water quality. Living shorelines:

- Encompass a range of shoreline stabilization techniques along estuaries, bays, tributaries, and other sheltered shorelines.
- Use primarily natural materials, incorporating vegetation and other living, natural "soft" elements. For added stability, these may be used alone or combined with "harder" structures, such as oyster reefs, rock sills, or anchored large wood (Table 11.4).
- Provide a connection between land and water, helping to stabilize the shoreline, reduce erosion, and enhance ecosystem functions.[29]

Case 11.4: Franklin-98 Living Shoreline Project, Florida

In Florida's Apalachicola Bay, the Franklin-98 Living Shoreline Project is restoring a six-mile stretch of eroding shoreline along U.S. Highway 98 between Eastpoint and Yent Bayou. This area is chronically impacted by wave energy and storm surge, threatening a critical evacuation route. Previous hard infrastructure measures, including seawalls and riprap, proved insufficient against repeated coastal damage. In response, the Apalachee Regional Planning Council, in partnership with WSP, Franklin's Promise Coalition, and North Florida Construction, Inc., initiated a living shoreline project combining natural and structural elements. The project includes the installation of nearshore reefs composed of dolomitic limestone. Approximately 4,470 tons of reef material were placed between October 2023 and January 2024. These reefs are designed to attenuate wave energy through their elevation and roughness, creating calmer conditions along the shore. Over time, they are expected to be colonized by oysters and other marine organisms, enhancing habitat value and serving as a larval source for the broader bay. Inland from the reef structures, intertidal marshes are being reestablished using Spartina plantings. Clumps of vegetation, spaced in a checkerboard pattern approximately one meter apart, have been planted since September 2023 and have withstood multiple tropical and winter storms. The reduced wave energy allows these marshes to stabilize sediment and support further vegetation growth. The project combines habitat restoration and shoreline stabilization, involving local labor and respecting traditional knowledge. It aligns with broader goals of coastal resilience and supports the ecological and economic recovery of Apalachicola Bay.[30]

Apalachee Regional Planning Council. "Franklin 98: Protecting Community. Conserving the Coast." (n.d.). https://arpc.org/franklin-98.

Coastal Wetlands

Coastal wetlands, such as saltmarshes and mangroves, play a crucial role in stabilizing shorelines, controlling erosion, and attenuating wave energy.[31] Saltmarshes are intertidal plant communities comprising sedges, rushes, reeds, grasses, succulent herbs, and low-growing shrubs that can tolerate high soil

Table 11.5 Guiding Principles for Restoring Coastal Wetlands

Principle	*Description*
Conserve and protect aquatic resources	Preserving existing wetlands is more effective and cost-efficient than attempting rehabilitation after degradation.
Aim for ecological integrity	Rehabilitation efforts should restore the natural structure, composition, and processes of ecosystems as fully as possible.
Restore natural structure and function	Reinstating the original structure of wetlands enhances their ecological functions and supports objectives such as improved water quality.
Work with the catchment and broader landscape context	Wetland rehabilitation must be planned at the catchment level, accounting for upstream and surrounding influences that affect long-term outcomes.
Understand the natural potential of the catchment and anticipate future changes	Set realistic goals by understanding historic conditions and forecasting future ecological, social, and physical changes.
Address ongoing concerns of degradation	Rehabilitation will be unsuccessful if sources of degradation persist; identify and address these issues to ensure lasting results.
Involve all relevant stakeholders	Ensure transparency and inclusivity by engaging local communities and stakeholders throughout the planning and implementation process.

salinity and periodic inundation with saltwater. They typically occupy the high-tide zone on sheltered, soft-substrate foreshores of coastal lakes and estuaries, often found behind mangroves. Mangroves provide natural coastal protection by absorbing the energy of tidal currents, storm surges, and waves, acting as a natural breakwater that reduces erosion. As climate change is expected to increase the frequency and intensity of cyclones, storm surges, and sea-level rise, mangroves will play an even more crucial role in safeguarding coastal areas.[32] Restoring coastal wetlands, including saltmarshes and mangroves, can be guided by key principles outlined in Table 11.5.[33]

Case 11.5: Wetland Restoration in the Barataria-Terrebonne Estuary, Louisiana

Located between the Mississippi and Atchafalaya Rivers, the Barataria-Terrebonne National Estuary represents one of Louisiana's most ecologically productive and culturally significant coastal regions. It is also one of the areas most impacted by wetland loss, with factors such as subsidence, levee construction, canal dredging, and sea-level rise contributing to land degradation. Since 1990, wetland restoration in this region has been supported through the Coastal Wetlands Planning, Protection and Restoration Act (CWPPRA) and coordinated with the Coastal Protection and Restoration Authority (CPRA), which oversees state-level planning and funding allocation. One of the earliest CWPPRA-supported restoration techniques implemented in this area was marsh creation, notably at Bayou Dupont. Sediment dredged from the Mississippi River was pumped via pipeline to restore degraded marsh, initially creating 471 acres of new wetland. The infrastructure was later reused for additional projects, increasing cost efficiency and expanding the area of restored habitat. The Fourchon Maritime Ridge project combined ridge and marsh restoration using dredged material from port expansion. Native trees and marsh grasses were planted to stabilize the ridge and enhance bird habitat. This effort reflects the importance of reestablishing elevated features that historically supported forested habitats and human development. Additionally, the West Bay sediment diversion project used an uncontrolled channel cut into the riverbank to reintroduce sediment into marshes, mimicking natural deltaic processes. This approach has created new land while supporting the growth of downstream marshes.[34]

Barataria-Terrebonne National Estuary Program & Coastal Wetlands Planning, Protection and Restoration Act. "Coastal Wetlands Restoration: Residents' Guide." (2015). https://lacoast.gov/new/Ed/BTNEP_book.pdf.

Bio-shields

Bio-shields, which utilize coastal vegetation, are an eco-DRR approach that protects ecosystems and livelihoods from storms and cyclones by absorbing wave energy and reducing wind speeds. Compared to hard engineering structures, bio-shields are a cost-effective and less intrusive solution that preserves the natural aesthetic of coastal landscapes. They offer a range of environmental benefits, including reducing soil erosion by lowering wind velocity, minimizing water loss through soil evaporation, providing habitats for aquatic life, and mitigating coastal pollution. Additionally, bio-shields support local communities by providing resources such as fruits, livestock fodder, and wood for fire or construction.

Several key factors should be considered when planning bio-shield projects:

- *Plant selection*: The choice of plants depends on site-specific conditions, including wind exposure, soil type and moisture, topographical changes from shifting sand, storm frequency, and wave exposure. Native, salt-tolerant species are recommended as they are better adapted to coastal conditions and require less maintenance.
- *Protecting plants*: Vegetation is most vulnerable before its root systems are established. Protective measures, such as ground blankets to stabilize soil and fences to shield young plants from wind, help ensure successful growth.
- *Invasive plants*: Invasive species should not be planted in coastal areas, as they often outcompete native species. If invasive plants are present, they should be replaced with native species that are appropriate for the area. However, replacing invasive plants may temporarily destabilize sandbanks, requiring careful planning.
- *Slope stability*: For bio-shields along steep banks, slope stability is critical. Steep slopes are prone to slumping or collapse, which can be mitigated by filling the lower part of the bank with soil, similar to the upper portion. If the slope extends to the high tide line, sediment may erode, requiring careful grading to stabilize the bank.[35]

Case 11.6: Coastal Bioshield Development in Ramanathapuram, Tamil Nadu, India

In Ramanathapuram district, Tamil Nadu, the Forest Department has undertaken large-scale tree-planting efforts to increase green cover and enhance coastal resilience. As part of the Green Tamil Nadu Mission, over 250,000 trees were planted in two weeks in late 2023. The initiative aimed to restore vegetative cover in a region where forest cover stood at 6.25%, significantly below the 33% target outlined in India's National Forest Policy. The district included approximately 3,150 hectares of coastal shelterbelt, with over 450 hectares planted with *Casuarina equisetifolia* and the remaining area consisting of native tree species. These shelterbelts functioned as bioshields – natural barriers designed to reduce the impacts of cyclones, storm surges, and saltwater intrusion. To further strengthen these bioshields, the department planted an additional 200,000 casuarina trees along the coastline. Simultaneously, efforts were made to restore *Borassus flabellifer* (palmyra) populations, a drought-resistant native species known for its deep roots and contribution to groundwater recharge. Approximately 50,000 palmyra nuts were dibbled across forest areas, and 5,500 saplings were cultivated in nurseries for future planting. A total of 20,000 palmyra trees – half seedlings and half saplings – were scheduled for planting during the 2023–24 period. Altogether, the Forest Department set a target of more than 550,000 tree plantings for the year, aiming to establish denser green cover that would serve both ecological and protective functions across the district's vulnerable coastal zones.[36]

The New Indian Express. "2.5l Trees Planted So Far in Tn's Ramanathapuram to Improve Green Cover." (n.d.). https://newindianexpress.com/states/tamil-nadu/2023/Nov/18/25l-trees-planted-so-far-in-tns-ramanathapuram-to-improve-green-cover-2633860.html.

Managed Realignment

Managed realignment involves modifying flood defenses to allow controlled flooding of previously protected land. This typically requires relocating actively maintained defenses to a new position further inland, ideally to rising ground. The primary benefit of managed realignment is the creation of intertidal habitats, which effectively attenuate wave energy, reduce offshore sediment transport, and help mitigate erosion.[37] There are two primary methods for realignment: breaching sections of the existing defense structure or removing the entire structure. The choice of method depends on site-specific conditions. While full removal interferes with natural processes the least, it is more expensive and, in some cases, infeasible. Once partial inundation begins, the risks associated with continuing removal efforts may become too high.

Determining whether managed realignment is appropriate requires careful assessment of various factors, including environmental conditions, economic viability, and potential impacts on surrounding land use. Table 11.6 outlines key considerations for evaluating the suitability of managed realignment.[38]

Table 11.6 Factors to Consider in Managed Realignment Projects

Positive Factors	*Negative Factors*
Naturally occurring higher ground exists (removing the need to build flood protection structures further inland)	Presence of contaminated land such as landfill sites or reuse-filled seawalls
Existing flood or coastal protection structures are in poor condition and/or reaching the end of design life	Presence of infrastructure not designed for saline conditions that are very costly to move (e.g. electric or gas cables)
Existing structures were built for land reclamation purposes (e.g. agricultural embankment) and structural integrity, design life or levels of protection provided are unknown	Presence of scheduled monuments or sites of historic importance
Landowners are willing to lose some land to the sea	Presence of protected areas that would not benefit from managed realignment
Potential to restore valuable intertidal habitat and enhance biodiversity	Poor site access

Saltmarshes and Mudflat Management and Restoration

Saltmarshes and mudflats play a crucial role in reducing wave and tidal energy in front of flood defenses, while providing valuable natural habitats that support fish, shellfish, and other marine life.[39] These ecosystems offer various ecosystem services, including coastal protection, biodiversity support, and carbon sequestration. Saltmarshes and mudflats are often found together, with mudflats typically located in front of saltmarshes. The formation of these habitats requires fine-grained sediments, such as silts and clays, to settle from the water column, which can only occur in areas with extremely low water velocities (less than 0.0002 centimeter/second). Consequently, saltmarshes are usually found in estuaries or sheltered coastal areas, such as bays, where the wind fetch distance, defined as the unobstructed length of water over which the wind can travel in a constant direction, is less than 2,000 meters.[40,41] The tidal range is a key factor in determining the extent of saltmarsh habitat that can be established. To increase the available space for saltmarsh, mudflats can be extended seawards, or coastal infrastructure can be moved landwards using a managed realignment approach. Most saltmarsh and mudflat restoration projects have relied on managed realignment, which allows these habitats to develop naturally while enhancing coastal resilience.[42,43]

Case 11.7: Essex Wildlife Trust's Saltmarsh Restoration Project

In 2018, Essex Wildlife Trust (EWT), in partnership with the Environment Agency, launched a saltmarsh restoration project focused on the

Blackwater and Colne estuaries. The initiative, aiming to restore vital coastal ecosystems, employed a cost-effective and low-impact method involving the use of "saltmarsh sausages" – coir rolls made of biodegradable materials. These structures were installed at Abbotts Hall and Moverons Farm to trap sediment, promote vegetation growth, and enhance the natural resilience of the marshes. The saltmarsh sausages were strategically placed in low-energy tidal channels and secured with chestnut stakes and hessian rope. Over time, the rolls were designed to accumulate sediment and support the establishment of native saltmarsh vegetation such as common sea lavender (*Limonium vulgare*). These restored habitats contribute to coastal biodiversity, providing habitat for invertebrates, fish, and bird species, while also acting as natural carbon sinks. By 2021, with additional funding from The Wildlife Trust's Nature-based Solutions Fund and the People's Postcode Lottery, EWT expanded the project. A collaboration with the University of Essex introduced drone monitoring, mapping, and data analysis to evaluate the progress of the restoration efforts. Results confirmed that the coir roll structures had successfully integrated into the saltmarsh environment, supporting vegetative growth and stabilizing sediment, thereby enhancing carbon sequestration. EWT's ongoing partnership with the University of Essex is generating new data to further refine and adapt the restoration methodology. The initiative is part of EWT's broader commitment to coastal and marine conservation, positioning saltmarsh restoration as a key natural solution to the climate emergency. To support wider application of this approach, EWT developed a Saltmarsh Restoration Project Toolkit aimed at practitioners exploring ecological restoration methods. The toolkit consolidates lessons learned and practical guidance from the project to assist others in adopting similar techniques.[44]

Essex Wildlife Trust. "Saltmarsh Restoration Project." (n.d.). https://essexwt.org.uk/what-we-do/protecting-wildlife/projects/saltmarsh.

Coastal Setbacks

Coastal setbacks are designated buffer zones along coastlines where development is restricted or prohibited to reduce exposure to coastal hazards. These setbacks are measured from a fixed coastal feature, such as a permanent vegetation line, and serve to protect infrastructure, ecosystems, and communities from erosion, flooding, and storm surges. Setbacks can be defined in two main ways. Lateral setbacks establish a minimum distance from the shoreline to prevent new buildings or infrastructure from being placed too close to areas at risk of coastal erosion. Elevation setbacks, on the other hand, specify a minimum height above sea level for development, ensuring structures are less vulnerable to coastal flooding. By implementing these setbacks, coastal resilience can be strengthened, reducing long-term risks associated with sea-level rise and extreme weather events.[45]

Oyster Reef Restoration

Oyster reefs develop as successive generations of larvae settle and attach to hard surfaces such as existing oyster reefs, rocks, pilings, downed trees, and recycled oyster shells. Over time, these layers of oysters accumulate, forming large reef structures within estuarine systems. Oyster reefs provide numerous essential ecosystem services, including stabilizing sediments, improving water quality through filtration, and creating crucial habitats for fish and other marine organisms. They also serve as natural wave barriers, helping dissipate wave energy before it reaches the shoreline. This function reduces coastal erosion and enhances shoreline protection, particularly during storm events. By restoring and conserving oyster reefs, coastal resilience can be improved while also supporting biodiversity and maintaining water quality in estuarine environments.[46]

Case 11.8: Community-Driven Oyster Reef Restoration in New York Harbor

The Billion Oyster Project is a large-scale initiative that restores oyster reefs in New York Harbor through collaboration between scientists, educators, and local communities. Oyster reefs provide natural coastal protection, reducing wave energy, minimizing flooding, and stabilizing shorelines. Since its inception, the project has restored over 150 million oysters across 17 sites, enhancing marine biodiversity and improving water quality. A key aspect of the initiative is its educational outreach, engaging over 100 New York City schools in hands-on STEM learning. Students participate in oyster monitoring, reef construction, and environmental data collection, integrating science with real-world applications. The project also partners with the Urban Assembly New York Harbor School, where students train for maritime careers through specialized programs in aquaculture, marine biology, and ocean engineering. The project operates an oyster shell recycling program, collecting over 2.8 million pounds of shells from local restaurants to use as reef substrate. These recycled shells provide a surface for young oysters to attach and grow, accelerating reef formation. Community engagement is at the heart of the initiative, with nearly 15,000 volunteers contributing to restoration efforts. Beyond ecological benefits, the project emphasizes long-term stewardship by involving residents in conservation efforts. Oyster reefs serve as living laboratories, demonstrating how urban restoration can enhance resilience against climate change. The initiative is supported by government grants, private donations, and research partnerships, ensuring continued expansion and innovation in habitat restoration.[47]

Billion Oyster Project. "Restoring Oyster Reefs to New York Harbor through Public Education Initiatives." (n.d.). https://billionoysterproject.org.

Coral Reef Restoration

Coral reef structures protect shorelines by absorbing wave energy and reducing the impact of storms, cyclones, and floods, helping to prevent loss of life, property damage, and coastal erosion.[48,49] The health of coral reef structures can be enhanced through the following measures:

Manage Land-Based and Near-Shore Water Fluxes and Quality

Improving water quality is essential for maintaining the ecosystem services that coral reefs provide. Excess nutrients and municipal wastewater contribute to ocean acidification. In contrast, reducing inorganic nutrient loading and organic matter enhances water quality, supports coral calcification, and minimizes calcium carbonate dissolution. Three key interventions for improving water quality on reefs include reducing point-source pollution and nutrient inputs, establishing natural barriers or filters, such as mangroves, between land-based pollution sources and the reef, and redirecting or diverting pollutants away from reef ecosystems, preferably through industrial or natural treatment facilities.[50]

Reduce Unsustainable Fishing Practices

Unsustainable fishing practices threaten reef resilience, while effective management can help maintain fish stocks and balance the ecosystem. Removing herbivorous fish and apex predators disrupts reef dynamics, often leading to excessive macroalgal growth that can outcompete corals. To support sustainable fishing, it is essential to identify and protect critical habitats for fish spawning and reproduction while reducing pressure on pre-productive fish to aid stock recovery. Strategies to enhance coral reef fish populations include establishing a hierarchy of fishing rights for stakeholders, expanding successful marine protected areas, implementing measures to protect juveniles and apex predators, safeguarding spawning and nursery areas, and eliminating destructive fishing practices that harm reef ecosystems.[51]

Explore and Test New Ecological Engineering Approaches

Coral reef restoration can be achieved through coral gardening and selective harvesting. Coral gardening involves cultivating coral colonies in nurseries before transplanting them onto the reef. This process typically uses asexual reproduction (fragmentation), where coral fragments are grown in land-based aquaria or designated marine nurseries until they are ready for transplantation. On the other hand, selective harvesting focuses on designing fishing practices that enhance reef resilience. For example, on macroalgal-dominated reefs, selective fishing targets predators of herbivorous fish, thereby increasing herbivore populations, which helps control algal overgrowth and promotes coral recovery.[52]

Spatial Planning

Marine spatial planning helps reduce stress on reef ecosystems. It supports long-term restoration efforts by designating marine protected areas, preserving key ecosystem services, and zoning coastal waters to balance conservation, food security, and livelihoods. When informed by local ecological, habitat, and hydrological knowledge, marine spatial planning can be tailored to enhance reef recovery and resilience. For example, identifying areas with favorable water quality and carbonate chemistry can help locate refuges from ocean acidification. Ensuring connectivity between habitats can facilitate gene flow, larval transport, and the recruitment of corals, invertebrates, and fish at restoration sites.[53]

Case 11.9: Hybrid Coral Reef Restoration for Coastal Protection in the Reefense Project

The Reefense project is a United States initiative funded by the Defense Advanced Research Projects Agency to develop hybrid structures that mimic coral reefs for coastal protection and restoration purposes. The project, known as X-REEFS (neXt generation Reef Engineering to Enhance Future Structures), is led by the University of Miami and includes contributions from several institutions, including the University of California, Santa Cruz (UCSC), Penn State University, and SECORE International. The objective of X-REEFS is to design and test engineered reef-like structures that combine physical wave protection with biological coral restoration. These structures are intended to provide immediate shoreline defense while supporting the growth and recovery of coral species. The project integrates engineered materials – such as novel concrete formulations and structural modules – with ecological and biological methods to encourage coral colonization and resilience. Researchers at UCSC are involved in hydrodynamic modelling and performance assessment to quantify how the structures influence wave energy and sediment transport. The University of Miami's facility is being used to study the interaction of waves, wind, and shorelines with the reef prototypes under controlled conditions. The project also involves testing adaptive coral biology, including faster-growing and thermally resilient coral strains, to ensure survival and growth under changing environmental conditions. X-REEFS includes field deployment and monitoring of prototype structures in Florida and the Caribbean. The aim is to develop reef systems that are modular, self-repairing, and capable of maintaining structural and ecological function under long-term coastal and climate stress.[54]

UC Santa Cruz. "Reefense Project Aims to Meld Biology and Engineering in Novel Structures for Coastal Protection." (n.d.). https://news.ucsc.edu/2022/06/reefense-grant/.

Droughts

Eco-DRR can help mitigate droughts by facilitating the gradual release of water from natural storage features, ensuring water availability during dry periods.

Constructed Wetlands

During drought, wetlands act as natural retention basins, gradually releasing stored water to sustain water availability. Constructed wetlands can be designed to mimic the hydrological processes of natural wetlands, providing an effective solution for drought regulation.[55] These artificial wetlands are typically established on uplands, outside floodplains, or within designated floodways to prevent damage from excess stormwater. Their construction involves excavation, backfilling, grading, diking, and installing water control structures to create optimal hydraulic flow patterns. In areas with highly permeable soils, an impervious liner, such as compacted clay, is often installed before covering it with the original soil. Wetland vegetation is then planted or allowed to establish naturally, supporting water retention and enhancing ecosystem resilience.[56]

Case 11.10: Constructed Wetlands for Water Reuse and Habitat in North Texas

As part of Texan by Nature's 2018 Conservation Wrangler program, constructed wetlands were supported as an innovative solution to meet the growing water demand in North Texas. With the region's population expected to double by 2060, the Tarrant Regional Water District (TRWD) and North Texas Municipal Water District (NTMWD), in collaboration with Plummer Associates, have developed large-scale constructed wetland systems to treat wastewater for natural reuse. The TRWD's George W. Shannon Wetlands Project and NTMWD's East Fork Water Reuse Project each divert up to 90–95 million gallons of treated effluent per day from the Trinity River. Water is directed into engineered wetland systems, which contain sedimentation basins and aquatic vegetation. These systems remove sediment and reduce nutrient concentrations, achieving up to 95% sediment removal and 50%–80% reductions in nitrogen and phosphorus. The treated water is then transferred to nearby reservoirs, Lavon Lake and Richland Chambers, for subsequent treatment to potable standards. Collectively, these projects span more than 6,500 acres and provide a raw water supply for approximately 3.8 million people. Beyond water treatment, the wetlands support over 200 avian species and offer recreational opportunities, including birdwatching, fishing, and hunting. They also serve as educational hubs, particularly through the John Bunker Sands Wetland Center, which offers programs for students and the broader community. Compared to new reservoir construction, these wetlands offer a cost-effective and spatially efficient alternative, delivering comparable water yields at approximately one-third the cost and within shorter timeframes.[57]

Texan by Nature. "Constructed Wetlands: 2018 Conservation Wrangler." https://texanbynature.org/projects/constructed-wetlands/.

Rainwater Harvesting

Rainwater harvesting involves collecting and storing rainwater at its source for later use. Water butts, commonly used for small-scale applications, collect runoff from rooftops through a connection to a downspout and are primarily used for watering household gardens and other non-potable purposes. Storage tanks are designed to capture and retain larger volumes of stormwater for use in more extensive applications. These tanks are more effective when implemented at a community or building-scale than at individual properties. Collected rainwater from storage tanks is typically used for irrigation or non-potable indoor purposes, such as toilet flushing, clothes washing, and evaporative cooling.[58,59]

Case 11.11: Cape Town's Rainwater and Greywater Integration System

In Cape Town, the integration of harvested rainwater with greywater systems has been promoted as a viable strategy for residential water conservation and irrigation. By combining these two sources, households can significantly reduce their reliance on potable water for garden use. Rainwater is collected from rooftops – preferably metal surfaces due to their higher runoff efficiency – and stored in tanks with capacities ranging from 1,500 to 5,000 liters. These tanks can be installed either aboveground along house walls or underground for aesthetic purposes. System suppliers offer configurations where a slow trickle of rainwater is fed into a greywater system. This integration not only extends the usability of greywater but also helps maintain the necessary flow within the system to prevent blockages or odor buildup. Suppliers typically assess roof area, roof material, household water usage, and desired application to determine optimal system size and layout. A rainwater harvesting system often includes filters to remove debris, and water is generally pumped for distribution to the garden. This approach also has cost benefits. As sewerage charges in Cape Town are based on municipal water meter readings, using alternative water sources reduces these charges. However, rainwater is not recommended for human consumption due to the potential presence of contaminants, such as airborne particles and pollutants, which can increase acidity levels. Households are advised to consult their system suppliers for maintenance requirements and available support. By adopting such systems, residents contribute to water conservation efforts in a region facing significant water scarcity.[60]

City of Cape Town. "Rainwater Harvesting: Alternative Water Resources (Pamphlet No. 4 of 4)." (n.d.). https://resource.capetown.gov.za/documentcentre/Documents/Graphics%20and%20educational%20material/Alternative_Water_Resources_Rainwater_English.pdf.

Soakaways

Soakaways are underground chambers designed to store surface water and gradually release it into the surrounding soil. They are typically constructed as square or circular excavations filled with rubble or lined with brickwork, precast concrete, or a perforated storage structure surrounded by granular backfill. Soakaways enhance soil moisture levels, support groundwater recharge, and attenuate stormwater by reducing surface runoff. Additionally, they contribute to stormwater treatment by filtering out sediments and pollutants before the water infiltrates the ground.[61]

Infiltration Trenches

Infiltration trenches are shallow excavations filled with rubble or stone that facilitate water infiltration into the surrounding soil through their bottom and sides. They help reduce surface runoff rates and volumes, support groundwater recharge, and maintain base flow in rivers. These trenches are versatile and can be integrated into various settings, including residential and commercial areas, playing fields, recreational spaces, and public open areas. Infiltration trenches are designed to blend into the landscape and require minimal land use while providing effective stormwater management.[62]

Community-Scale Practices for Water Retention and Infiltration

Forests play a crucial role in water retention and infiltration. Similarly, smaller-scale techniques can achieve comparable benefits. Well-vegetated surfaces function like forests by enhancing water retention and promoting infiltration. In agricultural settings, practices such as low- or zero-tillage, mulching, crop rotation, and maintaining vegetation cover contribute to a healthy soil structure, improve water retention, and increase resilience to drought conditions.[63]

Case 11.12: Australia's Climate-Smart Agriculture Program

Launched under the Australian Government's Natural Heritage Trust (NHT), the $302.1 million Climate-Smart Agriculture Program spans five years from 2023–24. It is structured around multiple investment streams, each designed to improve the sustainability, resilience, and productivity of the agricultural sector while reducing emissions and conserving natural resources. The *Partnerships and Innovation* stream allocates $45 million over four years to fund medium- and large-scale projects that trial and implement climate-smart tools and practices. This stream aims to promote sector-wide adoption of sustainable solutions that enhance profitability and improve environmental outcomes. The *Capacity Building* stream includes two open-competitive rounds distributing $25.5 million from 2024–25 to 2027–28. These grants focus on expanding access to knowledge and skills for community organizations, First Nations groups, and farmers. Round 2, launched in early 2025, specifically targets the broad-scale adoption of sustainable agriculture practices through educational initiatives, information resources, and support networks. *Small Grants* provide $2.5 million over two years to support grassroots organizations implementing on-ground projects that strengthen climate resilience and improve land management practices. These grants enable local, often community-led, action aligned with broader climate adaptation goals. The program also funds *Soil Capacity Building* through $6.26 million for Regional Soils Coordinators and $2.14 million for a National Soils Community of Practice, fostering collaboration between soil scientists, producers, and communities. Regional Delivery Partners and National Landcare organizations are further supported through $161.9 million in operational and project-based funding to deliver climate-smart agriculture outcomes across 52 regions, incorporating traditional knowledge and supporting First Nations participation.[64]

Australian Government, Department of Agriculture, Fisheries and Forestry. "Climate-Smart Agriculture Program." (n.d.). https://agriculture.gov.au/agriculture-land/farm-food-drought/natural-resources/landcare/climate-smart.

Restoration of Natural Infiltration to Groundwater

The restoration of natural infiltration to groundwater reduces runoff from surrounding land, enhancing the condition of groundwater aquifers and increasing water availability. This measure is also known as 'artificial groundwater recharge', and mechanisms to restore or enhance natural infiltration capacity include:

- Surface structures to facilitate/augment recharge, such as soakaways and infiltration basins
- Subsurface indirect recharge, where infiltration capacity is enhanced through wells drilled within the unsaturated zone
- Subsurface direct recharge, where infiltration and recharge of the groundwater aquifer is done through wells reaching the saturated zone[65]

Other Natural Disasters: Tsunamis

Well-managed ecosystems can protect people from various hazards, including tsunamis. Coastal forests, such as mangroves, can mitigate tsunami damage by dissipating wave energy, capturing drifting debris, and delaying the arrival time of tsunamis. In addition to their protective function, coastal forests provide multiple benefits, including shielding against strong winds and blown sand, shaping coastal landscapes, creating habitats for wildlife, and enhancing recreational opportunities.[66,67,68] The extent to which coastal forests can mitigate tsunami impacts depends on several key factors outlined in Table 11.7.[69]

Table 11.7 Parameters that Determine the Magnitude of Tsunami Mitigation by Coastal Forests

Parameter	Description
Forest width	Tsunami energy is gradually dissipated across the width of the forest due to drag created by tree trunks, branches, foliage, and undergrowth.
Forest density	Defined by tree spacing (horizontal density) and vertical structure. Denser forests reduce tsunami force through increased drag and energy absorption.
Forest age and tree diameter	Older forests with larger trees are more effective in tsunami mitigation. Larger tree diameters enhance structural resistance to toppling and improve energy dissipation.
Forest height	Taller trees offer a greater frontal area to resist and reflect tsunami waves, reducing the likelihood of overtopping.
Species composition	Diverse tree species contribute to variation in forest structure and drag resistance, improving the forest's capacity to disrupt and weaken tsunami flow.

Case 11.13: Japan's Great Forest Wall Project

Following the Great East Japan Earthquake and tsunami, Japan launched a large-scale initiative to establish "life-saving forests" along its coastal regions to mitigate future tsunami impacts. The project involves planting native trees using the Miyawaki Method, which prioritizes potential natural vegetation and high-density planting of diverse native species. Approximately 70,000 trees have been planted by residents and volunteers so far. Before planting, toxic debris such as crushed concrete and bricks is removed, ensuring healthy growth conditions. These forests are designed to develop into stable ecosystems without ongoing maintenance. The Miyawaki Method promotes rapid growth through close planting and species diversity, yielding multi-layered, resilient forests. These life-saving forests offer several benefits: they reduce tsunami energy, trap debris, require no maintenance once mature, fertilize nearby aquatic ecosystems, help mitigate global warming through carbon sequestration, and withstand environmental stresses such as wind, rain, and fire. The fire resistance is attributed to canopy structures that retain water, reducing flammability. Implementation strategies vary by setting. In Iwanuma, Miyagi Prefecture, small hills near residential areas are planted with trees to serve as evacuation zones. In Tagajo city, dense tree belts protect individual buildings, including temples and shrines. Trees are also planted along embankments and slopes to form natural breakwaters, reducing wave energy and stabilizing the land. Additionally, vegetation on tide embankments strengthens their resistance to storm surges. These different applications demonstrate the adaptability of the forest approach across urban and rural coastal landscapes, enhancing overall resilience to coastal hazards through ecological means.[70]

Great Forest Wall Project. "Great Forest Wall Project: Your Donation Will Help Us Create Forests." (n.d.). https://morinoproject.com/wp/wp-content/themes/gfw/images/home/panf-eng.pdf.

Case Summary

Various Eco-DRR strategies have been implemented to reduce disaster risk and improve community resilience (Table 11.8).

Table 11.8 Ecosystem-Based Disaster Risk Reduction Improving Disaster Risk Management and Resilience

Habitat	*Project Name*	*Activities*	*Co-Benefits*
Urban blue-green networks	Wairau Blue-Green Network – Auckland	Restores natural water flows in the Wairau catchment through wetland and detention basin development, stream daylighting, and flow path improvements.	Reduces flood risk for homes and infrastructure, enhances green public space, and improves climate resilience.
Wetlands	New Forest Wetland Restoration – United Kingdom	Re-meanders streams, fills drainage ditches, and raises channel beds across 5,000+ hectares to restore mires, bogs, and wet woodlands.	Supports biodiversity, improves flood regulation, enhances carbon storage, and increases drought resilience.
Coastal management	Gold Coast Beach Nourishment – Australia	Uses dredging and sand bypass systems to replenish beaches and restore sediment transport disrupted by coastal infrastructure.	Reduces coastal erosion, protects infrastructure, and maintains recreational and ecological beach value.
Living shorelines	Franklin-98 Living Shoreline – Florida	Constructs limestone reef breakwaters and reestablishes intertidal marshes using checkerboard plantings to stabilize shoreline.	Dampens wave energy, restores oyster and marsh habitats, and enhances storm resilience.
Estuary restoration	Barataria-Terrebonne Wetlands – Louisiana	Applies dredged sediment for marsh and ridge creation, sediment diversions, and native vegetation planting.	Rebuilds lost wetlands, supports bird and fish habitats, and mimics deltaic processes for coastal resilience.
Coastal bioshields	Ramanathapuram Coastal Bioshield – India	Plants over 550,000 trees including casuarina and palmyra to form shelterbelts against storms and saltwater intrusion.	Improves green cover, groundwater recharge, and disaster resilience in vulnerable coastal areas.
Saltmarsh restoration	Essex Saltmarsh Project – United Kingdom	Installs coir roll 'saltmarsh sausages' in estuaries to trap sediment and support native plant growth.	Enhances carbon storage, biodiversity, and natural flood defense in coastal marshes.

Table 11.8 (Continued)

Habitat	*Project Name*	*Activities*	*Co-Benefits*
Oyster reefs	Billion Oyster Project – New York Harbor	Restores oyster reefs with recycled shells and community planting; integrates education, aquaculture, and maritime training.	Improves water quality, protects coastlines, and engages citizens in ecosystem stewardship.
Coral reefs	Reefense Project – Hybrid Coral Reefs	Develops modular reef structures combining engineered materials and coral biology for wave protection and reef recovery.	Reduces erosion, fosters coral resilience, and advances scalable reef restoration technologies.
Constructed wetlands	North Texas Constructed Wetlands	Treats wastewater through engineered wetlands, providing potable-quality supply to reservoirs and supporting habitats.	Delivers cost-effective water reuse, supports biodiversity, and offers recreation and education.
Urban water reuse	Cape Town Rainwater and Greywater Integration	Promotes household systems combining rainwater harvesting with greywater reuse for garden irrigation, using rooftop collection and filtration.	Reduces potable water use, lowers utility charges, supports climate adaptation in water-scarce regions.
Climate-smart agriculture	Australia's Climate-Smart Agriculture Program	Funds national and local projects supporting climate-resilient practices, soil health, traditional knowledge, and emissions reduction across the agricultural sector.	Improves productivity, strengthens land management, supports First Nations participation, and builds sector-wide resilience.
Coastal forests	Japan's Great Forest Wall Project	Uses the Miyawaki Method to plant native forests along coastlines as life-saving barriers against tsunamis and climate hazards.	Mitigates tsunami and storm surge impacts, restores biodiversity, absorbs carbon, and supports community resilience.

Notes

1 United Nations Office for Disaster Risk Reduction and Centre for Research on the Epidemiology of Disasters, "The Human Cost of Weather-Related Disasters 1995-2015," (2015), https://unisdr.org/we/inform/publications/46796.

2 F.G. Renaud, K. Sudmeier-Rieux, and M. Estrella, *The Role of Ecosystems in Disaster Risk Reduction* (Tokyo: United Nations University Press, 2013).

3 IUCN, "About Ecosystem-Based Disaster Risk Reduction (Eco-Drr)," (n.d.), https://iucn.org/theme/ecosystem-management/our-work/environment-and-disasters/about-ecosystem-based-disaster-risk-reduction-eco-drr.
4 IUCN, "About Ecosystem-Based Disaster Risk Reduction (Eco-Drr)."
5 UNEP, "Ecosystem-Based Disaster Risk Reduction," (n.d.), https://unenvironment.org/explore-topics/disasters-conflicts/what-we-do/risk-reduction/ecosystem-based-disaster-risk.
6 Belinda Reyers et al., "Navigating Complexity through Knowledge Coproduction: Mainstreaming Ecosystem Services into Disaster Risk Reduction," *Proceedings of the National Academy of Sciences* 112, no. 24 (2015).
7 Reyers et al., "Navigating Complexity through Knowledge Coproduction: Mainstreaming Ecosystem Services into Disaster Risk Reduction."
8 U. Nehren, K. Sudmeier-Rieux, S. Sandholz, M. Estrella, M. Lomarda, and T. Guillén, "The Ecosystem-Based Disaster Risk Reduction: Case Study and Exercise Source Book," (2014), http://pedrr.org/pedrr/wp-content/uploads/2013/09/Eco-DRR-case-study-source-book_final.pdf.
9 IUCN, "Environmental Guidance Note for Disaster Risk Reduction: Healthy Ecosystems for Human Security and Climate Change Adaptation," (2013), https://iucn.org/content/environmental-guidance-note-disaster-risk-reduction-healthy-ecosystems-human-security-and-climate-change-adaptation.
10 Convention on Biological Diversity, "Voluntary Guidelines for the Design and Effective Implementation of Ecosystem-Based Approaches to Climate Change Adaptation and Disaster Risk Reduction," (2018), https://cbd.int/doc/c/3f7a/4589/5cc1b7058bf52427fa9bae84/sbstta-22-inf-01-en.pdf.
11 IUCN, "Five Reasons Why Ecosystems Are Central to Disaster Risk Reduction," (n.d.), https://iucn.org/theme/ecosystem-management/our-work/environment-and-disasters/about-ecosystem-based-disaster-risk-reduction-eco-drr/five-reasons-why-ecosystems-are-central-disaster-risk-reduction.
12 Convention on Biological Diversity, "Voluntary Guidelines for the Design and Effective Implementation of Ecosystem-Based Approaches to Climate Change Adaptation and Disaster Risk Reduction."
13 Brenden Jongman, "Effective Adaptation to Rising Flood Risk," *Nature Communications* 9, no. 1 (2018).
14 Bregje K. van Wesenbeeck, Stephanie IJff, Brenden Jongman, Simone A.B. Balog-Way, Stefanie M. Kaupa, Lauren V. Bosche, Glenn-Marie Lange, Niels B. Holm-Nielsen, Henk Nieboer, Yusuke Taishi, Pradeep H. Kurukulasuriya, and Imen Meliane, "Implementing Nature Based Flood Protection: Principles and Implementation Guidance (English)," (2017), http://documents.worldbank.org/curated/en/739421509427698706/Implementing-nature-based-flood-protection-principles-and-implementation-guidance.
15 WWF, "Natural and Nature-Based Flood Management: A Green Guide," (2017), https://worldwildlife.org/publications/natural-and-nature-based-flood-management-a-green-guide.
16 WWF, "Natural and Nature-Based Flood Management: A Green Guide."
17 Environment Agency, "Working with Natural Processes to Reduce Flood Risk," (2017), https://assets.publishing.service.gov.uk/government/uploads/system/uploads/attachment_data/file/654440/Working_with_natural_processes_one_page_summaries.pdf.
18 Robert Brears, "Integrated Flood Risk Management, Lessons from the Rhine and Danube for South Asia," *New Water Policy & Practice* 1, no. 2 (2015).

19 Brears, "Integrated Flood Risk Management, Lessons from the Rhine and Danube for South Asia."
20 Auckland Council, "Next Steps to Improve Flood Resilience for the Wairau Community," (n.d.), https://ourauckland.aucklandcouncil.govt.nz/news/2025/03/next-steps-to-improve-flood-resilience-for-the-wairau-community/.
21 Johannes A. Belle, Nacelle Collins, and Andries Jordaan, "Managing Wetlands for Disaster Risk Reduction: A Case Study of the Eastern Free State, South Africa," *Jàmbá: Journal of Disaster Risk Studies* 10, no. 1 (2018).
22 WWF, "Natural and Nature-Based Flood Management: A Green Guide."
23 HLS New Forest Higher Level Stewardship Scheme, "Wetland Restoration," (n.d.), https://hlsnewforest.org.uk/about-us/.
24 Udo Nehren et al., "Sand Dunes and Mangroves for Disaster Risk Reduction and Climate Change Adaptation in the Coastal Zone of Quang Nam Province, Vietnam," in *Land Use and Climate Change Interactions in Central Vietnam: Lucci*, ed. Alexandra Nauditt and Lars Ribbe (Singapore: Springer Singapore, 2017).
25 WWF, "Natural and Nature-Based Flood Management: A Green Guide."
26 Environment Agency, "Working with Natural Processes to Reduce Flood Risk."
27 Scottish Environmental Protection Agency, "Natural Flood Management Handbook," (2015), https://sepa.org.uk/media/163560/sepa-natural-flood-management-handbook1.pdf.
28 City of Gold Coast, "Beach Nourishment," (n.d.). https://goldcoast.qld.gov.au/Environment-sustainability/Protecting-our-environment/Coastal-management/Beach-nourishment.
29 NOAA Habitat Blueprint, "Living Shorelines," (n.d.), https://habitatblueprint.noaa.gov/living-shorelines/.
30 Apalachee Regional Planning Council, "Franklin 98: Protecting Community. Conserving the Coast," (n.d.), https://arpc.org/franklin-98.
31 Z. Sebesvari et al., "Opportunities for Considering Green Infrastructure and Ecosystems in the Sendai Framework Monitor," *Progress in Disaster Science* 2 (2019).
32 Department of Primary Industries, "Mangroves," (2008), http://dpi.nsw.gov.au/__data/assets/pdf_file/0020/236234/mangroves.pdf.
33 Department of Primary Industries, "Mangroves."
34 Protection and Restoration Act Barataria-Terrebonne National Estuary Program & Coastal Wetlands Planning, "Coastal Wetlands Restoration: Residents' Guide," (2015), https://lacoast.gov/new/Ed/BTNEP_book.pdf.
35 Ravishankar Thupalli and Tariq A. Deen, "An Investment Strategy for Reducing Disaster Risks and Coastal Pollution Using Nature Based Solutions," in *Multifunctional Wetlands: Pollution Abatement and Other Ecological Services from Natural and Constructed Wetlands*, ed. Nidhi Nagabhatla and Christopher D. Metcalfe (Cham: Springer International Publishing, 2018).
36 The New Indian Express, "2.5l Trees Planted So Far in Tn's Ramanathapuram to Improve Green Cover," (n.d.). https://newindianexpress.com/states/tamil-nadu/2023/Nov/18/25l-trees-planted-so-far-in-tns-ramanathapuram-to-improve-green-cover-2633860.html.
37 M.M. Linham and R.J. Nicholls, "Technologies for Climate Change Adaptation – Coastal Erosion and Flooding –," (2010), https://tech-action.unepdtu.org/wp-content/uploads/sites/2/2019/04/tnahandbook-coastalerosionflooding.pdf.
38 Scottish Environmental Protection Agency, "Natural Flood Management Handbook."

39 Environment Agency, "Working with Natural Processes – Evidence Directory," (2017), http://evidence.environment-agency.gov.uk/FCERM/Libraries/FCERM_Project_Documents/Working_with_natural_processes_evidence_directory.sflb.ashx.
40 Scottish Environmental Protection Agency, "Natural Flood Management Handbook."
41 U.S. Geological Survey, "Application of Wind Fetch and Wave Models for Habitat Rehabilitation and Enhancement Projects," (2008), https://pubs.usgs.gov/of/2008/1200/pdf/ofr2008-1200_web.pdf.
42 Scottish Environmental Protection Agency, "Natural Flood Management Handbook."
43 Environment Agency, "Working with Natural Processes – Evidence Directory."
44 Essex Wildlife Trust, "Saltmarsh Restoration Project," (n.d.), https://essexwt.org.uk/what-we-do/protecting-wildlife/projects/saltmarsh.
45 Linham, "Technologies for Climate Change Adaptation – Coastal Erosion and Flooding –."
46 Florida Department of Environmental Protection, "Oyster Reef Restoration," (n.d.), https://floridadep.gov/rcp/rcp/content/oyster-reef-restoration.
47 Billion Oyster Project, " Restoring Oyster Reefs to New York Harbor through Public Education Initiatives," (n.d.), https://billionoysterproject.org.
48 M.V.W. Cuttler et al., "Response of a Fringing Reef Coastline to the Direct Impact of a Tropical Cyclone," *Limnology and Oceanography Letters* 3, no. 2 (2018).
49 Filippo Ferrario et al., "The Effectiveness of Coral Reefs for Coastal Hazard Risk Reduction and Adaptation," *Nature Communications* 5 (2014).
50 Andreas J. Andersson et al., "Ecological and Socioeconomic Strategies to Sustain Caribbean Coral Reefs in a High-Co_2 World," *Regional Studies in Marine Science* 29 (2019).
51 Andersson et al., "Ecological and Socioeconomic Strategies to Sustain Caribbean Coral Reefs in a High-Co_2 World."
52 Andersson et al., "Ecological and Socioeconomic Strategies to Sustain Caribbean Coral Reefs in a High-Co_2 World."
53 Andersson et al., "Ecological and Socioeconomic Strategies to Sustain Caribbean Coral Reefs in a High-Co_2 World."
54 UC Santa Cruz, "Reefense Project Aims to Meld Biology and Engineering in Novel Structures for Coastal Protection," (n.d.), https://news.ucsc.edu/2022/06/reefense-grant/.
55 UN Environment and IUCN UN Environment-DHI, "Nature Based Solutions for Water Management: A Primer," (2018), http://unepdhi.org/-/media/microsite_unepdhi/publications/documents/unep_dhi/web_unep-dhi_nbs-primer-2018.pdf?la=en.
56 R.C. Brears, *Blue and Green Cities: The Role of Blue-Green Infrastructure in Managing Urban Water Resources* (London: Palgrave Macmillan UK, 2018).
57 Texan by Nature, "Constructed Wetlands: 2018 Conservation Wrangler," (2018), https://texanbynature.org/projects/constructed-wetlands/.
58 Natural Water Retention Measures, "Rainwater Harvesting," (2015), http://nwrm.eu/sites/default/files/nwrm_ressources/u2_-_rainwater_harvesting.pdf.
59 US EPA, "Rainwater Harvesting. Conservation, Credit, Codes, and Cost Literature Review and Case Studies," (2013), https://epa.gov/sites/production/files/2015-11/documents/rainharvesting.pdf.
60 City of Cape Town, "Rainwater Harvesting: Alternative Water Resources (Pamphlet No. 4 of 4)," (n.d.), https://resource.capetown.gov.za/documentcentre/Documents/Graphics%20and%20educational%20material/Alternative_Water_Resources_Rainwater_English.pdf.

61 Natural Water Retention Measures, "Soakaways," (2015), http://nwrm.eu/sites/default/files/nwrm_ressources/u7_-_soakaways.pdf.
62 "Infiltration Trenches," (2015), http://nwrm.eu/sites/default/files/nwrm_ressources/u8_-_infiltration_trenches.pdf.
63 UN Environment-DHI, "Nature Based Solutions for Water Management: A Primer."
64 Department of Agriculture Australian Government, Fisheries and Forestry, "Climate-Smart Agriculture Program," (n.d.), https://agriculture.gov.au/agriculture-land/farm-food-drought/natural-resources/landcare/climate-smart.
65 Natural Water Retention Measures, "Restoration of Natural Infiltration to Groundwater," (2015), http://nwrm.eu/sites/default/files/nwrm_ressources/n13_-_restoration_of_natural_infiltration_to_groundwater_0.pdf.
66 Ministry of the Environment, "Ecosystem-Based Disaster Risk Reduction in Japan: A Handbook for Practitioners," (2016), http://env.go.jp/nature/biodic/eco-drr/pamph04.pdf.
67 Iyyappan Jaisankar, Ayyam Velmurugan, and T.P. Swarnam, "Chapter 24 – Bioshield: An Answer to Climate Change Impact and Natural Calamities?," in *Biodiversity and Climate Change Adaptation in Tropical Islands*, ed. Chandrakasan Sivaperuman et al. (Academic Press, 2018).
68 Daniel M. Alongi, "Mangrove Forests: Resilience, Protection from Tsunamis, and Responses to Global Climate Change," *Estuarine, Coastal and Shelf Science* 76, no. 1 (2008).
69 FAO, "The Role of Coastal Forests in the Mitigation of Tsunami Impacts," (2007), http://fao.org/forestry/19290-09bf06569b748c827dddf4003076c480c.pdf.
70 Great Forest Wall Project, "Great Forest Wall Project: Your Donation Will Help Us Create Forests," (n.d.), https://morinoproject.com/wp/wp-content/themes/gfw/images/home/panf-eng.pdf.

References

Alongi, Daniel M. "Mangrove Forests: Resilience, Protection from Tsunamis, and Responses to Global Climate Change." *Estuarine, Coastal and Shelf Science* 76, no. 1 (2008/01/01): 1–13.

Andersson, Andreas J., Alexander A. Venn, Linwood Pendleton, Angelique Brathwaite, Emma F. Camp, Sarah Cooley, Dwight Gledhill, et al. "Ecological and Socioeconomic Strategies to Sustain Caribbean Coral Reefs in a High-Co_2 World." *Regional Studies in Marine Science* 29 (2019/05/01): 100677.

Belle, Johannes A., Nacelle Collins, and Andries Jordaan. "Managing Wetlands for Disaster Risk Reduction: A Case Study of the Eastern Free State, South Africa." *Jàmbá: Journal of Disaster Risk Studies* 10, no. 1 (2018).

Brears, R.C. *Blue and Green Cities: The Role of Blue-Green Infrastructure in Managing Urban Water Resources.* London: Palgrave Macmillan UK, 2018.

Brears, Robert. "Integrated Flood Risk Management, Lessons from the Rhine and Danube for South Asia." *New Water Policy & Practice* 1, no. 2 (2015/03/01): 59–67.

Convention on Biological Diversity. "Voluntary Guidelines for the Design and Effective Implementation of Ecosystem-Based Approaches to Climate Change Adaptation and Disaster Risk Reduction." (2018). https://cbd.int/doc/c/3f7a/4589/5cc1b7058bf52427fa9bae84/sbstta-22-inf-01-en.pdf.

Cuttler, M.V.W., J.E. Hansen, R.J. Lowe, and E.J.F. Drost. "Response of a Fringing Reef Coastline to the Direct Impact of a Tropical Cyclone." *Limnology and Oceanography Letters* 3, no. 2 (2018/04/01): 31–38.

Department of Primary Industries. "Mangroves." (2008). http://dpi.nsw.gov.au/__data/assets/pdf_file/0020/236234/mangroves.pdf.

Environment Agency. "Working with Natural Processes – Evidence Directory." (2017). http://evidence.environment-agency.gov.uk/FCERM/Libraries/FCERM_Project_Documents/Working_with_natural_processes_evidence_directory.sflb.ashx.

Environment Agency. "Working with Natural Processes to Reduce Flood Risk." (2017). https://assets.publishing.service.gov.uk/government/uploads/system/uploads/attachment_data/file/654440/Working_with_natural_processes_one_page_summaries.pdf.

FAO. "The Role of Coastal Forests in the Mitigation of Tsunami Impacts." (2007). http://fao.org/forestry/19290-09bf06569b748c827dddf4003076c480c.pdf.

Ferrario, Filippo, Michael W. Beck, Curt D. Storlazzi, Fiorenza Micheli, Christine C. Shepard, and Laura Airoldi. "The Effectiveness of Coral Reefs for Coastal Hazard Risk Reduction and Adaptation." *Nature Communications* 5 (2014/05/13): 3794.

Florida Department of Environmental Protection. "Oyster Reef Restoration." (n.d.). https://floridadep.gov/rcp/rcp/content/oyster-reef-restoration.

IUCN. "About Ecosystem-Based Disaster Risk Reduction (Eco-Drr)." (n.d.). https://iucn.org/theme/ecosystem-management/our-work/environment-and-disasters/about-ecosystem-based-disaster-risk-reduction-eco-drr.

IUCN. "Environmental Guidance Note for Disaster Risk Reduction: Healthy Ecosystems for Human Security and Climate Change Adaptation." (2013). https://iucn.org/content/environmental-guidance-note-disaster-risk-reduction-healthy-ecosystems-human-security-and-climate-change-adaptation.

IUCN. "Five Reasons Why Ecosystems Are Central to Disaster Risk Reduction." (n.d.). https://iucn.org/theme/ecosystem-management/our-work/environment-and-disasters/about-ecosystem-based-disaster-risk-reduction-eco-drr/five-reasons-why-ecosystems-are-central-disaster-risk-reduction.

Jaisankar, Iyyappan, Ayyam Velmurugan, and T.P. Swarnam. "Chapter 24 – Bioshield: An Answer to Climate Change Impact and Natural Calamities?" In *Biodiversity and Climate Change Adaptation in Tropical Islands*, edited by Chandrakasan Sivaperuman, Ayyam Velmurugan, Awnindra Kumar Singh, and Iyyappan Jaisankar, 667–98. Cambridge, United States: Academic Press, 2018.

Jongman, Brenden. "Effective Adaptation to Rising Flood Risk." *Nature Communications* 9, no. 1 (2018/05/29): 1986.

Linham, M.M., and Nicholls, R.J. "Technologies for Climate Change Adaptation – Coastal Erosion and Flooding –." (2010). https://tech-action.unepdtu.org/wp-content/uploads/sites/2/2019/04/tnahandbook-coastalerosionflooding.pdf.

Ministry of the Environment. "Ecosystem-Based Disaster Risk Reduction in Japan: A Handbook for Practitioners." (2016). http://env.go.jp/nature/biodic/eco-drr/pamph04.pdf.

Natural Water Retention Measures. "Infiltration Trenches." (2015). http://nwrm.eu/sites/default/files/nwrm_ressources/u8_-_infiltration_trenches.pdf.

Natural Water Retention Measures. "Rainwater Harvesting." (2015). http://nwrm.eu/sites/default/files/nwrm_ressources/u2_-_rainwater_harvesting.pdf.

Natural Water Retention Measures. "Restoration of Natural Infiltration to Groundwater." (2015). http://nwrm.eu/sites/default/files/nwrm_ressources/n13_-_restoration_of_natural_infiltration_to_groundwater_0.pdf.

Natural Water Retention Measures. "Soakaways." (2015). http://nwrm.eu/sites/default/files/nwrm_ressources/u7_-_soakaways.pdf.

Nehren, U., K. Sudmeier-Rieux, S. Sandholz, M. Estrella, M. Lomarda, and T. Guillén. "The Ecosystem-Based Disaster Risk Reduction: Case Study and Exercise Source Book." (2014). http://pedrr.org/pedrr/wp-content/uploads/2013/09/Eco-DRR-case-study-source-book_final.pdf.

Nehren, Udo, Hoang H.D. Thai, N.D. Trung, Claudia Raedig, and S. Alfonso. "Sand Dunes and Mangroves for Disaster Risk Reduction and Climate Change Adaptation in the Coastal Zone of Quang Nam Province, Vietnam." In *Land Use and Climate Change Interactions in Central Vietnam: Lucci*, edited by Alexandra Nauditt, and Lars Ribbe, 201–22. Singapore: Springer Singapore, 2017.

NOAA Habitat Blueprint. "Living Shorelines." (n.d.). https://habitatblueprint.noaa.gov/living-shorelines/.

Renaud, F.G., K. Sudmeier-Rieux, and M. Estrella. *The Role of Ecosystems in Disaster Risk Reduction*. Tokyo: United Nations University Press, 2013.

Reyers, Belinda, Jeanne L. Nel, Patrick J. O'Farrell, Nadia Sitas, and Deon C. Nel. "Navigating Complexity through Knowledge Coproduction: Mainstreaming Ecosystem Services into Disaster Risk Reduction." *Proceedings of the National Academy of Sciences* 112, no. 24 (2015): 7362–68.

Scottish Environmental Protection Agency. "Natural Flood Management Handbook." (2015). https://sepa.org.uk/media/163560/sepa-natural-flood-management-handbook1.pdf.

Sebesvari, Z., J. Woelki, Y. Walz, K. Sudmeier-Rieux, S. Sandholz, S. Tol, V. Ruíz García, K. Blackwood, and F.G. Renaud. "Opportunities for Considering Green Infrastructure and Ecosystems in the Sendai Framework Monitor." *Progress in Disaster Science* 2 (2019/07/01): 100021.

Thupalli, Ravishankar, and Tariq A. Deen. "An Investment Strategy for Reducing Disaster Risks and Coastal Pollution Using Nature Based Solutions." In *Multifunctional Wetlands: Pollution Abatement and Other Ecological Services from Natural and Constructed Wetlands*, edited by Nidhi Nagabhatla, and Christopher D. Metcalfe, 141–70. Cham: Springer International Publishing, 2018.

UN Environment-DHI, UN Environment and IUCN. "Nature Based Solutions for Water Management: A Primer." (2018). http://unepdhi.org/-/media/microsite_unepdhi/publications/documents/unep_dhi/web_unep-dhi_nbs-primer-2018.pdf?la=en.

UNEP. "Ecosystem-Based Disaster Risk Reduction." (n.d.). https://unenvironment.org/explore-topics/disasters-conflicts/what-we-do/risk-reduction/ecosystem-based-disaster-risk.

United Nations Office for Disaster Risk Reduction and Centre for Research on the Epidemiology of Disasters. "The Human Cost of Weather-Related Disasters 1995-2015." (2015). https://unisdr.org/we/inform/publications/46796.

US EPA. "Rainwater Harvesting. Conservation, Credit, Codes, and Cost Literature Review and Case Studies." (2013). https://epa.gov/sites/production/files/2015-11/documents/rainharvesting.pdf.

U.S. Geological Survey. "Application of Wind Fetch and Wave Models for Habitat Rehabilitation and Enhancement Projects." (2008). https://pubs.usgs.gov/of/2008/1200/pdf/ofr2008-1200_web.pdf.

van Wesenbeeck, Bregje K., Stephanie IJff, Brenden Jongman, Simone A.B. Balog-Way, Stefanie M. Kaupa, Lauren V. Bosche, Glenn-Marie Lange, Niels B. Holm-Nielsen, Henk Nieboer, Yusuke Taishi, Pradeep H. Kurukulasuriya, and Imen Meliane. "Implementing Nature Based Flood Protection: Principles and Implementation

Guidance (English)." (2017). http://documents.worldbank.org/curated/en/739421509427698706/Implementing-nature-based-flood-protection-principles-and-implementation-guidance.

WWF. "Natural and Nature-Based Flood Management: A Green Guide." (2017). https://worldwildlife.org/publications/natural-and-nature-based-flood-management-a-green-guide.

12 Adaptive Management and Nature-Based Solutions

Introduction

Ecosystems are dynamic, self-organizing systems that evolve through complex interactions across multiple scales, making the outcomes of nature-based solutions (NBS) inherently uncertain. Managing this uncertainty requires a structured framework that promotes reflection, experiential learning, and adaptive management. Long-term monitoring is essential for the success of NBS, as their stability depends on managers' ability to assess the effectiveness of interventions and make necessary adjustments. Monitoring programs should be institutionalized within organizations and stakeholder groups to ensure that NBS provide lasting social and environmental benefits. Ultimately, the effectiveness of NBS is strengthened by an adaptive management approach that enables continuous learning and improvement.[1,2,3,4]

Adaptive Management

Adaptive management is a continuous process of learning and adaptation, where knowledge accumulates over time, and management strategies are adjusted accordingly. This iterative cycle enhances understanding of resources and leads to more effective management decisions. A defining feature of adaptive management is the feedback loop between learning and decision-making, where learning informs decisions, and management actions serve as opportunities to gain insights into resource dynamics. Unlike trial-and-error approaches, adaptive management follows a structured framework that includes clearly defined objectives, identification of alternative strategies, predictions of outcomes, acknowledgment of uncertainties, systematic monitoring, and informed decision-making based on observed responses.[5,6] The key elements of adaptive management include:

- *Regularly review and revise management objectives*: Stakeholder differences and competing scientific perspectives are inevitable in adaptive management. Open discussions about these differences should be part of the learning process. However, for adaptive management to be effective,

DOI: 10.4324/9781003517948-12

there must be a shared understanding of goals and a clear definition of progress. Without this, the process becomes ineffective.

- *Establishing a model of the system being managed*: A clearly defined baseline understanding and assumptions about the system provide a foundation for learning. Models help test hypotheses by defining key variables that characterize the system's state, rates of change, and potential future trajectories.
- *Developing a range of management options*: Even when objectives are agreed upon, uncertainty often exists about the effectiveness of different management actions. Since existing data rarely indicate an optimal policy, a range of possible management options should be considered from the outset. Each option should be evaluated based on its likelihood of achieving the objectives and potential to generate new information or limit future choices.
- *Monitoring and evaluating outcomes*: Adaptive management requires a structured approach to track and compare the results of management decisions. Systematic data collection and evaluation enable testing alternative hypotheses and improve system understanding over time.
- *Incorporating learning into future decisions*: Adaptive management aims to improve decision-making through active learning. To achieve this, mechanisms must be in place to integrate new information into ongoing management processes, ensuring adjustments are made based on observed outcomes.
- *Maintaining a collaborative structure for stakeholder participation and learning*: Involving stakeholders affected by management actions is a widely recognized principle of ecosystem management. Meaningful participation that fosters active learning and consensus-building is essential for adaptive management to succeed, though it remains a significant challenge.[7]

The adaptive management framework is built on four core principles:

1. Identifying competing hypotheses to explain observed patterns or processes in the system being managed.
2. Developing models that incorporate these hypotheses to predict how the system will respond to different management interventions.
3. Monitoring resource responses to assess changes and gather data on system behavior.
4. Comparing actual outcomes with predicted responses to refine understanding and improve future decision-making.[8]

Conditions for Using Adaptive Management

Not all management decisions require an adaptive approach. In some cases, there may be minimal uncertainty about the best course of action or little opportunity to apply learning over time. Therefore, the need for adaptive management should be assessed at the start of a project. The decision to use adaptive management can be based on whether the following five conditions are met:

1 *Management is necessary despite uncertainty*: Action is required to address a significant issue, even though its outcomes cannot be predicted with certainty.
2 *Clear and measurable objectives guide decision-making*: Well-defined goals and metrics are essential for evaluating performance and determining the most effective actions.
3 *Learning can be applied to future management*: Multiple management options are available, and the framework allows for adjustments as new insights emerge.
4 *Monitoring can reduce uncertainty*: Data collection and analysis improve the understanding of system processes, leading to more informed decision-making.
5 *Sustained stakeholder commitment*: Stakeholders are engaged throughout the process, from setting objectives to recognizing uncertainty and contributing to data collection and analysis.[9]

Adaptive Management and Climatic Challenges

Adaptive management is a decision-making process that acknowledges and operates within the realm of uncertainty. Decision-makers must act despite imperfect information, as uncertainty is inherent in managing complex systems. Recognizing uncertainty does not prevent decisions from being made; instead, it ensures that adaptive capacity and resilience strategies are appropriately designed and implemented to mitigate the impact of uncertainty. The goal is not to eliminate uncertainty and understand its significance in relation to the decision. If a decision remains unchanged despite identified uncertainties in the supporting evidence, further refinement of the analysis may not be necessary.[10] There are three types of uncertainty in managing climate change risks:

1 Natural uncertainty refers to the inherent randomness in natural systems. It is addressed by considering a range of return periods for events such as storms, allowing for calculating of potential damage distributions and expected annual damage. While it is impossible to predict precisely when or where a major event will occur, planning based on probabilistic assessments helps mitigate risk.
2 Knowledge uncertainty relates to the limitations in our understanding of systems, our ability to measure and model them, and our capacity to predict future changes. Gaps in data, evolving scientific knowledge, and technological limitations contribute to this uncertainty.

3 Decision uncertainty arises from doubts about the best course of action. It is shaped by how knowledge uncertainty influences decision-making, the trade-offs society is willing to accept, and the level of investment deemed necessary to mitigate future risks. This includes acceptable risk levels, the balance between social equity and environmental protection, and the willingness to invest in precautionary measures.[11]

In the face of climate uncertainties, prioritizing no-regret actions can be crucial in maximizing benefits and minimizing risks. No-regret actions are policies and activities that support economic, environmental, or social development goals, even if the impacts of climate change do not materialize as expected. These actions reduce vulnerability to climate change and provide immediate benefits, such as improving quality of life, enhancing service provision, and strengthening community resilience. In many cases, no-regret actions help communities manage climate variability, address non-climate risk factors, reduce overall vulnerability, and build resilience to future shocks while delivering multiple co-benefits.[12,13]

Adaptive management seeks to enhance scientific understanding and develop flexible management strategies for future scenarios, including those involving unexpected events. In the context of climate change, continuous documentation and monitoring of decisions and their outcomes contribute to the advancement of scientific knowledge and support iterative policy adjustments. This long-term, structured decision-making process is designed to:

- Reduce uncertainty through monitoring, modelling, and implementing no-regret actions that minimize risk by identifying key vulnerabilities to extreme events.
- Address uncertainties and extreme climate events.
- Ensure policies achieve their intended objectives and can be modified as needed.
- Base decisions on the best available information to improve resilience and effectiveness over time.[14]

Adaptive Management and Non-Climatic Challenges

In addressing non-climatic challenges that lead to environmental degradation, adaptive management involves forming hypotheses about ecosystem functioning, monitoring outcomes, comparing results with expectations, and adjusting management strategies to achieve conservation goals. This approach acknowledges uncertainty in ecosystem management by treating policies and practices as experiments, allowing continuous learning and refinement.

Adaptive management enables managers to assess their progress toward both short-term and long-term ecosystem goals. It benefits three key groups. First, agencies and organizations gain better information and utilize resources more efficiently, thereby improving their outreach to the public and policymakers. Second, the public benefits from enhanced ecosystem services. Third, ecosystems benefit, as improved data enable informed decision-making, allowing for timely adjustments to projects and programs that lead to better conservation and restoration outcomes.[15]

Adaptive management is a practical approach for restoring or enhancing ecosystems affected by societal challenges such as urbanization. It acknowledges that ecosystems are only partially understood and emphasizes the importance of monitoring conditions and applying lessons learned during management. This approach is beneficial when ecosystems are responsive to interventions, but uncertainty exists about their impacts. Key characteristics of adaptive management in ecosystem management include:

- Ecosystems are dynamic, changing over time in response to environmental conditions and management actions that also evolve.
- Environmental variability is only partially predictable, with some changes remaining unrecognized. Ecological processes can be unpredictable, leading to system behaviors that are difficult to predict.
- Management interventions occur periodically and influence ecosystems either directly or indirectly.
- Effective management is constrained by uncertainty regarding ecosystem processes and the effects of interventions. Reducing this uncertainty through ongoing learning can enhance decision-making and outcomes.[16]

Uncertainty in Adaptive Management

Uncertainty is an inherent challenge in ecosystem management, often limiting the effectiveness of interventions. Four main types of uncertainty influence ecosystem management:

1. *Environmental variation*: This is the most significant and uncontrollable source of uncertainty, often driven by factors such as climate variability, which can significantly impact ecosystem dynamics.
2. *Partial observability:* This refers to uncertainty about the actual state of an ecosystem resulting from limitations in monitoring and data collection.
3. *Partial controllability*: This occurs when the intended outcomes of management actions differ from actual results, often due to indirect methods being used to achieve objectives.
4. *Structural or process uncertainty*: This arises from gaps in understanding the biophysical processes that regulate ecosystems and how management interventions influence these processes.[17]

Adaptive Management and Nature-Based Solutions

In the context of NBS, adaptive management relies on continuous monitoring, investigation, and research to deepen understanding of ecosystems and the services they provide. This approach supports the assessment of changing environmental conditions, the evaluation of management outcomes, and improved comprehension of climate change impacts. Effective adaptive management strategies should:

- Ensure that short-term decisions align with and support long-term objectives.
- Focus on dynamic pathways with successive decision points, rather than pursuing a fixed end state.
- Emphasize flexibility in both individual measures and overarching strategies, allowing for timely adjustments to avoid underperformance or overinvestment.
- Promote synergies with public and private sector goals, as well as development initiatives, to maximize overall impact.[18]

Developing NBS to enhance climate resilience and mitigate environmental degradation requires an adaptive management framework that includes planning, designing, implementing, and monitoring NBS to achieve social, environmental, and economic objectives.

1 *Developing a plan for implementing NBS*:
   a Define a clear vision that aligns with local priorities and has broad support.
   b Determine the geospatial scope of the project, unrestricted by political or administrative boundaries.
   c Establish a cross-sectoral steering group with strong leadership and representation from key stakeholders and the community.
   d Encourage collaboration across political, organizational, and land-ownership boundaries, integrating multiple disciplines and scales.
   e Identify the multifunctional benefits of existing and potential NBS.
   f Assess local challenges, risks, and community needs.
   g Secure necessary resources for successful implementation and long-term sustainability.
2 *Designing NBS*:
   a Develop and communicate a draft plan, strategy, or design incorporating the vision and objectives.
   b Refine and improve the plan based on feedback and stakeholder engagement.
   c Ensure the final design meets functionality, durability, and aesthetic requirements.
3 *Implementing NBS*:
   a Establish design and management standards with locally relevant criteria.
   b Secure sustainable funding mechanisms for ongoing maintenance.
   c Implement the project following the agreed plan.
   d Set clear milestones, targets, and performance indicators.

4. *Managing and maintaining NBS*:
   a. Continuously evaluate progress against objectives using key performance indicators and stakeholder feedback.
   b. Ensure effective aftercare management and oversee maintenance activities.[19]

Notes

1 Emmanuelle Cohen-Shacham et al., "Core Principles for Successfully Implementing and Upscaling Nature-Based Solutions," *Environmental Science & Policy* 98 (2019).

2 Nadja Kabisch et al., "Nature-Based Solutions to Climate Change Mitigation and Adaptation in Urban Areas: Perspectives on Indicators, Knowledge Gaps, Barriers, and Opportunities for Action," *Ecology and Society* 21, no. 2 (2016).

3 Cohen-Shacham et al., "Core Principles for Successfully Implementing and Upscaling Nature-Based Solutions."

4 Melbourne Water, "Healthy Waterways and Stormwater Strategies," (2013), http://melbournewater.com.au/aboutus/reportsandpublications/key-strategies/pages/healthy-waterways-and-stormwater-strategies.aspx.

5 U.S. Department of the Interior, "Adaptive Management: The U.S. Department of the Interior Applications Guide," (2012), https://doi.gov/sites/doi.gov/files/migrated/ppa/upload/DOI-Adapative-Management-Applications-Guide.pdf.

6 U.S. Department of the Interior, "Adaptive Management: The U.S. Department of the Interior Applications Guide."

7 National Research Council, *Adaptive Management for Water Resources Project Planning* (Washington, DC: National Research Council, 2004).

8 U.S. Department of the Interior, "Adaptive Management: The U.S. Department of the Interior Applications Guide."

9 U.S. Department of the Interior, "Adaptive Management: The U.S. Department of the Interior Applications Guide."

10 Y. Li, P. Sayers, G. Galloway, E. Penning-Rowsell, F. Shen, K. Wen, Y. Chen, and T. Le Quesne, "Flood Risk Management: A Strategic Approach," (2013), https://adb.org/sites/default/files/publication/30246/flood-risk-management.pdf.

11 Rasmus Heltberg, Paul B. Siegel, and Steen L. Jorgensen, "Addressing Human Vulnerability to Climate Change: Toward a 'No-Regrets' Approach," *Global Environmental Change* 19, no. 1 (2009).

12 Heltberg, Siegel, and Jorgensen, "Addressing Human Vulnerability to Climate Change: Toward a 'No-Regrets' Approach."

13 Ontario Centre for Climate Impacts and Adaptation Resources, "Adaptive Management in Climate Change Adaptation," (2012), http://climateontario.ca/doc/factsheets/AdaptiveManagement-final.pdf.

14 Ontario Centre for Climate Impacts and Adaptation Resources, "Adaptive Management in Climate Change Adaptation."

15 Thomas M. Franklin, Ronald H. Helinski, and Andrew M. Manale, "Using Adaptive Management to Meet Conservation Goals," in *U.S. Environmental Protection Agency Papers*, ed. U.S. EPA (2007).

16 U.S. Department of the Interior, "Adaptive Management: The U.S. Department of the Interior Applications Guide."

17 U.S. Department of the Interior, "Adaptive Management: The U.S. Department of the Interior Applications Guide."

18 Frans Klijn et al., "Adaptive Flood Risk Management Planning Based on a Comprehensive Flood Risk Conceptualisation," *Mitigation and Adaptation Strategies for Global Change* 20, no. 6 (2015).

19 Landscape Institute, "Green Infrastructure: An Integrated Approach to Land Use," (2013), https://landscapewpstorage01.blob.core.windows.net/www-landscapeinstitute-org/2016/03/Green-Infrastructure_an-integrated-approach-to-land-use.pdf.

References

City of Ottawa. "Putting down Roots for the Future. City of Ottawa Urban Forest Management Plan 2018-2037 " (2017). https://documents.ottawa.ca/sites/documents/files/final_ufmp_en.pdf.

Cohen-Shacham, Emmanuelle, Angela Andrade, James Dalton, Nigel Dudley, Mike Jones, Chetan Kumar, Stewart Maginnis, *et al.* "Core Principles for Successfully Implementing and Upscaling Nature-Based Solutions." *Environmental Science & Policy* 98 (2019/08/01/ 2019): 20–29.

Franklin, Thomas M Franklin, Ronald Helinski Helinski, and Andrew Manale Manale. "Using Adaptive Management to Meet Conservation Goals." In *U.S. Environmental Protection Agency Papers*, edited by U.S. EPA, 2007.

Government of the Netherlands. "Delta Programme Themes and Regions." https://government.nl/topics/delta-programme/delta-programme-themes-and-regions.

Heltberg, Rasmus, Paul Bennett Siegel, and Steen Lau Jorgensen. "Addressing Human Vulnerability to Climate Change: Toward a 'No-Regrets' Approach." *Global Environmental Change* 19, no. 1 (2009): 89–99.

Kabisch, Nadja, Niki Frantzeskaki, Stephan Pauleit, Sandra Naumann, McKenna Davis, Martina Artmann, Dagmar Haase, *et al.* "Nature-Based Solutions to Climate Change Mitigation and Adaptation in Urban Areas: Perspectives on Indicators, Knowledge Gaps, Barriers, and Opportunities for Action." *Ecology and Society* 21, no. 2 (2016).

Klijn, Frans, Heidi Kreibich, Hans De Moel, and Edmund Penning-Rowsell. "Adaptive Flood Risk Management Planning Based on a Comprehensive Flood Risk Conceptualisation." *Mitigation and Adaptation Strategies for Global Change* 20, no. 6 (2015): 845–64.

Landscape Institute. "Green Infrastructure: An Integrated Approach to Land Use." (2013). https://landscapewpstorage01.blob.core.windows.net/www-landscapeinstitute-org/2016/03/Green-Infrastructure_an-integrated-approach-to-land-use.pdf.

Melbourne Water. "Healthy Waterways and Stormwater Strategies." (2013). http://melbournewater.com.au/aboutus/reportsandpublications/key-strategies/pages/healthy-waterways-and-stormwater-strategies.aspx.

Ministry of Infrastructure and Water Management, the Ministry of Agriculture, Nature, and Food Quality, and the Ministry of the Interior and Kingdom Relations,. "Delta Programme 2019. Continuing the Work on the Delta: Adapting the Netherlands to Climate Change in Time." (2018).https://english.deltaprogramma.nl/site/binaries/site-content/collections/documents/2018/09/18/dp2019-en-printversie/DP2019+EN+printversie.pdf.

National Research Council. "Adaptive Management for Water Resources Project Planning." Washington D.C.: National Research Council, 2004.

Ontario Centre for Climate Impacts and Adaptation Resources. "Adaptive Management in Climate Change Adaptation." (2012). http://climateontario.ca/doc/factsheets/AdaptiveManagement-final.pdf.

PBL Netherlands Environmental Assessment Agency. "A Reflexive Monitoring and Evaluation Framework for the Delta Programme." https://pbl.nl/en/news/newsitems/2017/a-reflexive-monitoring-and-evaluation-framework-for-the-delta-programme.

Sayers, P., Y. Li, G. Galloway, E. Penning-Rowsell, F. Shen, K. Wen, Y. Chen, and T. Le Quesne. "Flood Risk Management: A Strategic Approach." (2013). https://adb.org/sites/default/files/publication/30246/flood-risk-management.pdf.

U.S. Department of the Interior. "Adaptive Management: The U.S. Department of the Interior Applications Guide,." (2012). https://doi.gov/sites/doi.gov/files/migrated/ppa/upload/DOI-Adapative-Management-Applications-Guide.pdf.

13 Financing Nature-Based Solutions

Introduction

An expanding range of public and private funding sources is available for implementing nature-based solutions (NBS) to address societal challenges while enhancing human well-being and biodiversity. However, financing NBS does not follow a universal approach, as solutions must be tailored to specific contexts. Factors such as climatic risks, ecosystem types, geographical scale, implementation levels, governance structures, socio-economic beneficiaries, and project timelines all influence the most suitable financing mechanisms.

Sources of Finance for Nature-Based Solutions

Sources of financing for NBS can be categorized into domestic and international funds, originating from either public or private sources. These funding streams are summarized as follows:

International Public Financial Sources

Multilateral funds are financial mechanisms that support projects, policy initiatives, and technical assistance for international cooperation. Some funds explicitly incorporate NBS approaches. These funds are financed by national governments and typically managed by a dedicated secretariat, such as the Green Climate Fund.[1]

Multilateral Development Banks

Multilateral development banks are supranational institutions established by sovereign states, which serve as their shareholders. Their mandates align with these states' development aid and cooperation policies. These banks support economic growth and social progress in developing countries by financing various projects incorporating NBS.[2] Examples include the African Development Bank, the Asian Development Bank, the European Investment

DOI: 10.4324/9781003517948-13

Bank, the International Finance Corporation, the Islamic Development Bank, and the World Bank.

Bilateral Cooperation

Bilateral cooperation sources facilitate technical and financial exchanges between the two governments to implement policies, projects, or specific measures. These initiatives are typically funded through bilateral development banks and national development organizations. Often grant-based, bilateral cooperation supports piloting innovative concepts, including NBS, while reducing financial risk exposure.[3] Examples of bilateral cooperation agencies include Kreditanstalt für Wiederaufbau (KfW), the Nordic Development Fund, and the UK International Climate Fund.

Private Finance Sources

Various national and international market mechanisms, certification schemes, and foundations provide financial support for implementing NBS, particularly in forests, peatlands, and coastal ecosystems such as mangroves. For instance, voluntary carbon certification schemes generate revenue by selling carbon credits from avoided emissions. Other financial sources include certification for ecotourism operations, sustainable forest and agricultural products, and responsibly managed aquaculture and fisheries.[4]

Domestic Public Sources of Finance

Domestic funding for NBS is often generated through fiscal instruments, such as taxes, levies, fees, and bond issuance. These mechanisms allow governments to mobilize resources to support large-scale environmental restoration, climate resilience projects, and sustainable land-use practices. Carbon emissions, plastic waste, or resource extraction taxes can provide a steady revenue stream for financing NBS. Levies and fees, such as water use charges or tourism-related environmental fees, can also be directed towards conservation efforts and ecosystem restoration.[5]

Combination of Financing Instruments

Actors seeking funding for NBS must identify financing options and instruments that align with their specific needs and the broader context of their activities. Often, a combination of different financial instruments and sources is required to ensure continuous support throughout the entire planning, implementation, and long-term operation of NBS. This comprehensive approach helps secure sustainable funding supporting initial investments and ongoing management.[6] Two main cost categories need to be considered:

- *Investment costs*: These are the initial expenses required for developing infrastructure, building capacities, and acquiring technologies necessary for NBS implementation. Investment costs typically include research and development, capacity-building and coordination, construction materials, and specialized equipment. These costs are usually highest during the early development phase of an NBS project. External financing is often required to cover these expenses, which may come from grants, public funding, or private investments expecting returns from the benefits generated by the NBS.
- *Operating costs*: These are the ongoing expenses necessary to maintain and manage NBS over its lifetime. They include costs related to staffing, equipment maintenance, communication, and administrative functions such as financial oversight, monitoring, evaluation, and adaptive management of projects and initiatives. Securing long-term funding sources, such as revenue from ecosystem services, public-private partnerships, or government budgets, is essential to ensure the sustained effectiveness of NBS.[7]

By integrating investment and operating cost considerations into financial planning, actors can develop more resilient funding strategies that ensure the long-term viability and success of NBS initiatives.

Financial Instruments for Implementing Nature-Based Solutions

NBS initiatives are typically financed through a combination of financial instruments, which may include the following:

Blended Concessional Finance

Blended concessional finance for private sector projects is a tool used by multilateral development banks and development finance institutions, collaborating with donors and other development partners, to increase investment in critical private sector activities, facilitate sustainable development, and attract private capital. This approach offers two key advantages. First, it offers more affordable borrowing rates, which lower annual financing costs or extend repayment periods. Second, it helps reduce risk perception for lenders by involving multilateral development banks and donors, whose participation and due diligence enhance investor confidence.[8,9]

A set of principles has been established to guide development finance institutions in implementing blended concessional finance initiatives:

1. *Additionality and justification for blended finance*: Development finance institutions should only provide concessional finance when it fills a market gap that would otherwise remain unaddressed. It should not compete with or replace private sector financing. Concessionality should not be the sole

reason for financial support, as this could distort markets by undercutting commercial finance providers.

2 *Crowding-in and minimum concessionality*: Blended finance should help attract private investment rather than displacing it. Concessional finance is most effective when it provides the missing element that makes a private project commercially viable while demonstrating its potential for commercial replication.
3 *Commercial sustainability*: Development finance should aim to support projects that can eventually operate without long-term concessional assistance. Financial support must contribute to the commercial viability of the sector or activity, ensuring projects do not become permanently dependent on subsidies and avoiding incentives for rent-seeking behavior.
4 *Market reinforcement*: Blended concessional finance should address market failures without distorting competition. It should complement broader policy measures that remove market barriers rather than delaying necessary regulatory or institutional reforms. The goal is to enhance market function and encourage new private sector entrants.
5 *Promotion of high standards*: Private sector projects benefiting from concessional finance should meet high standards in corporate governance, environmental and social responsibility, transparency, integrity, and disclosure. Poorly designed projects or unnecessary concessional funding can damage market development and create reputational risks. Given that some projects may fail, concessional finance should be deployed with rigorous assessment and strong governance to ensure its effectiveness and accountability.[10,11]

Debt-for-Nature Swaps

A debt-for-nature swap is an agreement in which a country's debt is reduced in exchange for a commitment by the debtor government to invest in environmental conservation. In this voluntary transaction, a creditor country either forgives a portion or the entirety of the debt or sells the outstanding debt to a third party – typically a conservation organization – at a discounted rate. In the latter case, the indebted country's government agrees to a structured repayment plan, typically through its central bank in the local currency or via bonds. The funds generated from these repayments are then directed towards conservation initiatives. The success of debt-for-nature swaps depends on several key factors:

- Strong political leadership is needed to ensure government commitment and policy alignment.
- Effective internal government coordination between financial and environmental authorities.
- Realistic conditions that align with national environmental policies and programs to ensure feasibility and long-term sustainability.

- Stakeholder engagement and good governance are crucial for ensuring transparency and accountability in the management of conservation funds.
- Availability of technical assistance to support the implementation and monitoring of conservation initiatives linked to the debt swap.[12,13]

Green Credit Lines

For many years, public financial institutions have provided credit lines to local financial institutions for on-lending to end-borrowers. More recently, these institutions have adapted this approach to support green lending. Green credit lines are financial intermediation tools extended by public financial institutions to participating local financial institutions – such as commercial banks, specialized financial institutions, and non-banking credit organizations – which then on-lend to end-borrowers, including households, small and medium enterprises, industrial enterprises, and project developers, for eligible green projects.

Typically, borrowing financial institutions pay interest only on the amount borrowed. Green credit lines may offer special financial conditions, such as reduced interest rates, longer repayment periods, extended grace periods, or incentive payments. They may also include technical assistance, funded by public financial institutions, to help local banks develop the capacity to finance green investments and assist end-borrowers in structuring viable proposals.

Green credit lines are becoming a key tool for national development banks and multilateral financial institutions to promote long-term investment in projects with positive environmental impacts. Major providers of green credit lines include the European Investment Bank, the European Bank for Reconstruction and Development, the Inter-American Development Bank, KfW, the International Bank for Reconstruction and Development, Agence Française de Développement (AFD), the International Finance Corporation, and the Japan International Cooperation Agency.[14]

Green Loans

Green loans are financing instruments specifically designed to fund or refinance, either partially or entirely, new or existing green projects. These loans follow four key principles:

1. *Use of proceeds*: The primary requirement of a green loan is that its proceeds must be used exclusively for green projects, which should be clearly defined in the finance documents. These projects must deliver clear environmental benefits, which should be assessed and, where possible, quantified, measured, and reported by the borrower.
2. *Project evaluation and selection*: The borrower must transparently communicate to lenders:

Table 13.1 Categories of Green Projects

Category	*Project Example*
Pollution prevention and control	Projects focused on reducing emissions and contamination, including wastewater treatment, greenhouse gas mitigation, air pollution control, and soil and groundwater remediation.
Environmentally sustainable management of natural resources and land	Initiatives promoting sustainable agriculture, climate-smart irrigation (e.g. drip systems), organic farming, eco-friendly aquaculture and fisheries, reforestation, afforestation, and the protection of natural landscapes.
Terrestrial and aquatic biodiversity conservation	Projects that safeguard biodiversity in coastal zones, marine ecosystems, inland watersheds, forests, and grasslands through habitat protection and restoration.
Sustainable water and wastewater management	Development of resilient infrastructure for clean drinking water, wastewater recycling, sustainable urban drainage systems, and flood risk reduction.
Climate change adaptation	Environmental interventions that enhance resilience to climate extremes, such as green infrastructure for cooling, flood barriers, and nature-based storm protection.
Green buildings	Construction or retrofitting of buildings that comply with green certification standards (e.g. LEED, BREEAM), focusing on energy efficiency, water conservation, and reduced environmental impact.

a Their environmental sustainability objectives.
b The process determines project eligibility within established green project categories (examples are listed in Table 13.1).
c The methods used to identify and manage material environmental risks associated with the proposed projects.

3 *Management of proceeds*: Loan proceeds should be credited to a dedicated account or tracked appropriately to ensure transparency and maintain the integrity of the financing.
4 *Reporting*: Borrowers must keep up-to-date information on how the loan proceeds are used. Where full disclosure is limited due to confidentiality agreements, competitive considerations, or many projects, borrowers can report the expected impact through qualitative performance indicators and, where feasible, quantitative measures (e.g., reduction or avoidance of greenhouse gas emissions).[15]

Green Bonds

Green bonds are financial instruments where proceeds are used exclusively to finance or refinance, either partially or entirely, new or existing eligible green projects. These projects, which include various environmental initiatives (examples listed in Table 13.2), may also encompass related expenditures

Table 13.2 Green Project Categories Supported by Green Bonds

Category	*Description*
Renewable energy	Projects related to the generation, transmission, and application of renewable energy, including clean energy products and technologies.
Energy efficiency	Initiatives that improve energy use in buildings, appliances, infrastructure, and systems such as smart grids, energy storage, and district heating.
Pollution prevention and control	Efforts to reduce air emissions and greenhouse gases, remediate soil, prevent waste, and promote recycling and energy-efficient waste-to-energy technologies.
Sustainable management of natural resources and land use	Environmentally sustainable practices in agriculture, forestry, fisheries, and aquaculture, including afforestation, reforestation, and the protection or restoration of natural landscapes.
Terrestrial and aquatic biodiversity conservation	Protection and restoration of coastal, marine, freshwater, and terrestrial ecosystems and habitats.
Sustainable water and wastewater management	Development of infrastructure for safe drinking water, efficient wastewater treatment, urban drainage systems, and flood control.
Climate change adaptation	Projects that enhance resilience to climate risks through planning, monitoring, and adaptive infrastructure and information systems.
Green buildings	Construction or renovation of buildings that meet recognized regional, national, or international green building standards or certifications.

such as research and development and can contribute to multiple environmental objectives.[16] There are three main types of green bonds:

1 *Classic bond*: Issued by an investment-grade entity to finance a portfolio of green assets. These bonds carry the same credit rating as the issuer's other debt instruments since they have full recourse to the borrower.
2 *Project bond*: Specifically ring-fenced for a particular green project. Investors are directly exposed to the risks and returns of one or more designated green projects.
3 *Asset-backed bond*: Collateralized by a portfolio of green assets or receivables, which may change over time. The total value of the underlying assets must always be equal to or greater than the bond's value.[17]

Payments for Ecosystem Services

Payments for Ecosystem Services (PES) are a key mechanism for creating markets that incentivize the protection and enhancement of ecosystem services. These payments are made to landowners or resource managers to encourage actions that improve the quantity and quality of ecosystem services, benefiting specific users or society, often in distant locations.

A PES is a voluntary transaction where at least one buyer purchases a well-defined ecosystem service, or a land-use practice that secures that service, from at least one provider, under the condition that the service continues to be supplied. A fundamental aspect of PES is that it provides financial incentives for individuals or communities to adopt environmentally beneficial practices beyond their legal obligations, allowing them to do so voluntarily. These practices ensure the sustained provision of quantifiable ecosystem services that might otherwise be economically unviable.[18] There are four primary types of PES markets, detailed in Table 13.3.[19]

Table 13.3 Types of Markets for Payments for Ecosystem Services

Type of Market	*Purpose*	*Description*
Public payment schemes for private landowners	To maintain or enhance ecosystem services	Government-led programs where public institutions make direct payments to landowners or land managers to support the provision of ecosystem services. These schemes are typically country-specific and involve targeted incentives.
Regulated/ mandated markets	To meet legally required ecosystem service levels	Created through legislation, these markets impose caps or thresholds on environmental impacts (e.g. pollution, deforestation). Entities must meet these standards directly or by purchasing credits from others who can do so at a lower cost.
Voluntary markets	To enhance brand value, meet stakeholder expectations, or prepare for future regulations	Companies or organizations engage in ecosystem service transactions by choice – often as part of corporate social responsibility (CSR), to improve reputation, or to anticipate future regulatory requirements.
Self-organized private deals	To secure or maintain specific ecosystem services	Direct, often informal agreements between beneficiaries and providers of ecosystem services, typically with minimal government involvement. These deals are common in regions without formal regulatory markets and may involve businesses or individuals paying landowners to modify land use practices.
Public payment schemes for private landowners	To maintain or enhance ecosystem services	Government-led programs where public institutions make direct payments to landowners or land managers to support the provision of ecosystem services. These schemes are typically country-specific and involve targeted incentives.

Conservation Easements

Conservation easements are a legal mechanism for protecting ecosystem services on private land when outright land acquisition is not feasible or desirable. These are voluntary, legally binding agreements between a landowner and a third-party entity, such as a government agency or conservation organization, that restrict certain land uses to preserve its ecological value. The agreement specifies the limitations placed on the land, how they will be enforced and monitored, and the rights the landowner retains. It may also include financial compensation for the landowner. Easements can be established through purchase, donation, or a combination of both, where the landowner sells the easement at a price below its market value.[20]

Case 13.1: Virginia Department of Forestry's Conservation Easement Program

The Virginia Department of Forestry (DOF) administers a conservation easement program designed to safeguard forestland across the Commonwealth through voluntary legal agreements with private landowners. These easements limit future development while allowing landowners to retain ownership, use, and control of their property, including the ability to engage in activities such as agriculture, forest management, and recreation. Easements are perpetual, ensuring that land remains protected for future generations. The DOF prioritizes large, intact tracts of working forest that provide environmental, economic, and social benefits. Easements under this program are tailored to protect forests, wetlands, and water resources while guiding long-term natural resource stewardship. Most easements are donated by landowners and may qualify as charitable contributions, which can provide federal income tax deductions and state tax credits. In select cases, the DOF collaborates with applicants to pursue competitive grant funding that offsets a portion of the easement's appraised value and related transaction costs. The application process includes both donated and grant-funded tracks. Donated easement applications are reviewed triannually, while grant-funded easement proposals must be submitted in advance of external grant deadlines. Final decisions depend on the conservation value of the property and the department's capacity to support additional easements. As of late 2021, the DOF held more than 196 easements, covering over 87,500 acres across more than 60 jurisdictions. These agreements protect a substantial portion of Virginia's forestland and waterways, including areas within the Chesapeake Bay watershed. Easements are enforced by the DOF or partner organizations, ensuring compliance across generations of ownership.[21]

Virginia Department of Forestry. "Conservation Easements." https://dof.virginia.gov/forest-management-health/forestland-conservation/conservation-easements/.

Tradable Permits

Tradable permits are market-based instruments that allow the buying and selling of pollution rights in regulated markets. Unlike environmental taxes, which set a price for pollution, tradable permits set a fixed limit on total emissions or environmental impacts, allowing entities to trade allowances and ensuring cost-effective pollution reduction. These permits can be based on either 'bads' (such as pollution and emissions) or 'goods' (such as ecosystem restoration).

There are two main types of tradable permit systems. In 'cap and trade' schemes, an overall pollution limit (the cap) is set, and companies exceeding it face penalties. Companies that emit below their limit can sell unused allowances to others, creating a financial incentive for emission reduction. In 'baseline and credit' schemes, companies earn credits for reducing emissions below a set baseline, which can then be sold to other companies needing to offset their excess emissions. Trading allows firms to find the lowest-cost option for compliance, promoting efficiency. Companies that can reduce pollution at a lower cost will sell permits to those facing higher reduction costs, benefiting both parties.[22,23,24]

Several key features shape the design of tradable permit markets:

- *Definition of the tradable commodity*: Permits are typically measured in physical units (e.g., tons of emissions). The choice of where to measure emissions – at the source or further down the supply chain – impacts the cost and ease of enforcement.
- *Allocation of allowances*: Permits can be distributed through auctions, which promote efficiency, transparency, and revenue generation. Alternatively, free allocation methods, such as 'grandparenting' (based on historical emissions) or 'benchmarking' (based on performance standards), can ease the transition for industries.
- *Rules of participation*: Some systems allow open-market trading, while others restrict participation to regulated firms or require a minimum trade size.
- *Pre-approval requirements*: Some schemes require regulatory approval before trades occur, while others encourage open trading to increase market activity.
- *Compliance regimes*: Most schemes have an annual deadline for companies to meet their emission limits, often resulting in a surge of trading activity toward the compliance date.

- *Penalty regimes*: Non-compliance typically results in financial penalties or additional permit requirements in future periods. Fixed penalties create a price ceiling for allowances, whereas extra compliance obligations deter firms from exceeding limits.[25,26]

Case 13.2: Wisconsin's Statewide Water Quality Trading Program

Wisconsin's water quality trading (WQT) framework allows point source permit holders to comply with Wisconsin Pollutant Discharge Elimination System (WPDES) permit limits by purchasing pollutant reduction credits from other sources within the same watershed. Trading is a voluntary, market-based approach designed to offer flexibility in meeting water quality-based effluent limitations, particularly for phosphorus and total suspended solids. Initially piloted in the Red Cedar, Fox-Wolf, and Rock River basins, the program expanded following the 2010 adoption of statewide numeric phosphorus criteria. In 2011, the Wisconsin Department of Natural Resources (WDNR) formalized guidance and permitting procedures for WQT. By 2020, legislation authorized the creation of a centralized trading clearinghouse, now operated by Wisconsin Clearinghouse LLC, to streamline credit transactions. Trading plans are incorporated into WPDES permits and require the prior submission of a notice of intent, a credit generation strategy, and an implementation timeline. Credits may be generated through point or nonpoint sources, with best management practices such as prairie conversion, streambank stabilization, and buffer strips used to reduce pollutants. Trade ratios, typically averaging 2:1, account for uncertainty, delivery distance, downstream effects, and the form of pollutants. As of 2023, over 40 permit holders had indicated intent to trade, with 23 submitting approved plans. On average, permittees purchase 400 pounds of phosphorus credits annually. WDNR oversees credit validation, including best management practices inspections and annual reporting. While total maximum daily load constraints limit trading in some areas, interim credits from nonpoint sources have increased participation. The program provides a cost-effective alternative to facility-based treatment upgrades.[27]

Wisconsin Department of Natural Resources. "Wisconsin's Water Quality Trading." https://dnr.wisconsin.gov/topic/Wastewater/WaterQualityTrading.html.

Subsidies

Subsidies can have either positive or negative environmental impacts, depending on the activities they support. Environmentally beneficial subsidies

Table 13.4 Benefits of Using Subsidies for Nature-Based Solutions

Benefit	*Description*
Immediate effectiveness	Subsidies have an instant impact once implemented, enabling rapid action to promote resource efficiency with minimal delay.
Competitive advantage for enterprises	National or regional subsidies can strengthen a country's industrial base, enhance international competitiveness, and help prevent companies from relocating to lower-cost regions.
Supports innovation at an early stage	By lowering the initial cost of new technologies, subsidies help pioneering products enter the market sooner and accelerate consumer awareness and adoption.
Promotes green growth	Green subsidies are crucial for scaling clean technologies and expanding environmental infrastructure, fostering a low-carbon economy.
Supports small and medium-sized enterprises (SMEs)	Subsidies provide essential financial support to SMEs, which often lack the internal resources to invest in new products or technologies without external funding.

promote public goods, such as encouraging the adoption of NBS by reducing costs for businesses and households. These subsidies can take three primary forms:

- *Direct subsidies*: Financial grants or credits that help cover part of the costs associated with resource-efficient technologies or conservation efforts.
- *Indirect subsidies*: Tax exemptions, reduced tax rates, or the provision of essential services such as water and electricity at below-market prices.
- *Direct price support*: Financial assistance provided to producers or consumers to stabilize prices or incentivize specific behaviors.

Subsidies are used across various sectors, including brownfield redevelopment following hazardous contamination, agricultural grants for erosion control, low-interest loans for small farmers, and land conservation initiatives. When applied to NBS, subsidies offer numerous advantages, such as promoting sustainable land use, enhancing climate resilience, and improving biodiversity conservation (see Table 13.4).[28,29]

Case 13.3: The Netherlands' MIA and Vamil Tax Incentives for Environmental Investments

In the Netherlands, businesses investing in environmentally friendly technologies or assets may access tax benefits through two national schemes: the Environmental Investment Allowance (Milieu-investeringsaftrek, MIA) and the Random Depreciation of Environmental

Investments scheme (Willekeurige afschrijving milieu-investeringen, Vamil). Both are administered by the Netherlands Enterprise Agency (RVO) and aim to incentivize private sector contributions to environmental and climate goals. The MIA scheme allows businesses to deduct 27%, 36%, or 45% of eligible investment costs from taxable profit, depending on the asset's classification on the annually updated Environmental List. This deduction is in addition to regular investment allowances. Vamil enables companies to randomly depreciate up to 75% of investment costs at a time of their choosing, providing cash flow and liquidity advantages. The remaining 25% follows standard depreciation rules. To qualify, companies must be registered in the Netherlands or associated territories, invest a minimum of €2,500 per application, and select assets listed on the Environmental List. The schemes can be combined, though some restrictions apply, particularly where other forms of government support are involved. Disinvestment surcharges may apply if the asset is sold or decommissioned within a five-year period. Eligible costs include acquisition, production, installation, and, in some cases, environmental consultancy for SMEs. Applications must be submitted to RVO within three months of the investment obligation.[30]

Netherlands Enterprise Agency (RVO). "Tax Schemes for Environmentally-Friendly Investments (Mia and Vamil)," (n.d.).

Environmental Taxes and Charges

Without government intervention, economic actors have little incentive to account for environmental damage, as its costs are widely distributed and do not directly affect the polluter. Traditionally, environmental policy relied heavily on command-and-control regulations, which were highly prescriptive, such as banning certain substances or mandating specific technologies for pollution control. However, in recent decades, there has been a shift toward environmental taxes and charges, which provide more flexibility and efficiency in addressing environmental harm.

An environmental tax is levied on a physical unit (or a proxy) of an activity or product with a proven negative environmental impact. Common examples include energy taxes, transport taxes, pollution taxes, and resource taxes. In contrast, environmental charges are fees paid for the right to use environmental resources or services to reduce harm and encourage responsible environmental practices, such as waste disposal fees or water extraction charges.[31,32]

Using environmental taxes and charges instead of command-and-control regulations offers several key benefits:

- *Internalizing environmental costs*: By incorporating environmental harm into market prices, taxes and charges ensure that polluters bear the cost of

Table 13.5 Multiple Benefits of Environmental Taxes and Charges

Benefit	*Description*
Ongoing incentives to abate	Unlike fixed targets or technology standards, environmental taxes and charges provide a continuous financial incentive to reduce emissions – even beyond the initial compliance level.
Improved competitiveness of low-polluting alternatives	By increasing the cost of polluting activities, taxes make cleaner alternatives more attractive, boosting demand and allowing these alternatives to benefit from economies of scale – often without the need for direct subsidies.
Strong incentives to innovate	Taxes raise the cost of pollution, encouraging firms to invest in new technologies and adapt existing ones to lower their environmental impact and reduce tax liabilities.
Transparency	Environmental taxes and charges can be designed to clearly show who is taxed, how much is charged, and what pollution levels trigger costs – ensuring accountability and clarity for all stakeholders.
Environmental certainty	Taxes directly link the cost of polluting products or activities to their environmental harm, making it easier to forecast the environmental benefits of policy interventions.

their actions. This corrects market failures by making environmental externalities visible in economic decisions.

- *Providing flexibility in compliance*: Unlike rigid regulatory mandates, taxes and charges allow businesses and individuals to choose how to reduce their environmental impact. This avoids government overreach in dictating specific solutions and promotes cost-effective mitigation strategies.

Additional benefits of environmental taxes and charges, including their ability to drive innovation, enhance economic efficiency, and generate government revenue, are summarized in Table 13.5.[33,34]

Environmental Performance Bonds

Governments often require firms to obtain environmental performance bonds – insurance ensuring they can cover the cost of any environmental damage they cause. This acts as both a financial safeguard and an economic deterrent, as firms that violate environmental regulations may face increased insurance costs or even be unable to obtain coverage in the future. Additionally, bonding companies are unlikely to issue performance bonds to firms that are likely to violate environmental standards. Performance bonds are typically applied to industries with a higher-than-average risk of significant environmental harm, such as oil transport, forestry, mining, and heavy industry, where the additional cost of the bond is proportionate to the potential environmental risk.[35]

Case 13.4: Financial Assurances for Environmental Risk Management in Victoria, Australia

In Victoria, Australia, the Environment Protection Authority (EPA) utilizes financial assurances as a regulatory tool to ensure that businesses involved in high-risk environmental activities possess the financial means to address potential contamination or site remediation. These assurances serve as a form of security deposit, protecting the public from incurring cleanup costs if a duty holder fails to meet its environmental responsibilities. Under the Environment Protection Act 2017 and associated Regulations, EPA may require a financial assurance as a condition of a development or operating license, environmental action notice, or site management order. Activities that typically require assurances include reportable priority waste management, large-scale waste and resource recovery operations, bulk storage, and the containment of contaminated sites. Operators must submit a financial assurance proposal during the licensing process, detailing the estimated costs of remediation or restoration. EPA evaluates these proposals using published calculation methodologies but may request tailored estimates supported by evidence where activities differ materially from standard cases. Bank guarantees are the most accepted form of assurance, though alternative forms may be considered in specific circumstances. EPA may claim against the assurance if the operator fails to fulfil cleanup obligations and can require the assurance to be replenished following a claim. This mechanism supports proactive risk management by ensuring that entities engaging in environmentally hazardous activities can meet the cost of remediating environmental harm. It also incentivizes compliance by directly linking financial responsibility to the scale and risk of the permitted activity.[36]

Environment Protection Authority Victoria. "Meeting Your Obligations." https://epa.vic.gov.au/meeting-your-obligations.

Public-Private Partnerships

Public-private partnerships (PPPs) involve the private sector financing, planning, designing, constructing, operating, and maintaining infrastructure and related services under a contractual agreement. While the government typically retains responsibility for delivering the core service, private sector costs are recovered over time through payments from the government or, in some cases, partially or entirely from users. Payments commence once operations begin and usually include both a capital component and an ongoing service delivery component.[37,38]

There are two types of green PPP projects: brownfield and greenfield. Brownfield projects involve private sector participation in existing infrastructure, whereas greenfield projects focus on developing new infrastructure. While both can incorporate environmentally friendly characteristics, greenfield PPPs offer greater flexibility in integrating sustainability from the outset.

PPPs can be crucial in raising public awareness, diversifying stakeholders in NBS, and fostering corporate social responsibility. Additionally, they can create markets for green products by improving the investment environment. The success of NBS-related PPPs depends on several key factors:

- *Effective partnerships*: Unlike traditional procurement contracts, which are shorter-term, NBS-related PPPs can last from 15 to over 90 years. Establishing a cooperative and credible partnership is crucial, and public authorities must develop expertise to define contract terms effectively.
- *Interaction and negotiation*: Engaging with private operators during the bidding phase helps clarify partnership objectives and encourages innovative technological solutions that the public sector may not have considered, particularly in projects involving rapidly evolving technologies.
- *Clear environmental objectives*: Environmental targets must be well-defined, measurable, and integrated into project specifications. Retroactively adding green requirements after PPP design can be costly and may conflict with prior technological decisions.
- *Flexibility*: Given the evolving nature of environmental challenges and technological advancements, NBS-related PPPs must remain adaptable. Contracts should be structured to allow efficient responses to changing requirements and emerging innovations.[39]

Case 13.5: Sustain Our Great Lakes Program

The Sustain Our Great Lakes (SOGL) program is a public–private partnership administered by the National Fish and Wildlife Foundation (NFWF) that supports on-the-ground restoration efforts in the Great Lakes Basin. Backed by funding and support from federal agencies, corporate sponsors, and philanthropic foundations – including the U.S. Environmental Protection Agency, U.S. Fish and Wildlife Service, U.S. Forest Service, Cleveland-Cliffs, the Ralph C. Wilson Jr. Foundation, and the Caerus Foundation – the program is designed to restore and enhance fish and wildlife populations, improve water quality, and increase ecological resilience across one of the most vital freshwater systems on the planet. SOGL operates by awarding competitive grants for habitat restoration, green stormwater infrastructure, invasive species control, and community activation of restored green spaces. Between 2006 and 2024, the program awarded $142.4 million in grant funding across 529 projects. Grantees matched this investment with an additional $180.9 million, resulting in a combined conservation impact of over $323.3 million. Notable achievements include reconnecting over 2,630 miles of stream for fish passage, restoring 58,888 acres of wetland and upland habitat, and adding 389.7 million gallons of stormwater storage capacity. A key strength of SOGL is its alignment with the federally funded Great Lakes Restoration Initiative, ensuring that investments target high-priority ecological outcomes. The program leverages diverse partnerships to build conservation capacity and scale effective practices, reflecting a collaborative model that integrates public accountability with private innovation.[40]

National Fish and Wildlife Foundation. "Sustain Our Great Lakes Program." https://nfwf.org/programs/sustain-our-great-lakes-program?activeTab=tab-1.

Environmental Impact Bonds

Environmental impact bonds use a pay-for-success model, where investors purchase a bond, and repayment is linked to achieving specific environmental outcomes. A performance payment is issued if the NBS meets a pre-established, mutually defined target, such as preventing land degradation. Both investors and contractors share this payment, which is provided by local asset owners who benefit directly from earlier, high-quality restoration or conservation efforts. The benefits of environmental impact bonds include:

- *Aligning interests in NBS with better project performance*: A performance-based payment, verified by an independent third party, ensures that governments, asset owners, contractors, and investors are all committed to achieving measurable environmental results.
- *Attracting impact investors*: These bonds appeal to private investors, particularly those seeking financial returns and positive environmental or social impact. Investors interested in NBS are more likely to support this financial instrument.
- *Providing measurable proof of NBS benefits*: Environmental impact bonds demonstrate the value of investing in natural restoration by quantifying the environmental and economic benefits of NBS projects. This helps build support among stakeholders.
- *Enhancing government effectiveness*: By leveraging private investment, environmental impact bonds enable governments to accelerate nature restoration efforts, protect communities, create green jobs, and sustain tax rev enues.[41,42,43]

Case 13.6: Environmental Impact Bond for Green Infrastructure in Atlanta, Georgia

In 2019, the City of Atlanta's Department of Watershed Management (DWM) issued a $14 million environmental impact bond to finance green infrastructure improvements in the Proctor Creek Watershed. Supported by the Rockefeller Foundation and developed in partnership with Quantified Ventures and Neighborly, this marked the first publicly issued environmental impact bond in the United States. The bond enables the city to fund a portfolio of nature-based projects aimed at reducing stormwater runoff, alleviating flooding, and improving water quality in historically underserved Westside neighborhoods. The environmental impact bond incorporates a Pay-for-Success mechanism, where investor returns are partially tied to measurable environmental outcomes – in this case, the volume of stormwater captured. If the projects exceed a threshold of 6.5 million gallons of stormwater capture capacity, investors receive a performance payment. Projects funded through the bond include stream restorations, constructed wetlands, rain gardens, and stormwater planters integrated into parks and streetscapes. Beyond the hydrological benefits, the initiative aims to deliver co-benefits, including the creation of green spaces, improved air quality, and local workforce development. Construction began in 2021, with all projects scheduled for completion by 2024. Community engagement was a central feature of project design, with DWM conducting over 25 public meetings to ensure alignment with local needs and values.[44]

City of Atlanta Department of Watershed Management. "Environmental Impact Bond (Eib) for Green Infrastructure in Proctor Creek Watershed." https://atlantawatershed.org/environmental-impact-bond/.

Biodiversity Offsets

Biodiversity offsets compensate for economic losses caused by economic activities by requiring developers to fund environmental protection or restoration projects elsewhere. There are three main types of biodiversity offset markets:

- *Public agency model*: A state agency oversees valuation, transactions, and matching between impact and compensation sites. The government plans, conducts, and monitors the compensation project, while the developer covers the costs.
- *Mandatory markets*: In this system, developers can choose among multiple offset providers, and compensation prices are influenced by market competition. Governments establish property rights, define developer liabilities, set biodiversity valuation metrics, and regulate the trading of biodiversity credits. Private mitigation banks or compensation agencies

identify landowners, implement conservation actions, receive credits from government agencies, and sell these credits to developers, thereby fulfilling their compensation obligations.

- *Voluntary offsets*: Companies voluntarily fund biodiversity conservation without regulatory requirements as part of corporate social responsibility or risk management. Resource-intensive industries often undertake voluntary offsets, relying on guidance from environmental experts and NGOs. Benefits include securing a 'license-to-operate' by gaining social acceptance from local communities and environmental groups and fostering regulatory goodwill for future business operations.[45]

Environmental Corporate Social Responsibility

Environmental corporate social responsibility (CSR) minimizes a business's negative environmental impact through sustainable practices. These efforts may include reducing energy and water consumption, improving waste management, increasing recycling, cutting emissions, and adopting eco-friendly office and travel policies. Failing to integrate environmental CSR into planning, design, construction, and operations can lead to higher business risks, reputational damage, and increased costs (Table 13.6).[46,47]

Table 13.6 Costs of Environmental Corporate Social Responsibility Inaction

Cost	*Description*
Planning permission refused or delayed	Failure to incorporate NBS can result in planning applications being rejected or delayed due to non-compliance with policy or the need for additional conditions – leading to financial setbacks and potential loss of investor confidence.
Increased programme costs	Replacing existing natural systems with hard infrastructure often incurs higher costs during construction, including demolition, disposal, and engineered replacements.
Flooding	Inadequate adaptation measures can lead to increased flood risk, resulting in loss of revenue, damage to property, reduced property values, operational downtime, and higher insurance premiums.
Temperature increases and urban heat island effect	Rising temperatures can accelerate the deterioration of infrastructure, leading to higher repair and maintenance costs – particularly in areas lacking green infrastructure.
Higher maintenance costs	Hard-engineered solutions typically require more frequent and intensive maintenance compared to the lower, ongoing upkeep needed for most NBS.
Loss of client relationships/ investment	Stakeholders value environmental responsibility. A poor sustainability reputation may lead to reduced investor interest, damaged brand trust, and competitive disadvantage.
Staff well-being	Poorly designed work environments lacking nature can negatively impact employee health, increase absenteeism, and reduce talent retention.

Case 13.7: Nestlé's Global Reforestation Program

Nestlé has launched a comprehensive Global Reforestation Program (GRP) as part of its broader Net Zero Roadmap and Forest Positive Strategy. The initiative aims to plant and grow 200 million trees by 2030 across Nestlé's sourcing landscapes, including farms and supply sheds where the company sources its key ingredients. This effort directly supports climate change mitigation through Natural Climate Solutions, which help remove and store carbon over the long term. Reforestation and restoration under the GRP are explicitly targeted at degraded natural ecosystems. These efforts are designed not only to sequester carbon but also to restore biodiversity, enhance soil health, protect water resources, and support the livelihoods of Indigenous peoples and local communities. Nestlé's reforestation projects are developed in partnership with expert organizations, including One Tree Planted, PUR Projet, South Pole, and NatureCO, ensuring rigorous planning, verification, and monitoring. The GRP includes a robust carbon accounting and monitoring framework, with each project following a minimum 20-year lifecycle. Nestlé tracks both the number of trees planted and the amount of CO_2 sequestered, and it is also developing systems to measure additional non-carbon benefits. Advanced technologies such as satellite imagery are being piloted to monitor long-term tree growth and ecosystem recovery. By 2030, Nestlé forecasts the GRP will deliver around 2 million metric tons of CO_2-equivalent removals. This not only supports the company's science-based net-zero target but also contributes to reversing biodiversity loss and creating resilient, sustainable supply chains. An independent External Advisory Council guides the strategy, ensuring accountability and alignment with broader environmental and social goals.[48]

Nestlé. "Long-Term Forest Conservation and Restoration." https://nestle.com/sustainability/nature-environment/forests/conservation-restoration.

Voluntary Carbon Markets

Voluntary carbon markets facilitate buying and selling emission reduction credits (offsets) outside government-regulated systems. Demand for verified carbon credits comes from individuals, such as travelers offsetting their carbon footprint, and businesses aiming to exceed legal requirements, fulfil corporate social responsibility commitments, or prepare for potential future regulations.[49] Voluntary carbon markets often prioritize additional environmental and social benefits, unlike compliance markets. For example, a regional peatland restoration project contributes to climate mitigation while enhancing biodiversity, improving water quality, reducing wildfire risk, and providing recreational benefits. Regional voluntary carbon markets, which can operate at national or subnational levels, typically attract local investors supporting climate mitigation initiatives. To ensure credibility, projects must demonstrate that they would not have occurred without external investment, a principle known as 'additionality.'[50] This can be assessed using four tests:

- *Legal test*: A project is not considered additional if existing laws or regulations mandate carbon mitigation activities.
- *Global finance contribution test*: The project qualifies as additional only if it would not have been possible without sponsor investments. However, projects do not need to rely solely on this funding, so that a threshold may be set.
- *Investment test*: The project must show that, without carbon finance, it would be economically or financially unviable or less attractive than other land-use options.
- *Barrier test*: The project must face obstacles that prevent its implementation without external funding, such as a lack of community support, accessibility issues, limited skilled labor, or high upfront costs. If sponsorship can remove these barriers, the project is considered additional.[51]

Crowdfunding

Crowdfunding offers a convenient means of financing NBS by aggregating small contributions from numerous individuals. These funds can support ecosystem restoration, green infrastructure, and conservation initiatives that might otherwise struggle to secure traditional financing. Standard crowdfunding models include:

- *Debt-based contributions*: Individuals lend funds to NBS projects, expecting repayment over time, often with interest. Depending on the project's risk profile, these may be structured as either secured or unsecured loans.
- *Contingent or performance-based loans*: Also known as forgivable loans, repayment is only required if certain project milestones are achieved, such as measurable environmental outcomes or future revenue generation.
- *Pre-financing or advance purchase*: Supporters contribute upfront for future access to project benefits or outputs, such as ecosystem services, carbon credits, or symbolic items like tree sponsorships. This model often blends pre-sales with rewards-based incentives.[52]

Case 13.8: ForestFinance's Prepaid Tree Sponsorships and Forest Gifts

ForestFinance is a pioneer in sustainable forestry and agroforestry, operating for nearly three decades across more than 9,000 hectares worldwide. Headquartered in Germany, the company enables individuals and businesses to invest in long-term forest protection, afforestation, and carbon sequestration initiatives. ForestFinance has planted more than 10 million trees and sequestered around 462,625 tons of CO_2 in Panama alone (as of 2018), highlighting the role of forest-based carbon projects in contributing to climate mitigation, biodiversity protection, and local development. The organization's model integrates forest protection, organic farming, and agroforestry to generate environmental and socio-economic co-benefits. ForestFinance projects are certified under the Gold Standard, ensuring high-integrity climate outcomes. In addition to restoring degraded lands and supporting biodiversity through species-rich forest landscapes, their initiatives provide fair employment opportunities and promote local economic resilience. By blending afforestation with income-generating activities such as sustainable cocoa cultivation, ForestFinance contributes to long-term ecosystem stewardship. Their projects are designed not only to sequester carbon but also to serve as seed banks, preserve habitats, and sustain traditional knowledge. The organization's transparency and emphasis on fair wages, local stakeholder involvement, and biodiversity enhancement further strengthen its credibility. ForestFinance also engages the public through innovative initiatives, such as corporate forests and tree-gifting platforms, which allow individuals and businesses to symbolically "make forest" by sponsoring trees or forest parcels. These offerings directly contribute to ongoing restoration and conservation efforts, making climate action more accessible to a broader audience.[53]

ForestFinance. "We Are Building Forests: Sustainable Investments for a Greener Future." https://forestfinance.de/en/.

Green Bank Cards

Green cards are bank cards designed to support sustainable development. Each time a transaction is made – whether a withdrawal or payment – a small fee is applied to the interbank rate, contributing to a sustainability or biodiversity fund. These cards can function as either debit or credit cards, with the funds generated primarily used to maintain or restore biodiversity assets.[54]

Case 13.9: Tred's Reforestation-Linked Debit Card

Launched in November 2021, Tred is the UK's first green fintech to introduce a Mastercard debit card designed to help users track, reduce, and offset their carbon footprint with every transaction. The green debit card, made from recycled ocean plastic, is linked to a mobile app that calculates a user's personalized carbon footprint based on their spending habits. Through a tailored monthly subscription, users can offset their emissions by planting trees or contributing to other climate mitigation projects. Each purchase made with the Tred card directly supports reforestation efforts, aligning daily spending with environmental impact. To celebrate its launch, Tred committed to planting 10,000 trees in partnership with Mastercard's Priceless Planet Coalition – an initiative aiming to plant 100 million trees by 2025. Enabled by Nium's card-issuing platform, Tred offers a seamless way to integrate sustainability into everyday financial behavior. With the growing demand for eco-conscious financial services, Tred's innovation removes barriers to climate action, empowering individuals to make more informed, sustainable lifestyle choices.[55]

Mastercard. "Tred Launches Mastercard Green Debit Card to Help Track and Offset Users' Carbon Footprint." https://mastercard.com/news/europe/en-uk/newsroom/press-releases/en-gb/2021/november/tred-launches-mastercard-green-debit-card-to-help-track-and-offset-users-carbon-footprint/.

Conservation-Focused Insurance

Conservation-focused insurance products offer a way to finance NBS as a strategy for risk management.[56] The insurance industry can play a key role in supporting NBS, particularly for climate adaptation, by:

- Applying risk management expertise to assess vulnerabilities, quantify risks, and design risk reduction and transfer mechanisms.
- Prioritizing NBS by enhancing capacity and advising on the cost-effectiveness of nature-based measures.
- Incentivizing loss reduction by informing stakeholders about risks, recommending NBS solutions, and offering insurance products supporting loss prevention.
- Developing new insurance products to cover risks linked to climate change and environmental factors, such as human health impacts, crop failures, and animal diseases.
- Raising awareness among governments, regulators, businesses, civil society, and academia about climate risks, vulnerable communities' needs, and insurance's role in promoting NBS.[57]

Case 13.10: Expanded Coral Reef Insurance Policy Enhances Hawai'i's Post-Storm Recovery and Climate Resilience

In February 2024, The Nature Conservancy (TNC) and WTW announced the renewal and expansion of the first U.S. coral reef insurance policy, covering Hawai'i's coral reefs. This revised policy replaces the initial version launched in 2022 and significantly increases the coverage area by an additional 314,976 square miles. The maximum annual payout has been set at $2 million, with up to $1 million available per storm, and the minimum payout has doubled to $200,000. Coverage is triggered when sustained tropical storm winds reach 50 knots or greater within the defined zone. The policy was developed through a competitive process involving seven bids, with Munich Re selected as the insurer. The broadened interest from insurers reflects growing recognition of parametric insurance as a viable tool for protecting natural assets from climate-related events. Parametric insurance enables the rapid disbursement of funds post-disaster, based on pre-agreed triggers, ensuring timely response and restoration efforts. According to TNC, the increased minimum payout facilitates faster action, such as damage assessments and coral reattachment, which must occur within weeks to avoid permanent reef loss. The policy also builds on the foundation laid by the 2022 initiative, which helped establish response protocols and stakeholder coordination. Hawai'i's reefs are integral to the state's economy and environmental resilience, reducing storm-related wave energy by up to 97% and supporting tourism and coastal protection valued at over $2 billion. The upgraded insurance model marks an evolving commitment to nature-based solutions as climate-related risks intensify.[58]

WTW. "Major Upgrade to First U.S. Coral Reef Insurance Policy Increases Coverage and Enables More Robust Post-Storm Response." https://wtwco.com/en-gb/news/2024/02/major-upgrade-to-first-us-coral-reef-insurance-policy-increases-coverage-and-enables-more-robust.

Case Summary

A range of innovative financial tools has been developed to support the design, implementation, and sustainability of NBS (Table 13.7).

Table 13.7 Innovative Financial Tools for Nature-Based Solutions

Financial Tool	*Project Name*	*Mechanism*	*Co-Benefits*
Conservation finance	Virginia Forestry Conservation Easements	Uses voluntary easements to protect private forestland from development while allowing continued use; includes donated and grant-funded tracks.	Preserves working forests, protects water quality, supports tax incentives and long-term stewardship.
Water quality trading	Wisconsin Statewide WQT Program	Enables point sources to meet permit limits by buying pollutant reduction credits from others in the same watershed.	Offers cost-effective compliance, promotes watershed-scale pollution control, and supports conservation practices.
Green investment incentives	Netherlands MIA and Vamil Schemes	Provides tax benefits for investments in environmentally friendly technologies, including accelerated depreciation and investment deductions.	Stimulates private green investment, improves liquidity, and supports sustainability goals.
Environmental risk regulation	Victoria EPA Financial Assurances	Requires financial assurances for high-risk environmental activities to ensure remediation responsibilities can be fulfilled.	Protects the public, incentivizes compliance, and ensures remediation funding is available when needed.
Public-private restoration	Sustain Our Great Lakes Program	Awards competitive grants for restoration and green infrastructure in the Great Lakes Basin through multi-stakeholder partnerships.	Restores habitat, improves water quality, expands stormwater capacity, and fosters ecological resilience.
Impact bonds	Atlanta Green Infrastructure Environmental Impact Bond	$14 million bond ties investor returns to stormwater capture metrics in underserved neighborhoods using green infrastructure.	Supports equitable climate resilience, creates green jobs, and demonstrates pay-for-success financing.
Corporate forestry	Nestlé Global Reforestation Program	Targets planting of 200 million trees by 2030 to restore degraded lands across sourcing landscapes and track CO_2 removals.	Sequesters carbon, restores biodiversity, and supports Indigenous and community livelihoods.

Table 13.7 (Continued)

Financial Tool	*Project Name*	*Mechanism*	*Co-Benefits*
Carbon forestry & gifts	ForestFinance Tree Sponsorships	Enables individuals and companies to invest in certified forest and agroforestry projects for carbon and biodiversity goals.	Supports climate mitigation, sustainable livelihoods, and biodiversity via certified, transparent forestry.
Consumer reforestation	Tred Reforestation Debit Card	Links spending to tree-planting initiatives through transaction-based donations and recycled plastic cards.	Engages consumers in everyday climate action, offsets emissions, and supports local/global planting efforts.
Nature insurance	Hawai'i Coral Reef Insurance	Parametric insurance model covers coral reefs against storm damage, triggering payouts based on wind speeds.	Enables rapid reef recovery post-storm, builds climate resilience, and secures coastal protection services.

Notes

1. GIZ, "Finance Options and Instruments for Ecosystem-Based Adaptation: Overview and Compilation of Ten Examples," (2018), https://adaptationcommunity.net/wp-content/uploads/2018/06/giz2018-en-eba-finance-guidebook-low-res.pdf.
2. GIZ, "Finance Options and Instruments for Ecosystem-Based Adaptation: Overview and Compilation of Ten Examples".
3. GIZ, "Finance Options and Instruments for Ecosystem-Based Adaptation: Overview and Compilation of Ten Examples" .
4. GIZ, "Finance Options and Instruments for Ecosystem-Based Adaptation: Overview and Compilation of Ten Examples".
5. GIZ, "Environmental Fiscal Reform: Case Studies," (2014), http://greenfiscalpolicy.org/wp-content/uploads/2013/08/Environmental-Fiscal-Reform-Case-Studies.pdf.
6. GIZ, "Finance Options and Instruments for Ecosystem-Based Adaptation: Overview and Compilation of Ten Examples".
7. GIZ, "Finance Options and Instruments for Ecosystem-Based Adaptation: Ov erview and Compilation of Ten Examples".
8. EDFI, "Dfi Working Group on Blended Concessional Finance for Private Sector Projects," (2017), https://edfi.eu/wp/wp-content/uploads/2017/10/DFI-Blended-Concessional-Finance-for-Private-Sector-Operations_Summary-R....pdf.
9. OECD, "Easing the Transition to Commercial Finance for Sustainable Water and Sanitation," (2017), http://oecd.org/water/Background-Document-OECD-GIZ-Conference-Easing-the-Transition.pdf.
10. EDFI, "Dfi Working Group on Blended Concessional Finance for Private Sector Projects."

11 OECD, "Easing the Transition to Commercial Finance for Sustainable Water and Sanitation."
12 IUCN, "Coastal "Blue" Carbon. A Revised Guide to Supporting Coastal Wetland Programs and Projects Using Climate Finance and Other Financial Mechanisms," (2015).
13 UNDP, "Financing Solutions for Sustainable Development," (2017), https://undp.org/content/dam/sdfinance/doc/Debt%20for%20Nature%20Swaps%20_%20UNDP.pdf.
14 Institute for Climate Economics, "Using Credit Lines to Foster Green Lending: Opportunities and Challenges," (2017), https://cbd.int/financial/2017docs/ice-creditlines2017.pdf.
15 Loan Market Association, "Green Loan Principles. Supporting Environmentally Sustainable Economic Activity," (2018), https://lma.eu.com/application/files/9115/4452/5458/741_LM_Green_Loan_Principles_Booklet_V8.pdf.
16 International Capital Market Association, "Green Bond Principles. Voluntary Process Guidelines for Issuing Green Bonds," (2018), https://icmagroup.org/green-social-and-sustainability-bonds/green-bond-principles-gbp/.
17 Australian Water Association, "Alternative Models for Financing Water Infrastructure," (2017), https://awa.asn.au/documents/Discussion_Paper_Alternative_Finance_for_Discussion.pdf.
18 M.S. Reed et al., "A Place-Based Approach to Payments for Ecosystem Services," *Global Environmental Change* 43 (2017).
19 R.C. Brears, *The Green Economy and the Water-Energy-Food Nexus* (London: Palgrave Macmillan UK, 2017).
20 Smart Prosperity Institute, "Economic Tools for Increasing Nature Conservation on Private Land," (2018), https://institute.smartprosperity.ca/library/publications/economic-tools-increasing-nature-conservation-private-land.
21 Virginia Department of Forestry, "Conservation Easements," (n.d.), https://dof.virginia.gov/forest-management-health/forestland-conservation/conservation-easements/.
22 OECD, "Tradeable Permits," (2004), https://oecd-ilibrary.org/environment/tradeable-permits_9789264015036-en.
23 Economics for the Environment Consultancy and Institute for European Environmental Policy, "The Use of Market-Based Instruments for Biodiversity Protection – the Case of Habitat Banking – Technical Report," (2010), https://ec.europa.eu/environment/enveco/pdf/eftec_habitat_technical_report.pdf.
24 R.C. Brears, *Natural Resource Management and the Circular Economy* (Cham: Springer International Publishing, 2018).
25 OECD, "Environmentally Related Taxes and Tradable Permit Systems in Practice," (2008), http://oecd.org/officialdocuments/publicdisplaydocumentpdf/?cote=COM/ENV/EPOC/CTPA/CFA(2007)31/FINAL&doclanguage=en.
26 International Carbon Action Partnership, "Allocation: How Emissions Permits Are Distributed," (2017), https://icapcarbonaction.com/zh/?option=com_attach&task=download&id=493.
27 Wisconsin Department of Natural Resources, "Wisconsin's Water Quality Trading," (n.d.), https://dnr.wisconsin.gov/topic/Wastewater/WaterQualityTrading.html.
28 US EPA, "Economic Incentives," (n.d.), https://epa.gov/environmental-economics/economic-incentives#subsidies.
29 GIZ, "Environmental Fiscal Reform: Case Studies."

30 Netherlands Enterprise Agency (RVO), "Tax Schemes for Environmentally-Friendly Investments (Mia and Vamil)," (n.d.).

31 Eurostat, "Environmental Taxes," (n.d.), http://ec.europa.eu/eurostat/web/environment/environmental-taxes.

32 Japan Center for a Sustainable Environment and Society, "What Are Environmental Taxes?" (n.d.), http://jacses.org/en/paco/envtax.htm.

33 Don Fullerton, Andrew Leicester, and Stephen Smith, *Environmental Taxes* (Cambridge, MA: National Bureau of Economic Research, 2008).

34 OECD, "Environmental Taxation: A Guide for Policy Makers," (2011), https://oecd.org/env/tools-evaluation/48164926.pdf.

35 UNEP, "Training Resource Manual: The Use of Economic Instruments for Environmental and Natural Resource Management," (2009), http://wedocs.unep.org/bitstream/handle/20.500.11822/7802/-Training%20Resource%20Manual_%20The%20Use%20of%20Economic%20Instruments%20for%20environmental%20and%20Natural%20Resource%20management-2009863.pdf?sequence=3&isAllowed=y.

36 Environment Protection Authority Victoria, "Meeting Your Obligations," (n.d.), https://epa.vic.gov.au/meeting-your-obligations.

37 World Bank and World Resources Institute, "Integrating Green and Gray: Creating Next Generation Infrastructure," (2019), https://wri.org/publication/integrating-green-gray.

38 Australian Water Association, "Alternative Models for Financing Water Infrastructure."

39 O. Merk, S. Saussier, C. Staropoli, E. Slack, and J.-H. Kim, "Financing Green Urban Infrastructure," in *OECD Regional Development Working Papers* (OECD, 2012).

40 National Fish and Wildlife Foundation, "Sustain Our Great Lakes Program," (n.d.), https://nfwf.org/programs/sustain-our-great-lakes-program?activeTab=tab-1.

41 EDF, "Environmental Impact Bonds Can Help Make Coastal Communities Safer, Sooner. Here's How," (n.d.), http://blogs.edf.org/growingreturns/2018/08/14/environmental-impact-bond-coastal-communities/.

42 "Financing Resilient Communities and Coastlines," https://edf.org/ecosystems/financing-resilient-communities-and-coastlines.

43 World Bank and World Resources Institute, "Integrating Green and Gray: Creating Next Generation Infrastructure."

44 City of Atlanta Department of Watershed Management, "Environmental Impact Bond (Eib) for Green Infrastructure in Proctor Creek Watershed,"(n.d.), https://atlantawatershed.org/environmental-impact-bond/.

45 Niak S. Koh, Thomas Hahn, and Wiebren J. Boonstra, "How Much of a Market Is Involved in a Biodiversity Offset? A Typology of Biodiversity Offset Policies," *Journal of Environmental Management* 232 (2019).

46 Invest NI, "Corporate Social Responsibility (Csr)," (n.d.), https://nibusinessinfo.co.uk/content/corporate-social-responsibility-environmental-impact.

47 UK Green Building Council, "Demystifying Green Infrastructure," (2015), https://ukgbc.org/wp-content/uploads/2017/09/Demystifying-Green-Infrastructure-report-FINAL.pdf.

48 Nestlé, "Long-Term Forest Conservation and Restoration," (n.d.), https://nestle.com/sustainability/nature-environment/forests/conservation-restoration.

49 IUCN, "Coastal "Blue" Carbon. A Revised Guide to Supporting Coastal Wetland Programs and Projects Using Climate Finance and Other Financial Mechanisms."

50 Aletta Bonn et al., "Investing in Nature: Developing Ecosystem Service Markets for Peatland Restoration," *Ecosystem Services* 9 (2014).
51 Bonn et al., "Investing in Nature: Developing Ecosystem Service Markets for Peatland Restoration."
52 FAO, "Sustainable Financing for Forest and Landscape Restoration: Opportunities, Challenges and the Way Forward. Discussion Paper," (2015), http://fao.org/3/a-i5174e.pdf.
53 ForestFinance, "We Are Building Forests: Sustainable Investments for a Greener Future," (n.d.), https://forestfinance.de/en/.
54 ForestFinance, "We Are Building Forests: Sustainable Investments for a Greener Future."
55 Mastercard, "Tred Launches Mastercard Green Debit Card to Help Track and Offset Users' Carbon Footprint," (n.d.), https://mastercard.com/news/europe/en-uk/newsroom/press-releases/en-gb/2021/november/tred-launches-mastercard-green-debit-card-to-help-track-and-offset-users-carbon-footprint/.
56 World Bank and World Resources Institute, "Integrating Green and Gray: Creating Next Generation Infrastructure."
57 UNEPFI, "Global Insurance Industry Statement on Adapting to Climate Change in Developing Countries," (2010), https://unepfi.org/publications/insurance-publications/global-insurance-industry-statement-adapting-to-climate-change-in-developing-countries/.
58 WTW, "Major Upgrade to First U.S. Coral Reef Insurance Policy Increases Coverage and Enables More Robust Post-Storm Response," (n.d.), https://wtwco.com/en-gb/news/2024/02/major-upgrade-to-first-us-coral-reef-insurance-policy-increases-coverage-and-enables-more-robust.

Bibliography

Australian Water Association. "Alternative Models for Financing Water Infrastructure." (2017). https://awa.asn.au/documents/Discussion_Paper_Alternative_Finance_for_Discussion.pdf.

Bonn, Aletta, Mark S. Reed, Chris D. Evans, Hans Joosten, Clifton Bain, Jenny Farmer, Igino Emmer, et al. "Investing in Nature: Developing Ecosystem Service Markets for Peatland Restoration." *Ecosystem Services* 9 (2014/09/01): 54–65.

Brears, R.C. *The Green Economy and the Water-Energy-Food Nexus*. London: Palgrave Macmillan UK, 2017.

Brears, R.C. *Natural Resource Management and the Circular Economy*. Cham: Springer International Publishing, 2018.

Economics for the Environment Consultancy and Institute for European Environmental Policy. "The Use of Market-Based Instruments for Biodiversity Protection – The Case of Habitat Banking – Technical Report." (2010). https://ec.europa.eu/environment/enveco/pdf/eftec_habitat_technical_report.pdf.

EDF. "Environmental Impact Bonds Can Help Make Coastal Communities Safer, Sooner. Here's How." (2018). http://blogs.edf.org/growingreturns/2018/08/14/environmental-impact-bond-coastal-communities/.

EDF. "Financing Resilient Communities and Coastlines." (2018). https://edf.org/ecosystems/financing-resilient-communities-and-coastlines.

EDFI. "Dfi Working Group on Blended Concessional Finance for Private Sector Projects." (2017). https://edfi.eu/wp/wp-content/uploads/2017/10/DFI-Blended-Concessional-Finance-for-Private-Sector-Operations_Summary-R....pdf.

Eurostat. "Environmental Taxes." (n.d.). http://ec.europa.eu/eurostat/web/environment/environmental-taxes.

FAO. "Sustainable Financing for Forest and Landscape Restoration: Opportunities, Challenges and the Way Forward. Discussion Paper." (2015). http://fao.org/3/a-i5174e.pdf.

Ferwerda, Willem H. "Chapter 4.6 – Four Returns, Three Zones, 20 Years: A Systemic Approach to Scale up Landscape Restoration by Businesses and Investors to Create a Restoration Industry." In *Land Restoration*, edited by Ilan Chabay, Martin Frick, and Jennifer Helgeson, 319–47. Boston, MA: Academic Press, 2016.

Fullerton, Don, Andrew Leicester, and Stephen Smith. *Environmental Taxes*. Cambridge, MA: National bureau of economic research, 2008.

GIZ. "Environmental Fiscal Reform: Case Studies." (2014). http://greenfiscalpolicy.org/wp-content/uploads/2013/08/Environmental-Fiscal-Reform-Case-Studies.pdf.

GIZ. "Finance Options and Instruments for Ecosystem-Based Adaptation: Overview and Compilation of Ten Examples." (2018). https://adaptationcommunity.net/wp-content/uploads/2018/06/giz2018-en-eba-finance-guidebook-low-res.pdf.

Institute for Climate Economics. "Using Credit Lines to Foster Green Lending: Opportunities and Challenges." (2017). https://cbd.int/financial/2017docs/ice-creditlines2017.pdf.

International Capital Market Association. "Green Bond Principles. Voluntary Process Guidelines for Issuing Green Bonds." (2018). https://icmagroup.org/green-social-and-sustainability-bonds/green-bond-principles-gbp/.

International Carbon Action Partnership. "Allocation: How Emissions Permits Are Distributed." (2017). https://icapcarbonaction.com/zh/?option=com_attach&task=download&id=493.

Invest NI. "Corporate Social Responsibility (Csr)." (n.d.). https://nibusinessinfo.co.uk/content/corporate-social-responsibility-environmental-impact.

IUCN. "Coastal "Blue" Carbon. A Revised Guide to Supporting Coastal Wetland Programs and Projects Using Climate Finance and Other Financial Mechanisms." (2015).

Japan Center for a Sustainable Environment and Society. "What Are Environmental Taxes?" (n.d.). http://jacses.org/en/paco/envtax.htm.

Koh, Niak S., Thomas Hahn, and Wiebren J. Boonstra. "How Much of a Market Is Involved in a Biodiversity Offset? A Typology of Biodiversity Offset Policies." *Journal of Environmental Management* 232 (2019/02/15): 679–91.

Loan Market Association. "Green Loan Principles. Supporting Environmentally Sustainable Economic Activity." (2018). https://lma.eu.com/application/files/9115/4452/5458/741_LM_Green_Loan_Principles_Booklet_V8.pdf.

Merk, O., S. Saussier, C. Staropoli, E. Slack, and J.-H. Kim. "Financing Green Urban Infrastructure." In *OECD Regional Development Working Papers*. OECD, 2012.

OECD. "Easing the Transition to Commercial Finance for Sustainable Water and Sanitation." (2017). http://oecd.org/water/Background-Document-OECD-GIZ-Conference-Easing-the-Transition.pdf.

OECD. "Environmental Taxation: A Guide for Policy Makers." (2011). https://oecd.org/env/tools-evaluation/48164926.pdf.

OECD. "Environmentally Related Taxes and Tradable Permit Systems in Practice." (2008). http://oecd.org/officialdocuments/publicdisplaydocumentpdf/?cote=COM/ENV/EPOC/CTPA/CFA(2007)31/FINAL&doclanguage=en.

OECD. "Tradeable Permits." (2004). https://oecd-ilibrary.org/environment/tradeable-permits_9789264015036-en.

Reed, M.S., K. Allen, A. Attlee, A.J. Dougill, K.L. Evans, J.O. Kenter, J. Hoy, et al. "A Place-Based Approach to Payments for Ecosystem Services." *Global Environmental Change* 43 (2017/03/01): 92–106.

Smart Prosperity Institute. "Economic Tools for Increasing Nature Conservation on Private Land." (2018). https://institute.smartprosperity.ca/library/publications/economic-tools-increasing-nature-conservation-private-land.

UK Green Building Council. "Demystifying Green Infrastructure." (2015). https://ukgbc.org/wp-content/uploads/2017/09/Demystifying-Green-Infrastructure-report-FINAL.pdf.

UNDP. "Financing Solutions for Sustainable Development." (2017). https://undp.org/content/dam/sdfinance/doc/Debt%20for%20Nature%20Swaps%20_%20UNDP.pdf.

UNEP. "Training Resource Manual: The Use of Economic Instruments for Environmental and Natural Resource Management." (2009). http://wedocs.unep.org/bitstream/handle/20.500.11822/7802/-Training%20Resource%20Manual_%20The%20Use%20of%20Economic%20Instruments%20for%20environmental%20and%20Natural%20Resource%20management-2009863.pdf?sequence=3&isAllowed=y.

UNEPFI. "Global Insurance Industry Statement on Adapting to Climate Change in Developing Countries." (2010). https://unepfi.org/publications/insurance-publications/global-insurance-industry-statement-adapting-to-climate-change-in-developing-countries/.

US EPA. "Economic Incentives." https://epa.gov/environmental-economics/economic-incentives#subsidies.

World Bank and World Resources Institute. "Integrating Green and Gray: Creating Next Generation Infrastructure." (2019). https://wri.org/publication/integrating-green-gray.

14 Best Practices and Conclusions on Nature-Based Solutions

Introduction

A set of best practices has been identified based on case studies of nature-based solutions (NBS) applied to climate change mitigation, climate adaptation, ecosystem restoration, urban sustainability, and disaster risk reduction. These practices offer practical guidance for other locations seeking to implement NBS to address societal and environmental challenges effectively.

Developing Climate Change Mitigation

From the case studies on NBS implemented for climate change mitigation, a diverse range of best practices has emerged that can be applied in other contexts:

- *Soils*: Incentivizing farmers to adopt conservation practices, such as cover cropping, compost application, and reduced tillage, enhances soil health and fertility while contributing to long-term carbon sequestration and reductions in agricultural emissions.
- *Soils/Agriculture*: Climate-smart agriculture can rehabilitate degraded pastures by using drought-resistant species, promoting silvopastoral systems, and delivering technical training to farmers. These interventions boost carbon storage, increase productivity, improve farmer incomes, and enhance resilience to drought.
- *Forests/Agriculture*: Integrated approaches that combine sustainable agricultural practices with forest rehabilitation and policy reform effectively reduce deforestation while supporting rural livelihoods. Key practices include supporting smallholder farmers, improving land tenure systems, and strengthening institutional coordination.
- *Grasslands*: Restoring grasslands through rotational grazing and effective water management can enhance soil carbon content and generate income through carbon credit schemes. Using digital monitoring tools effectively supports the tracking of soil carbon and the optimization of land use.

DOI: 10.4324/9781003517948-14

- *Peatlands*: Providing financial and technical support for restoring degraded peatlands, using methods such as ditch-blocking and rewetting, helps reduce greenhouse gas emissions while improving water quality and biodiversity.
- *Coastal Wetlands*: Supporting community-led restoration of mangroves, seagrasses, and saltmarshes enhances coastal resilience, carbon sequestration, and ecosystem health. Engagement of local and Indigenous communities ensures inclusive and sustainable project outcomes.
- *Mangroves*: Combining mangrove restoration with sustainable livelihoods and income-generating activities, such as small-scale aquaculture and natural product development, offers economic incentives while preserving critical carbon sinks.
- *Seagrass*: Community-driven seagrass restoration, utilizing traditional knowledge and hands-on techniques such as turf transplants, supports ecosystem regeneration, carbon capture, and youth engagement in conservation.
- *Livestock*: Reducing methane emissions from livestock through feed additives and dietary interventions provides a scalable solution for lowering agricultural emissions and increasing the sustainability of meat production.
- *Marine Protected Areas (MPAs)*: Strengthening marine conservation by protecting carbon-rich ecosystems, such as kelp forests and saltmarshes, contributes to climate mitigation and biodiversity preservation. Incorporating carbon valuation into marine policies helps align ecological and climate goals.

Developing Climate Change Adaptation

From the case studies of NBS implemented for climate adaptation, a variety of best practices have emerged that can be adopted in diverse geographies and contexts:

- *Agriculture*: Climate-smart agriculture initiatives, which integrate soil and water conservation, agroforestry, crop diversification, and small-scale irrigation, improve agricultural productivity and build resilience to climate extremes. Key enablers include climate-responsive advisory services and local capacity-building.
- *Forests*: Large-scale forest protection and restoration efforts that combine reforestation, invasive species removal, carbon credit generation, and water provisioning demonstrate the value of safeguarding fragile ecosystems. Supporting sustainable tourism and aligning forest management with climate adaptation goals are also essential.
- *Rivers*: Transforming urban rivers into resilient natural systems through river path reconfiguration, vegetation restoration, and integration of green

infrastructure enhances water quality, flood resilience, and urban livability while supporting economic regeneration.

- *Wetlands and Floodplains*: Utilizing engineered environmental infrastructure, such as flow regulators and channels, to mimic natural hydrological cycles can help regenerate wetlands and enhance ecosystem resilience in flood-prone areas.
- *Water Resources*: Managed aquifer recharge systems that capture and store excess surface water underground help replenish aquifers, stabilize groundwater levels, mitigate drought conditions, and prevent land subsidence – essential for maintaining water security in arid environments.
- *Coastal Wetlands*: Restoring tidal flows to degraded wetlands by modifying infrastructure, such as floodgates, enhances saltmarsh ecosystems, fish habitats, and resilience to sea-level rise. These interventions provide cost-effective methods for restoring natural functions.
- *Coastal Protection*: Nature-based coastal defense strategies that harness natural processes – like the redistribution of sand by wind and waves – strengthen shorelines, reduce maintenance needs, promote biodiversity, and provide recreational and tourism benefits.
- *Mangroves*: Community-led mangrove restoration, incorporating local nurseries, scientific planting methods, and community engagement, ensures optimal hydrological design, proper spacing, and effective monitoring – key factors for ecosystem recovery and long-term resilience.
- *Coral Reefs*: Reef restoration, utilizing selective propagation techniques and partnerships with local stakeholders, enhances coral resilience. Integrating education, ecosystem services, and restoration activities supports ecological recovery and community adaptation capacity.
- *Fisheries*: Community-based fisheries management that combines traditional knowledge with scientific approaches strengthens food security, biodiversity, and the adaptive capacity of coastal communities. Inclusive governance models also ensure equitable participation in resource management.

Restoring Degraded Ecosystems

From the case studies on restoring degraded ecosystems through NBS, a variety of best practices have emerged that can be replicated across other landscapes and seascapes:

- *Agriculture*: On-farm trials utilizing precision tools, such as GPS and variable-rate technology, can optimize nitrogen application, enhance nutrient use efficiency, reduce runoff, and inform sustainable nutrient management, ultimately increasing productivity.
- *Forests*: Long-term forest recovery can be supported through silvopastoral systems, community engagement initiatives, conservation agreements, and

water-harvesting infrastructure. These strategies promote soil and water retention, enhance biodiversity, and strengthen sustainable livelihoods.

- *Drylands*: Restoration in dryland regions benefits from integrated approaches that combine agroforestry, soil and water conservation practices, improved market access, and financial inclusion mechanisms. These efforts enhance food security, climate resilience, and institutional capacity.
- *Grasslands*: Protecting endangered grasslands requires designating conservation zones, regulating public access, removing invasive species, and adapting mowing practices. These measures maintain habitat connectivity and protect biodiversity.
- *Riparian Zones*: Ecological engineering along riparian corridors – such as implementing setback levees and planting native vegetation – enhances floodplain function, supports biodiversity, and improves agricultural outcomes. Combining flood control and ecological goals is essential.
- *Wetlands*: Large-scale wetland restoration through hydrological improvements, including reservoir construction, water treatment zones, and canal reconfiguration, can restore ecosystem function, improve water quality, and increase habitat resilience.
- *Saltmarsh*: Securing land tenure and fostering collaboration among public and private landowners, as well as involving the Indigenous community, are critical for successful saltmarsh restoration. Regular monitoring and stewardship are essential for maintaining long-term success.
- *Seagrass*: Manual seeding and long-term monitoring can regenerate aquatic habitats, enhance water quality, and support fisheries. These efforts also provide economic opportunities in coastal communities.
- *Coral Reefs*: Coral restoration using nursery cultivation, structural placement, and rope-based propagation techniques enhances reef biodiversity and resilience. Establishing land-based nurseries expands restoration potential and strengthens reef recovery.
- *MPAs*: Participatory zoning and inclusive governance of MPAs – led by communities in collaboration with non-governmental organizations and authorities – can protect key ecosystems, such as mangroves and marine nurseries, while supporting biodiversity and sustainable fisheries.

Enhancing Sustainable Urbanization

From the case studies of NBS implemented to enhance sustainable urbanization, a diverse range of best practices has emerged that other cities can adopt to improve livability, resilience, and biodiversity:

- *Green Roofs*: Financial incentives, integration into building codes, and scientific monitoring can significantly boost the adoption of green roofs. Prioritizing native vegetation, allowing for public use where appropriate, and contributing to thermal regulation and biodiversity are key goals. Removing regulatory barriers and offering clear technical guidance are also critical success factors.

- *Water Harvesting*: Promoting rainwater harvesting at scale is most effective when paired with financial incentives, public awareness efforts, and capacity-building activities. Pre-approval processes and hands-on training ensure systems are correctly installed and maintained for long-term water conservation and drought resilience.
- *Urban Biodiversity and Pollinator Support*: Everyday urban infrastructure, such as transit shelters and public furniture, can be adapted to support pollinators and urban biodiversity. Incorporating vegetated surfaces and sustainable materials helps improve air quality and fosters environmental stewardship.
- *Flood Management*: Urban parks designed as dual-purpose spaces for recreation and flood storage provide multiple co-benefits. These areas mimic natural wetlands, enhancing resilience to extreme weather events, supporting cooling, and improving local biodiversity in high-density environments.
- *Urban Green Space*: Establishing permanent green belts within urban areas supports biodiversity, offers space for urban agriculture, promotes cultural identity, and enhances recreational opportunities. These belts help cities maintain an ecological balance while supporting the well-being of residents.
- *Stormwater Wetlands*: Transforming underutilized urban spaces into stormwater wetlands creates multifunctional areas that improve water management, support biodiversity, and serve as educational resources for the public.
- *Rain Gardens*: Retrofitting sidewalks with vegetated rain gardens helps manage stormwater runoff, mitigate urban heat, and enhance habitat quality. Community engagement and stewardship are crucial for achieving long-term success and fostering local buy-in.
- *Floodplain Restoration*: Converting flood-prone properties into green infrastructure represents a long-term strategy for risk reduction. These spaces restore habitat, prevent future losses, and generate public value by creating greenways and community trails.
- *Green Streets and Urban Forestry*: Incorporating bioswales, permeable pavements, and tree trenches into urban street design enhances stormwater management, beautifies public spaces, and provides natural cooling, particularly beneficial in heat-vulnerable neighborhoods.
- *Green Corridors*: Integrating green corridors into the urban landscape by revitalizing waterways and connecting open spaces enhances access to nature, strengthens climate resilience, and improves ecological connectivity, while also reflecting community priorities and cultural identity.

Improving Disaster Risk Management and Resilience

From the case studies of NBS implemented in the context of disaster risk management and resilience, a variety of best practices have emerged that other regions can adopt to reduce hazard exposure, enhance recovery capacity, and deliver co-benefits:

- *Urban Blue-Green Networks*: Restoring natural water flows in urban areas through detention basins, stream daylighting, and flow path enhancement can reduce flood risks, expand green public spaces, and strengthen resilience to extreme rainfall events.
- *Wetland Restoration*: Re-meandering streams, blocking artificial drains, and rewetting wetlands across broad landscapes help regulate floodwaters, enhance drought resilience, store carbon, and restore ecological integrity.
- *Coastal Flood Protection*: Reestablishing sediment flow through sand bypass systems and dredging reduces coastal erosion, safeguarding shorelines and supporting tourism, ecosystems, and infrastructure.
- *Living Shorelines*: Combining natural materials like limestone breakwaters with marsh vegetation – planted in structured patterns – absorbs wave energy, stabilizes coastlines, and provides biodiversity and storm protection benefits.
- *Estuary Restoration*: Techniques such as sediment diversion, reuse of dredged material, and native vegetation planting help rebuild marshlands, restore natural delta processes, and protect vital coastal ecosystems.
- *Coastal Bioshields*: Planting shelterbelts of resilient tree species along coastlines reduces storm surge impacts, prevents saltwater intrusion, and enhances local livelihoods through ecosystem restoration and green cover.
- *Saltmarsh Restoration*: Utilizing natural materials to trap sediment and promote native plant growth in estuarine environments enhances carbon sequestration, improves habitat quality, and reinforces natural flood defenses.
- *Oyster Reef Restoration*: Restoring oyster reefs with recycled materials and involving local communities in planting efforts enhances water quality, reduces wave energy, and provides valuable habitats, while also offering educational and vocational opportunities.
- *Coral Reef Protection*: Deploying hybrid reef structures that integrate engineered components with live coral propagation improves shoreline defense, supports biodiversity, and enables scalable reef restoration in vulnerable coastal zones.
- *Constructed Wetlands*: Building wetlands for municipal wastewater treatment enhances water reuse for non-potable needs, increases drought resilience, supports wildlife habitats, and provides recreational and educational opportunities.
- *Water Harvesting and Reuse*: Integrating rooftop rainwater harvesting with greywater reuse systems for gardens and non-potable household uses

helps reduce potable water demand and strengthens climate adaptation in water-scarce regions.

- *Climate-Smart Agriculture*: Supporting soil health, reducing emissions, and promoting inclusive land stewardship practices enhances agricultural resilience, improves adaptive capacity, and fosters sustainable land management in diverse climates.
- *Coastal Forests*: Planting native trees using dense, multi-layered methods along coastlines creates long-term buffers against natural hazards, such as tsunamis, sequesters carbon, restores habitats, and fosters community involvement in ecosystem stewardship.

Innovative Financial Tools for Nature-Based Solutions

From the case studies of financial instruments supporting NBS, a variety of innovative tools have emerged that can be replicated to enable investment, scale impact, and ensure long-term sustainability:

- *Conservation Finance*: Voluntary conservation easements can protect working landscapes while allowing for continued land use and management. These agreements incentivize long-term stewardship, safeguard water resources, and may provide tax benefits to landowners through grant-supported or donation-based mechanisms.
- *Water Quality Trading*: Water quality trading schemes enable regulated entities to purchase pollutant reduction credits from others within the same watershed. These market-based mechanisms support regulatory compliance, enhance watershed health, and promote the uptake of conservation practices.
- *Green Investment Incentives*: Tax incentives, such as deductions or accelerated depreciation, can encourage corporate investment in green technologies. These mechanisms promote environmentally responsible business practices while improving liquidity and supporting broader sustainability goals.
- *Environmental Risk Regulation*: Regulatory frameworks requiring financial assurances for high-risk industrial activities ensure funds are available for site remediation. These measures protect the public, enforce accountability, and support compliance with environmental standards.
- *Public-Private Restoration*: Competitive grant programs funded through partnerships between public agencies and private foundations can finance ecosystem restoration and green infrastructure. These models deliver measurable ecological and social benefits while encouraging cross-sector collaboration.
- *Environmental Impact Bonds*: Pay-for-success financial instruments tie investor returns to environmental outcomes, such as stormwater capture or flood reduction. These bonds incentivize performance, generate green jobs, and direct resources toward underserved communities, enhancing equitable climate resilience.

- *Corporate Forestry*: Private sector investment in large-scale reforestation and landscape restoration initiatives can significantly contribute to biodiversity conservation, carbon sequestration, and sustainable livelihoods. These corporate-led efforts align environmental responsibility with long-term sustainability commitments.
- *Carbon Forestry and Gifts*: Impact-oriented investment tools, such as tree sponsorships and certified agroforestry programs, enable individuals and companies to support climate mitigation, biodiversity, and rural development while accessing transparent reporting on outcomes.
- *Consumer Reforestation Tools*: Fintech models linking consumer spending to automatic tree planting offer scalable solutions for engaging the public in climate action. These tools harness everyday behavior to fund reforestation and offset emissions.
- *Nature Insurance*: Parametric insurance products designed for ecosystem protection can trigger payouts based on predefined natural events (e.g. storms), enabling rapid recovery efforts. These instruments support long-term resilience and safeguard valuable ecological assets.

Conclusion

NBS offers a practical, cost-effective, and scalable approach to addressing various environmental, social, and economic challenges. Across diverse ecosystems – from forests and grasslands to coastlines and urban environments – case studies demonstrate that NBS can support climate change mitigation and adaptation, restore degraded landscapes, enhance urban sustainability, and reduce disaster risks.

The best practices identified throughout this book highlight the importance of integrated planning, cross-sector collaboration, community engagement, and long-term financing. Successful implementation often relies on aligning ecological outcomes with social benefits, supported by enabling policies, appropriate governance, and sustained investment.

As climate pressures intensify, NBS provide an opportunity to work with nature rather than against it. When designed and implemented effectively, these solutions contribute to environmental resilience and more inclusive and sustainable development pathways.

Index

For Product Safety Concerns and Information please contact our EU representative GPSR@taylorandfrancis.com
Taylor & Francis Verlag GmbH, Kaufingerstraße 24, 80331 München, Germany

www.ingramcontent.com/pod-product-compliance
Lightning Source LLC
LaVergne TN
LVHW021119300926
849160LV00010B/777

"This book makes a significant contribution to our understanding of the role of finance in the growth dynamics of the micro, small and medium enterprises in India, looking particularly into the adverse effect of credit constraints on growth of such enterprises, and how there is gender and caste discrimination in the credit market adversely impacting the performance of enterprises owned by women and socially disadvantaged entrepreneurs. Based on a careful, sophisticated and painstaking analysis of unit-level data on about 1.2 million enterprises collected in an official census of India's MSME sector, the book is timely and highly policy relevant, and will be quite useful for students, researchers, scholars of economics and policy makers".

Bishwanath Goldar, *Former Professor, Institute of Economic Growth, Delhi, India.*

"The book by Dr Rajesh Raj Natarajan and Dr Subash Sasidharan is a major quantitative study of the MSME sector in India. Using data from the Fourth census of MSMEs, and applying modern econometric techniques, the book examines gender-based and caste-based differential access to formal credit issues. It produces compelling empirical evidence with important policy implications and recommendations for the MSME sector which provides vast employment opportunities for workers with modest educational qualifications. Students of public policy will benefit a great deal from the book".

Professor K.L. Krishna, *Former Director, Delhi School of Economics; Former Chairman, Center for Economic and Social Studies, Hyderabad and Former Chairman, Madras Institute of Development Studies, Chennai.*

"This book provides a rich understanding of the role of finance in the development of small firms in India, combining carefully done empirical analysis with a comprehensive coverage of policy issues. The book will be valuable reading to academics and policy makers interested in India's economic development".

Kunal Sen, *Director, United Nations University-World Institute for Development Economics Research and Professor of Development Economics, University of Manchester, United Kingdom (UK).*

"I am glad to note that the book 'Small Firm Ownership and Credit Constraints in India' is released by an IIT-M Scholar and his co-author from Sikkim University. Generally, the MSME sector is a major contributor to the Global business which needs more documents for its past and future references.

With the limited availability of sources on MSMEs, this book will throw new light on financial availability to the particular sector of the society with the Indian context. I hope it will be useful for the Policy Makers, Executors, Associations and Researchers, relating to this sector".

R. Ramamurthy, *President, The Coimbatore District Small Industries Association (CODISSIA), Coimbatore.*